2024 IEEE Silicon Nanoelectronics Workshop (SNW 2024)

Honolulu, Hawaii, USA
15-16 June 2024

IEEE Catalog Number: CFP24SNW-POD
ISBN: 979-8-3503-9164-0

**Copyright © 2024 by the Institute of Electrical and Electronics Engineers, Inc.
All Rights Reserved**

Copyright and Reprint Permissions: Abstracting is permitted with credit to the source. Libraries are permitted to photocopy beyond the limit of U.S. copyright law for private use of patrons those articles in this volume that carry a code at the bottom of the first page, provided the per-copy fee indicated in the code is paid through Copyright Clearance Center, 222 Rosewood Drive, Danvers, MA 01923.

For other copying, reprint or republication permission, write to IEEE Copyrights Manager, IEEE Service Center, 445 Hoes Lane, Piscataway, NJ 08854. All rights reserved.

*** *This is a print representation of what appears in the IEEE Digital Library. Some format issues inherent in the e-media version may also appear in this print version.*

IEEE Catalog Number: CFP24SNW-POD
ISBN (Print-On-Demand):   979-8-3503-9164-0
ISBN (Online):   979-8-3503-9163-3
ISSN: 2161-4636

Additional Copies of This Publication Are Available From:

Curran Associates, Inc
57 Morehouse Lane
Red Hook, NY 12571 USA
Phone: (845) 758-0400
Fax: (845) 758-2633
E-mail: curran@proceedings.com
Web: www.proceedings.com

Table of Contents

Opening Remarks

Peide Ye & Zhihong Chen, Purdue University

Session 1: Plenary

Session Chair: *Pei-Wen Li, NYCU*

1.1 (Keynote) Future Perspectives of CMOS Logic Innovation Beyond 2nm, T. Hiramoto[1], H. Wakabayashi[2], *[1]University of Tokyo, [2]Tokyo Inst. of Tech.* — 1

1.2 (Keynote) Gate Oxide Reliability: Upcoming Trends, Challenges and Opportunities, B. Kaczer[1], R. Degraeve[1], J. Franco[1], T. Grasser[2], Ph. J. Roussel[1], E. Bury[1], P. Weckx[1], A. Chasin[1], S. Tyaginov[1], *[1]imec, [2]TU Vienna* — 3

Session 2: GAA/Nanosheet Transistors I

Session Chair: *Takahiro Shinada, Tohoku University*

2.1 (Invited) Heterogeneous CFET: A Pathway to CMOS Performance Balance Engineering, X.-R. Yu[1], W.-H. Lu[1], S.-W. Chang[1], W.-H. Chang[2], Darsen D. Lu[1], Y.-J. Lee[3,4], T. Maeda[3], W.-F. Wu[4], Y.-M. Li[3], Y.-H. Wang[1], *[1]Natl. Cheng Kung University, [2]AIST, [3]National Yang Ming Chiao Tung University, [4]TSRI* — 5

2.2 Enhanced Electrical Performance of Ultrathin Body Nanosheets, B.-W. Huang, Y.-R. Chen, T. Chou, H.-C. Lin, C.-T. Tu, Y.-C. Liu, W.-H. Hsieh, W.-J. Chen, M.-K. Lin, Y.-Q. Liu, L.-K. Wang, H.-C. Chou, Y. Huang, D.-W. Lin, C.-W. Liu, *National Taiwan University* — 7

Session 3: Quantum Computing Devices/Circuits

Session Chair: *Louis Hutin, CEA-Leti*

3.1 (Invited) Integration Technology for Superconducting Quantum Circuits, *Shiro Kawabata, AIST* — 9

3.2 Demonstration of 99.9% Single Qubit Control Fidelity of a Silicon Quantum Dot Spin Qubit Made in a 300 mm Foundry Process, N. D. Stuyck[1,2], F. Mengke[1,2], W. H. Lim[1,2], S. S. Ramirez[1,2], C. Escott[1,2], T. Botzem[1,2], T. Tanttu[1,2], H. Yang[1,2], A. Saraiva[1,2], A. Laucht[1,2], S. Kubicek[3], J. Jussot[3], S. Beyne[3], B. Raes[3], C. Godfrin[3], D. Wan, K. D. Greve[3,4], A. Dzurak[4]. *[1]Diraq, [2]UNSW, [3]imec, [4]KU-Leven* — 11

3.3 Tunneling Current Spectroscopy of Self-assembled Ge Double Quantum Dots with Hard-wall Barrier Engineering, *Y.-W. Chiu, I. H. Wang, T. Tsai, H.-C. Lin, P.-W. Li, National Yang Ming Chiao Tung University* — N/A

Session 4: Memory Technology

Session Chair: Jiezhi Chen, Shandong University

4.1 (Invited) IGZO Channel VCT (Vertical Channel Transistor) Technology for sub-10nm DRAM : Challenges and Opportunities, *Y. Lee, Daewon Ha, W. Lee, S. Yoo, J. H. Bae, M. H. Cho, K. Yoo, S. M. Lee, S. Lee, M. Terai, T. H. Lee, K. J. Moon, C. Sung, M. Hong, D. G. Cho, H. Kim, J. H. Seo, K. Park, B. J. Kuh, S. Hyun, S. J. Ahn, and J. H. Song, Samsung Electronics* — 15

4.2 A Novel 2-Tier Stacked Vertical Channel GAA Transistor Architecture with Area-Saving 3-Dimensional Local Interconnect to Achieve Gb-Density 6T SRAM with Acceptable Low Standby Chip Leakage (~1mA), *H.-T. Lue, W. Chen, T.-H. Yeh, K.-C. Wang, and C.-Y. Lu, Macronix* — 17

4.3 A Comparative Study of HBL and VBL 3D DRAM: Signal Margin, Bit-cell Density, and Scalability, *X. Wu[1], L. Upton[1], P.-K. Hsu[2], J. Chen[1], S. Yu[2], H.-S. P. Wong[1], [1]Stanford University, [2]Georgia Institute of Technology* — 19

4.4 Impact of Proton Irradiation on SiNx and Si–SiO$_2$ Interfaces in FLASH Memory, *H. Kim[1], J. Park[1], J. Im[2], H. Kim[2], Y. Jun Yoon[3], Y.-M. Kim[4], K. Park[5], S. Y. Woo[2], J.-H. Bae[1], [1]Kookmin University, [2]Kyungpook National University, [3]Andong National University, [4]Korea Atomic Energy Research Institute, [5]Yonsei University* — N/A

4.5 A Novel RRAM-based Ternary Strong-PUF with High Security Intensity Feasible for Low-earth Orbit Satellites in the 6G Era, *S. H. Lin, E. R. Hsieh, National Central University* — 23

Session 5: Ferroelectric Memory

Session Chair: K. Toprasertpong, The Univ. of Tokyo

5.1 (Invited) Outlook and Prospects for Ferroelectric Augmentation of Vertical NAND Flash Technology, *Asif Khan, Georgia Institute of Technology* — N/A

5.2 Investigating the Correlation Between Material Ferroelectricity and Silicon-Channel FeFET Performances: Insights for Material Engineering in Device Optimization, *X. Wang, L. Jiao, Z. Zhou, Z. Zheng, Y. Chen, R. Shao, C. Sun, D. Zhang, G. Liu, X. Gong, National University of Singapore*

N/A

5.3 Ferroelectricity Engineered AlScN Thin Films Prepared by Hydrogen Included Reactive Sputtering for Analog Applications, *S.-M. Chen[1], H. Nishida[1], T. Hoshii[1], K. Tsutsui[1], H. Wakabayashi[1], Edward Y. Chang[2], and K. Kakushima[1] Tokyo Inst. of Technology, National Yang Ming Chiao Tung Univ.*

29

5.4 Investigation of HfO_2 and ZrO_2 Nanolamination Thickness on Superlattice HZO FeRAMs Exhibiting Enhanced Remnant Polarization and Improved Endurance, *D. R. Hsieh, Z. Y. Hong, W. J. Yeh, J. C. Ni, H. E. Luo, Y.-K. Liang, S. L. Hsieh, K. L. Chen, T.-S. Chao, National Yang Ming Chiao Tung University*

31

5.5 Impact of Time Delay Schemes on Reliability Degradation During Program/Erase Cycling in HZO-based FeFETs, *X. Li[1], G. Zhao[1], L. Tai[1], P. P. Sang[1], X. Dou[1], X. Zhan[1], X. Wang[2], J. Wu[1], J. Chen[1], [1]Shandong University, [2]Institute of Microelectronics of Chinese Academy of Sciences*

33

5.6 Study on the Anomalous Characteristics of Random Telegraph Noise in FeFETs, *P. Cai[1], Y. Wu[2], Z. Sun[1], H. Li[1], X. Wang[3], Z. Ji[2], R. Wang[1], R. Huang[1], [1]Peking University, [2]Shanghai Jiao Tong University, [3]Institute of Microelectronics of Chinese Academy of Sciences*

35

Session 6: Neuromorphic Computing
Session Chair: Takahide Oya, Yokohama National University

6.1 Reservoir Computing Using Nonlinear Polarization and Charge Dynamics of Anti-ferroelectric HZO/Si FETs, *S.-Y. Min, K. Toprasertpong, E. Nako, R. Nakane, M. Takenaka, S. Takagi, The University of Tokyo*

37

6.2 Design Space Trade-offs in Networks of Coupled Injection-locked Ring Oscillators for Ising-based Optimum Search, *A. Bazzi, F. Badets, L. Hutin, CEA-Leti*

39

6.3 Background-Pattern-Dependency-Tolerant Weight Transfer Method for Accurate NAND-Flash Based Spiking Neural Networks,

41

B. Jeon, J. H. Chang, W. Y. Choi, Seoul National University

6.4 Artificial VO$_2$ Spiking Neurons with Protective Mechanism for Enhancing Resilience of Spiking Neural Network Against Adversarial Attacks, *C. Ban[1], L. Shan[1], G. Yang[1], J. Gao[1], L. Wu[1], Z. Wang[1,2], Y. Cai[1,2], R. Huang[1,2], [1]Peking University, [2]Beijing Advanced Innovation Center for Integrated Circuits*   43

6.5 Flash-based Computing-in-Memory (CiM) Towards Stochastic Computing with Low Power-consumption and High Noise-immunity, *H. Wang, Y. Feng, X. Zhan, M. Bai, P. P. Sang, J. Wu, Q. Wang, J. Chen, Shandong University*   45

Session 7: GAA/Nanosheet Transistors II
Session Chair: Woo Young Choi, SNU

7.1 (Invited) Multi-Vt Gate Stack Technologies for Nanosheet and CFET Devices, *H. Arimura[1], J. Franco[1], L.-Å. Ragnarsson[1], A. Vandooren[1], S. Brus[1], W. Maqsood[1], T. Conard[1], G. Alessio Verni[2], J. W. Maes[2], B. Kannan[2], M. Givens[2], N. Horiguchi[1], [1]imec, [2]ASM*   47

7.2 Investigation of Sheet Width Dependence on Hot Carrier Degradation in GAA Nanosheet Transistors, *Z. Sun, Z. Wang, R. Wang, R. Huang, Peking University*   49

Session 8: Process and Transistor Technology
Session Chair: Xiao Gong, NUS

8.1 Investigating HfO$_2$/ZrO$_2$ Superlattice Dielectric with High-κ Value of 52 via Annealing Temperature and Layer Thickness Modulation Beyond Moore's law, *Y. C. Wu[1], Y.-J. Yao[1], H. J. Chang[1], C.-Y. Wei[1], Y.-M. Fu[1], B.-X. Chen[1], T.-J. Lin[1], Y.-T. Fang[1], G.-L. Luo[2] and F.-J. Hou[2], [1]National Tsing Hua University, [2]Taiwan Semiconductor Research Institute*   N/A

8.2 First Demonstration of SRAM Transistor Based on 3-dimensional Stacked FET with Back Side Interconnection Structure Beyond 1nm Node, *M. Kim[1,2], J. Park[2], S. Park[2], J. Park[2], J. Kim[1], D. Ha[2], H. Shin[1,3], [1]Seoul National University, [2]Samsung, [3]Integra Semiconductor*   53

8.3 High Performance Te/ZnO CMOS Inverter Operated at 77K, *M. Kim, K. Kim, K. Kim, J. H. Jun, H.-W. Lee, B. H. Lee, Pohang University of Science and Technology*   N/A

8.4 Steep Slope Device N-type Gate-Controlled Carrier-Injection SOI-Transistor: Suppression of Hysteresis by Ar-ion Implantation and Possibility of CMOS, *H. Yonezaki, T. Mori, J. Ida, Kanazawa Institute of Technology*   57

Session 9: 2D Materials and Oxide Semiconductor Devices for 3D integration

Session Chair: Byoung Hun Lee, POSTECH

9.1 (Invited) Potential and Challenges of 2D Materials Based Electronics, *L.-J. Li, University of Hong Kong*   N/A

9.2 A Nanosheet Oxide Semiconductor FET Using ALD InZnOx Channel, *S.-H. Kim, K. Hikake, Z. Li, T. Saraya, T. Hiramoto, M. Kobayashi, University of Tokyo*   61

9.3 Investigation of In-Sn-Zn Composition on the Characterization of Submicron Channel Length Ultra-Thin Atomic Layer Deposited InSnZnO Channel Transistors, *Y.-K. Liang[1,2], L.-C. Peng[1], Y.-L. Lin[1], J.-Y. Zheng[1], D. R. Hsieh[1], T.-T. Chou[3], H.-Y. Huang[4], Y.-M. Lin[4], Y.-C. Tseng[1], T.-S. Chao[1], E. Y. Chang[1], K. Toprasertpong[2], S. Takagi[2], C.-H. Lin[1], [1]National Yang Ming Chiao Tung University, [2]The Univ of Tokyo, [3]Taiwan Instrument Research Institute, [4]Taiwan Semiconductor Manufacturing Company*   63

9.4 BEOL-Compatible In$_2$O$_3$ Thin-Film Transistor with Linear Dielectric ZrO$_2$ Achieving Dielectric Constant over 27 and Enhanced Field-Effect Mobility up to 89.3 cm^2/V·s, *Z. Lin, P. Ye, Purdue University*   65

9.5 First Investigation of Low-Frequency Noise in Scaled Atomic-Layer-Deposited IGZO FETs with Different Indium Ratio, *S. Lee, C. Niu, Z. Zhang, Y. Zhang, L. Long, Z. Lin, H. Wang, P. Ye, Purdue University*   N/A

9.6 A Novel Ternary Transistor with Nested Source Design Incorporating Hybrid Switching Mechanism for Low-Power and High-Performance Applications, *S. Xu[1], L. Tianyang[1], T. Luo[1], R. Huang[1,2], Q. Huang[1,2], [1]Peking University, [2]Beijing Advanced Innovation Center for Integrated Circuits*   69

Session 10: Poster Session 1

Session Chairs: Steve Chung, National Yang Ming Chiao Tung University

P1-1 Single Electron Charge Detection in Nanoscale Metal Double-dot: DC Single Electron Transistor vs. gate RF sensing, *M. I. Rahaman[1], G. Szakmany[1], X. Jehl[2], A. Orlov[1], G. Snider[1], [1]University of Notre Dame, [2]Univ. Grenoble Alpes* — 71

P1-2 Ferromagnetic Manganese Silicide Nanoparticles Formed by Ion Implantation in Silicon, *R. Ohsugi, M. Kawano, Y. K. Wakabayashi, Y. Krockenberger, H. Sumikura, J. Noborisaka, K. Nishiguchi , NTT* — 73

P1-3 The Simulation of Double Germanium Quantum Dots in a Ring-Shaped Quantum Structure, *C. Liang, Y.-T. E. Tang, National Central University* — 75

P1-4 Improved Uniformity and Excellent Endurance Characteristics of TaOx-Based RRAM by Laser-Mediated Interface Engineering, *L. Wu[1], Q. Wang[1], H. Liao[1], C. Ban[1], L. Shan[1], Z. Wang[1,2], Y. Wang[1,2], Y. Cai[1,2], [1]Peking University, [2]Beijing Advanced Innovation Center for Integrated Circuits* — 77

P1-5 Effects of Modulation Pulse on Resistive Switching Dynamics of RRAM Devices, *C.-H. Chien, Y.-H. Wang, National Cheng Kung University* — 79

P1-6 Innovative Switching Mechanism, Performance Investigations and Scaling Effects in a-V_2O_3 Based ReRAM devices, *K. Veyret[1,2], L. Laborie[1], R. Bon[1], G. Navaro[1], R. Hida[1], N. Castellani[1], C. Carabasse[1], P. Gonon[1] and E. Jalaguier[1], [1]CEA-Leti, [2]Univ. Grenoble Alpes* — 81

P1-7 Floating Gate Field Effect Transistors for Logic-In-Memory Application, *S. Jae Baik[1], S. Kim[2], M. Kang[3], and J. Jeon[2], [1]Samsung Electronics, [2]Sungkyunkwan University, [3]Korea National University of Transportation* — N/A

P1-8 Withdrawn

P1-9 High Performance HfLaO-based TiO_2-Channel FE V-NAND with High Consistency and Low Operation Voltage, *X. Song, S. Li, D. Sun, X. Liu, J. Kang, Peking University* — 85

P1-10 Accelerated Program Inhibition Failure by Trap-Mediated Tunneling in the Polycrystalline Floating-Channel 3-D NAND Flash Memory Array, *S. Kim[1], U. Lee[2], Y. Lee[1], C. Ahn[2], J. Sim[2], S. Cho[1], [1]Ewha Womans University, [2]SK hynix inc.* — 87

P1-11 Investigation of Row Hammer and Passing Gate Effects Based on the Work Functions of Dual Gates in DRAM Cells, *H. Kim[1], J. Im[1], J. Kim[1], S. Woo[1], T. Kwon[1], Y. J. Yoon[2], J.-H. Bae[3], S. Y. Woo[1], [1]Kyungpook National University, [2]Andong National University, [3]Kookmin University* — 89

P1-12 Analysis of Row Hammer and Passing Gate Effect in DRAM Cells by BCAT — 91

Structural Design, *J. Im[1], H. Kim[1], J. Kim[1], S. Woo[1], T. Kwon[1], Y. J. Yoon[2], J.-H. Bae[3], S. Y. Woo[1], [1]Kyungpook National University, [2]Andong National University, [3]Kookmin University*

P1-13 Controllable Polarization Switching of Hafnia-Based Ferroelectric Bilayers, *J. Jeong, H. Park, J. Woo, Kyungpook National University* 93

P1-14 Short-term and Long-term T-O Phase Transition Responsible for Two stages of Wake-up Process in Ferroelectric Hf$_{0.5}$Zr$_{0.5}$O$_2$ Film, *D. Chen, G. Qiang, F. Yuyan, Y. Zikang, J. Liu, S. Mengwei, L. Xiuyan, Shanghai Jiao Tong University* 95

P1-15 Enhancing Annealing Strategies for Back-End-of-Line-Compatible HfO$_2$-Based Ferroelectric Capacitors with Time Periods and Gas Species, *C.-H. Wu[1], T.-Y. Lin[2], C.-Y. Chiu[2], C.-J. Su[2], V. P.-H. Hu[1], [1]National Taiwan University, [2]National Yang Ming Chiao Tung University* N/A

P1-16 Plasma-enhanced Atomic Layer Deposition Based FEFETs, *C. Park[1,2], P. V. Ravindran[1], D. Das[3], P. G. Ravikumar[1], W. Chern[1], S. Yu[1], A. I. Khan[1], [1]Georgia Institute of Technology, [2]SK hynix, [3]NIT Silchar* N/A

P1-17 Investigation on the Local Polarization Scheme in Ferroelectric Tunnel Field-Effect Transistors for Ternary Content-Addressable Memory Applications, *M. Ryu, J. S. Woo, W. Y. Choi, Seoul National University* 101

P1-18 Study of Spacing-induced Fringing Effects on Emerging CFET Technology Nodes, *Y.-Z. Chen[1], N. Thoti[2], Y.-T. E. Tang[1], [1]National Central University, [2]University of Oulu* 103

Session 11: Poster Session 2

Session Chairs: K. Kakushima, Tokyo Institute of Technology

P2-1 Compact Model of Feedback Field-Effect Transistor Using Artificial Neural Network, *J. S. Su, J. H. Oh, Y. S. Yu, Hankyong National University* 105

P2-2 Neural Encoder Using "PN-Body Tied SOI-FET", *M. Kobayashi, J. Ida, T. Mori, Kanazawa Institute of Technology* 107

P2-3 Engineering Strategies in HfOx RRAM-Based Analog Synapses Toward Linear Weight Update for Neuromorphic Hardware Accelerators, *Y. Kim, H. Choi, J. Woo, Kyungpook National University* 109

P2-4 Enhancing Single-Electron Reservoir Computing Performance with Delay Function and Multiple-layer Reservoir Circuits, *S. Watanabe, T. Oya, Yokohama National University* 111

P2-5 Design of Single-Electron Circuit Representing Brownian Motion of Particles to 113

Implement Circuit Capable of Computing Diffusion Limited Aggregation Model, *R. Miyakoshi, T. Oya, Yokohama National University*

P2-6 A Machine-Learning-based Model for Emerging Memories Featuring Multiple States, *Z. Wang[1], Z. Rong[2], R. Wang[3], M. Chan[2], L. Zhang[1], [1]Peking University Shenzhen, [2]Hong Kong University of Science and Technology, [3]Peking University* — 115

P2-7 Automated Recipe Creation and Verification of Single Wafer Wet Etching: Ensemble Learning with Backcasting and Forecasting AIs Using Scarce Data, *K. Shibata[1], H. Horiguchi[2], C. Matsui[1], K. Takeuchi1, [1]University of Tokyo, [2]SCREEN Semiconductor Solutions Co., Ltd.* — 117

P2-8 Design of the 2-nm Nanosheet NAND-type TCAM with High Speed and Compat Cell-size: 45% Layout-reduction of 3-nm TCAM, *L.-A. Yu, C.-Y. Chou, C.-C. Lin, T.-Y. Tsai and E. R. Hsieh, NCU* — 119

P2-9 Design of Single-electron Information-processing Circuit for Particle Computation, *S. Mizuno, T. Oya, Yokohama National University* — 121

P2-10 Predicting the Retention Property of Scaled Cylindrical IGZO 2T0C DRAM Cells, *S.-M, Jeong, S. —M. Hong, GIST* — 123

P2-11 Withdrawn — N/A

P2-12 Withdrawn

P2-13 Simulation of Trap-Induced Noise Characteristics in 3-nm Complementary FET, *J. Xu, Z. Zhou, F. Liu, X. Liu, Peking University* — 127

P2-14 Simultaneously Improved Electrical Characteristics of Ge n/p-FinFETs by Using HfN Interface Layer with In-Situ Plasma Oxidation Treatments, *K.-S. Chang-Liao[1], Y.-Y. Chen[2], D.-B. Ruan[1], C.-H. Li[1], [1]National Tsing Hua University, [2]Fuzhou University* — N/A

P2-15 High Performance Ge FinFET CMOS Invertor with Post Plasma Oxidation and Nitridation Treatments before Supercritical Fluid Process, *K.-S. Chang-Liao[1], W.-C. Hung[2], D.-B. Ruan[1], K.-C. Yang[1], [1]National Tsing Hua University, [2]Fuzhou University* — N/A

P2-16 Characteristics of Aluminum-based Oxide with ALD SiO_2 Interfacial Layer as the Gate Dielectric of the Silicon Carbide (SiC) MOS Capacitor with RTA Annealing, *C.-L. Lin, B. X. Su, Y. L. Lee, Feng Chia University* — 133

P2-17 Electrical Characteristics of BEOL-Compatible Top-Gate In_2O_3 Transistors, *P. Hong J. Hao, X. Li, Huazhong University of Science and Technology* — 135

2024 IEEE Silicon Nanoelectronics Workshop

Hilton Hawaiian Village

Honolulu, HI USA

June 15-16, 2024

2024 IEEE Silicon Nanoelectronics Workshop

Welcome Message

The 2024 IEEE Silicon Nanoelectronics Workshop (SNW) is a satellite workshop of the 2024 Symposium on VLSI Technology and Circuits sponsored by the IEEE Electron Devices Society. This workshop is held alternately between Hawaii and Kyoto in an annual series, i.e., even year in Hawaii and odd year in Kyoto. It is the twenty-ninth workshop in the series, which showcases original work on semiconductor materials, devices, technologies, and applications that utilize silicon or which are based on novel materials on silicon substrates.

2020–2022 will be long remembered for the wide-spreading of COVID-19 pandemic. We are deeply sorry for those who have been suffering, and we sincerely wish you and your family are doing well, safe, and healthy. Fortunately, we come back with a full in-person event this year. We must largely acknowledge all those who made this event possible, including our organizing team, invited speakers, and all the authors. 2024 SNW would not have been delivered without the full support of all these people.

In spite of the dynamic changing of the world especially in the economy, this year we received a reasonably good number of submissions. The program includes 2 plenary talks, 6 invited talks, 27 oral presentations, and 36 poster papers, as well as is organized into 8 oral presentation sessions and 2 poster sessions. Contributions from researchers from over 11 countries around the world are featured. The workshop provides an excellent opportunity for engineers, researchers, professors, and students to share and discuss their recent work on the nanometer-scale devices and technologies.

We are very delighted to have two keynote speakers for the plenary session: Prof. T. Hiramoto of the University of Tokyo, speaking on "Future Perspectives of CMOS Logic Innovation Beyond 2nm" and Dr. Ben Kaczer of imec on "Gate Oxide Reliability: Upcoming Trends, Challenges and Opportunities". We also deeply appreciate all the invited speakers and authors with their enthusiastic support and participation.

This year's workshop will not be possible without the strong support from the members of the Technical Program Committee, the contributions from all keynote/invited speakers and authors, and the participation from the attendees. We also appreciate the support of *IEEE Electron Devices Society* along with its assistance in conference management. We hope that you enjoy the in-person meeting in Hawaii and do not forget a pleasant stay in Sunshine Hawaii !

Peide (Peter) Ye
General Chair
K. Kakushima
General Co-Chair

Zhihong Chen
Technical Program Chair
Daniel Moraru
Technical Program Co-Chair

2024 IEEE Silicon Nanoelectronics Workshop

Committee Members of the 2024 IEEE Silicon Nanoelectronics Workshop

General Chair
Peide Ye, Purdue University

General Co-Chair
K. Kakushima, Tokyo Institute of Technology

Technical Program Chair
Zhihong Chen, Purdue University

Technical Program Co-Chair
Daniel Moraru, Shizuoka University

Publication Chair
Sumeet Gupta, Purdue University

Treasurer
Vita Hu, National Taiwan University

Program Committee

Yang Chai (PolyU HK)	Byoung Hun Lee (POSTECH)
Mansun Chan (HKUST)	P. W. Li (NYCU)
K. S. Chang-Liao (NTHU)	Ming Liu (Fudan Univ.)
Jiezhi Chen (Shandong Univ.)	T. Marulame (Toshiba)
Woo Young Choi (SNU)	H. Mizuta (JAIST)
Steve Chung (NYCU)	Takahide Oya (YNU)
Kazuhiko Endo (Tohoku Univ.)	Wolfgang Porod (Univ. Notre Dame)
Jacopo Franco (imec)	Takahiro Shinada(Tohoku Univ.)
Xiao Gong (NUS)	T. Takahashi (Univ. Tokyo)
Ru Huang (Peking Univ.)	K. Toprasertpong (Univ. Tokyo)
Louis Hutin (CEA, Leti)	Tiwei Wei (Purdue Univ.)
T. Irisawa (AIST)	T. Yajima (Kyushu Univ.)
Jaehun Jeong (Samsung)	G. Yamahata (NTT)
Masaharu Kobayashi (Univ. Tokyo)	Shimeng Yu (GIT)

Future Perspectives of CMOS Logic Innovation Beyond 2nm

Toshiro Hiramoto[1], Hitoshi Wakabayashi[2]

[1]Institute of Industrial Science, The University of Tokyo, [2]Tokyo Institute of Technology

Email: hiramoto@nano.iis.u-tokyo.ac.jp

Abstract —**The historical trends of advanced CMOS in the past are reviewed and the future technological trends beyond 2nm forecasted by the roadmap [1] are discussed. Metrics for speed and energy tell us that although the progress may slow down, the CMOS advancements will not stop soon. One of the most serious issues to be addressed is a rapid increase in the power density.**

Keywords: CMOS logic, beyond 2nm technology

I. INTRODUCTION

The dramatic advancements of CMOS logic technology were mainly driven by the developments of microprocessors in 1990s and 2000s and by smart phones in 2010s. Recently, the technology driver has obviously been replaced by high performance computing for AI. Fig. 1(a) shows the historical trends of the technology generations (nodes) since 1980 reported in the international conferences (closed blue symbols) [2-23]. It is surprising to see that the technology node has shrunk at almost the same rate over 20 generations during last forty years. At present, the 3nm CMOS platform is available [22,24]. Thanks to increasing strong demands for higher performance, lower energy consumption, higher integration, and lower cost for the abovementioned technology derivers, the CMOS progresses will continue in the future. IRDS forecasted that the technology node will proceed at the same rate as in the past, as shown in Fig. 1(a) [1].

II. TECHNOLOGICAL TRENDS

Fig. 1(b) shows the actual numbers of the gate length (L_g, closed red symbols) and metal pitch (closed green symbols) along with the technology node. The introductions of new innovative technologies including strain silicon (2002), high-k/metal-gate (HKMG, 2007), FinFET (2012), and EUV (2019) are also shown. In 2000s, Lg shrunk very rapidly, and it was smaller than the node number, but the L_g shrink almost stopped by the introduction of HKMG. It again started to shrink by the FinFET introduction. Thus, the technology node does not coincide with L_g. On the other hand, the metal pitch has shrunk continuously as the technology node. This indicates that the node does not express the transistor size but the integration density of transistors. It is well understood that the CMOS technology has progressed not only by simple geometrical scaling but also by various technology boosters [1].

Fig. 1(b) also shows the future forecasts of L_g and metal pitch by IRDS [1] (open symbols). The shrink rates of L_g and metal pitch are slow and will stop in around 2028. The difference between the node and L_g will become larger and larger, and the technology node will not represent the integration level any more in the future. IRDS confronts us with the limitation of conventional two-dimensional scaling.

III. SPEED, ENERGY, AND POWER DENSITY

The intrinsic speed and operation energy consumption of a transistor are roughly given by CV/I and CV^2, respectively, where C is the parasitic capacitance, V the supply voltage, I the drive on-current. For simplicity, C is the gate capacitance including fringing capacitance [1]. Fig. 3(a) shows the trends of CV/I and CV^2 of high performance (HP) transistors [25]. The transistor speed and energy improved rapidly until 2000s mainly thanks to geometrical and V_{dd} scaling. However, the improvements became very slow recently. To look into the reasons, the trends of other parameters: saturation on-current (I_{on}), and off-current (I_{off}), supply voltage (V_{dd}), and oxide thickness (CET) are shown in Fig.3(b-c). The reason is obvious. In 1990s, the constant-field scaling worked perfectly, but V_{dd} was reduced below 1V and I_{off} is also reduced (Vth increased) to suppress dynamic and static power in 2010s, resulting in the temporary decrease in I_{on} and speed. By introducing HKMG, FinFET, and EUV, the performance has been still improved.

Further introductions of new innovative technologies are expected in next 15 years [1]. They include structure innovations such as gate-all-around (GAA) stacked nanosheets, CFET, and 3D stacked VLSI. However, one of the most serious issues in 3D is the increase in power density per area [26]. Fig. 3(d) shows the trends of the power density. Even by 2D scaling only, the power density has increased until 2020. The 3D stacked structure will lead to more drastic increase in power density. The backside interconnect technology will not be the fundamental solution. It is urgent to develop devices with less parasitics, higher mobility, and steep subthreshold swing, and high thermal conductivity. High mobility channels by atomic 2D materials that can be stacked in a 3D manner may be one of the candidates.

IV. SUMMARY

The CMOS technological trends in the past and the future perspectives are reviewed. Supported by increasing demands for higher speed and lower power/cost, the CMOS innovations will continue, but the increasing power density is serious issues to be solved.

REFERENCES

[1] International Roadmap for Devices and Systems (IRDS), 2022 Edition, https://irds.ieee.org/. [2] L. C. Parillo et al., IEDM, p. 706, 1980. [3] L. C. Parillo et al., IEDM, p. 752, 1982. [4] J. Agraz-Guerena et al., IEDM, p. 63, 1984. [5] R. A. Chapman et al., IEDM, p. 362, 1987. [6] J. Hayden et al., IEDM, p. 417, 1989. [7] R. A. Chapman et al., IEDM, p. 101, 1991. [8] M. Bohr et al., IEDM, p. 273, 1994. [9] M. Bohr et al., IEDM, p. 847, 1996. [10] S. Yang et al., IEDM, p. 197, 1998. [11] S. Tyagi et al., IEDM, p. 567, 2000. [12] S. Thompson et al., IEDM, p. 61, 2002. [13] P. Bai et al., IEDM, p. 657, 2004. [14] K. Mistry et al., IEDM, p. 247, 2007. [15] S. Natarajan et al., IEDM, p. 941, 2008. [16] C. Auth et al., VLSI, p. 131, 2012. [17] S. Natarajan et al., IEDM, p. 71, 2014. [18] C. Auth et al., IEDM, p. 673, 2017. [19] S.-Y. Wu et al., IEDM, p.43, 2016. [20] G. Yeap et al., IEDM, p.

979-8-3503-9164-0/24 $31.00 © 2024 IEEE

879, 2019. [21] B. Sell et al., VLSI, p. 282, 2022. [22] S.-Y. Wu, IEDM, p. 639, 2022. [23] J. Jeong et al., VLSI, T1-2, 2023. [24] S. Choi, IEDM, 1-1, 2023. [25] T. Hiramoto, VLSI-TSA, p.3, 2020. [26] H. Wakabayashi and T. Hiramoto, EDTM, 4D-1, 2024.

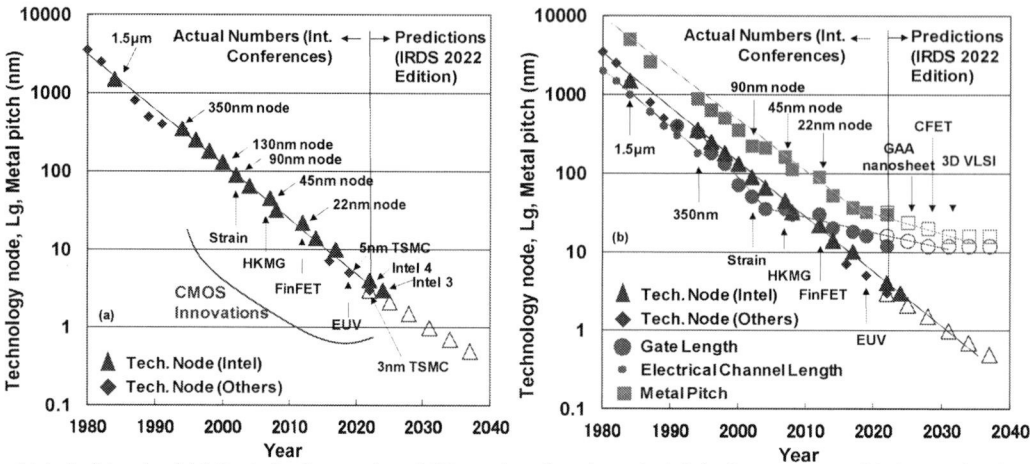

Fig. 1. The historical trends of (a) the technology node and (b) gate length and metal pitch in the past reported in international conferences [2-23]. The data from Intel are mainly taken because Intel reported detailed device parameters in most of generations (closed triangles). Other data are from Bell Lab (1980 and 1982), TI (1987 and 1991), Motorola (1989), and TSMC (2017, 2019, and 2022). The future nodes predicted by IRDS [1] are also shown.

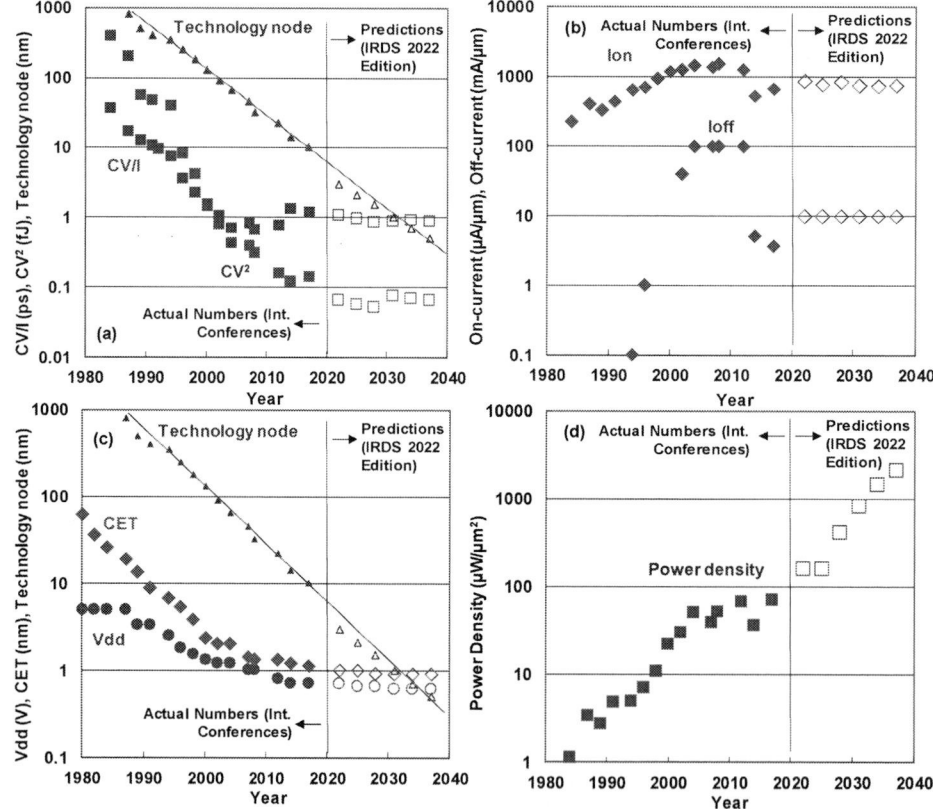

Fig. 3. Trends of (a) speed (CV/I) and energy (CV^2), (d) on-current (I_{on}) and off-current (I_{off}), (c) supply voltage (V_{dd}) and equivalent capacitance thickness (CET), and (d) power density of HP transistors in the past and future. The technology node is also shown in (a) and (c). It is assumed that gate width $W=10L_g$ in planar transistors, fin number reduced from 3 (2012) to 1 (2024), GAA stacked sheet number increased from 3 (2025) to 4 (2031), CFET introduced in 2028, stacked tier number increased from 2 (2031) to 6 (2037), and FO=1. I_{on} and I_{off} are normalized to effective width (W_{eff}), while layout footprint is used calculating the power density.

979-8-3503-9164-0/24 $31.00 © 2024 IEEE

Gate oxide reliability: upcoming trends, challenges, and opportunities

B. Kaczer[1], R. Degraeve[1], J. Franco[1], T. Grasser[2], Ph. J. Roussel[1], E. Bury[1], P. Weckx[1], A. Chasin[1], S. Tyaginov[1], M. Vandemaele[1], A. Grill[1], B. O'Sullivan[1], J. Diaz Fortuny[1], P. Saraza Canflanca[1], M. Waltl[2], P. Rinaudo[1,3], Y. Zhao[1], E. Kao[1,3], R. Asanovski[1], E. Catapano[1,3], A. Beckers[1], A. Vici[1,3], B. Truijen[1], Y. Higashi[1], S. Clima[1], Y. Xiang[1], D. Sangani[1,3], L. Panarella[1,3], Q. Smets[1], T. Knobloch[2], D. Waldhör[2], B. Van Troeye[1], Y. Guo[1,3], A. Kruv[1], K. Viswakarma[1,3], M. Gonzalez[1], D. Linten[1]

[1]imec, Leuven, Belgium; [2]TU Vienna, Austria; [3]KULeuven, Belgium

Any product that stops functioning before its declared useful lifetime will be deemed substandard at best, dangerous or fatal at worst. Integrated circuits (ICs) are typically a part of a larger product, which, depending on the target market, may be expected to operate reliably for up to multiple decades, often in extreme environments, and in mission-critical applications. It is therefore essential that reliability is engineered into the ICs and their individual components already during their development, and not ignored altogether, or merely evaluated as an afterthought.

In particular, the FET gate oxide is the source of multiple mechanisms degrading the FET characteristics over time (Fig. 1) [1,2]. These adverse mechanisms cannot be typically eliminated entirely but need to be adequately suppressed to make the device sufficiently reliable within the required operating specifications. Indeed, some technologies, "such as Ge or GaAs, were not selected [...] mainly due to their lack of a stable native oxide a low defect density interface" [3].

Gate oxide degradation can be traced to charge carrier i) trapping in and ii) leakage through preexisting and newly generated defect states in the oxide and its interfaces (Fig. 2) [4,5]. We argue that only full understanding the detailed properties of these defects, including their *chemistry* [5,6], is the only guaranteed approach to predict with confidence the overall time-to-failure of the device at any $\{V_G, V_D\}$ bias point and temperature T, determined by *circuit simulations* (see e.g. the path in Fig. 1). This process further requires the understanding of the *physics* of all the microscopic mechanisms in the device, which in turn depend on the *electrostatics* of the device, charge carrier and phonon *transport* [7,8], but also local *mechanical strain* [9,10]. Lastly, it requires the understanding of the *statistics* stemming from the distributed properties of the defects (e.g., bond strengths) as well as the stochasticity of the physical mechanisms. The challenge then lies in incorporating all into suitable models, with the sweet spot between accuracy and efficiency for each target user, ranging from full TCAD [11], Comphy/U [12,13], and reliability-enabled compact models [14].

Although multifaceted, this *defect-centric* approach to FEOL reliability reaps enormous benefits e.g. in development of low-T budget gate oxides for future upcoming VLSI technologies. By comparison of defect energies extracted from DFT calculations and from measurements and Comphy [12] simulations, hydroxyl-E' defects in the interfacial SiO_2 layer have been identified as the main culprit of NBTI and suppressed by a novel remote H plasma treatment (Fig. 3) [5]. H release from the gate oxide [15] and its incorporation in the channel also appears to be an additional BTI degradation component in IGZO based devices (Fig. 4) [16].

Temperature is a factor accelerating most degradation mechanisms. Heat is locally generated during FET operation, especially in the HCD regime (cf. Fig. 1). It affects charge carrier transport in the channel [17], as well as defect generation and oxide bulk trapping, and must be considered in TCAD simulations (Fig. 5a) [18]. The exact channel temperature depends on the heat balance, i.e., phonon transport inside the device and outward (Fig. 5b) [19].

At cryogenic temperatures, trapping in high-k oxide defects is observed to persist [20] due to nuclear tunneling (Fig. 6a) [21] through otherwise a sizeable barrier between defects states (~1 eV on average). Time-zero variability, HCD [20] and low-f noise are also observed to increase at cryogenic T's, the latter due to localized defect-like states in the channel (Fig. 6b) [22].

Stochastic variability of the physical degradation mechanisms, such as TDDB (Fig. 7a) [23] and gate oxide trapping (Fig. 7b) [4], is further enhanced by the ever-decreasing FET dimensions and the accompanying increase in as-fabricated variability (Fig. 7b). Characterizing all reliability and variability components is then greatly facilitated by our on-chip "SmartArrays" (Fig. 8) [2,24,25]. These complex arrays are also indispensable for efficient cryogenic measurements.

The fundamental learning from stochastic defect-assisted conduction developed for describing the TDDB time-to-breakdown distributions [26] has been re-applied to the description of so-called moving bits in Flash memories (Fig. 9) [27,28] and RRAM and OTS-based memory elements (Fig. 10) [29, 30].

Gate oxide trapping also plays a central role in InGaAs [31] and SiC power devices [32], in which it constitutes one of the remaining hurdles, as well as in FerroFETs with H(Z)O-based gate oxide, where trapping (Fig. 11) [33] in and near the interfacial layer [34] is crucial, and in fact essential for proper switching of the ferroelectric polarization [35]. Significant trapping in the gate oxide and its interfaces, but also likely in the channel itself (Fig. 12), presently hinders operation of 2D Transition Metal Dichalcogenide (TMD) FETs [36] and, more practically, mere extraction of the fundamental device parameters.

Thorough and systematic understanding of the degradation mechanisms and the defect properties, however, can be also used to our benefit in many novel applications. Our on-chip odometer (Fig. 13) [37] can potentially detect illicitly (re)used IC's, while our soft-BD based PUF solution (Fig. 14) can be used to generate unique IC "fingerprints" for challenge-response transactions [38]. Finally, trapping (NBTI) in a handful of FETs has been proposed, simulated, and tested in 28 nm FETs to enable an entire solution with one-shot learning and subject recognition (Fig. 15) [39].

[1] A. Chasin *et al.*, IEDM (2017); [2] E. Bury *et al.*, IRPS (2019); [3] M. L. Green *et al.*, JAP (2001); [4] B. Kaczer *et al.*, Microel. Rel. (2018); [5] J. Franco *et al.*, TED (2023); [6] T. Grasser, Microel. Rel. (2012); [7] M. Bina *et al.*, TED (2014); [8] R. Hussin *et al.*, SISPAD (2015); [9] A. Kruv *et al.*, IRPS (2020); [10] K. Lee *et al.*, EDL (2021); [11] M. Vandemaele *et al.*, IRPS (2019); [12] D. Waldhör *et al.*, Microel. Rel. (2023); [13] Z. Wu *et al.*, IRPS (2021); [14] P. Weckx *et al.*, IRPS (2017); [15] T. Grasser *et al.*, IEDM (2015); [16] A. Chasin *et al.*, IEDM (2021); [17] M. A. Alam *et al.*, TED (2019); [18] S. Tyaginov *et al.*, IRPS (2022); [19] E. Bury *et al.*, IEDM (2016); [20] A. Grill *et al.*, IRPS (2020); [21] J. Michl *et al.*, TED Part I & II (2021); [22] R. Asanovski *et al.*, TED (2023); [23] A. Vici *et al.*, TED (2023); [24] P. Saraza-Canflanca *et al.*, IRPS 2023; [25] P. Saraza-Canflanca *et al.*, IRPS (2024); [26] S. Sahhaf *et al.*, TED (2009); [27] R. Degraeve *et al.*, TED (2004); [28] D. Ielmini *et al.*, TED (2002); [29] N. Raghavan *et al.*, VLSI (2013); [30] R. Degraeve *et al.*, IRPS (2021); [31] V. Putcha *et al.*, IRPS (2017); [32] T. Grasser *et al.*, IRPS (2024); [33] M. N. K. Alam *et al.*, JEDS (2019); [34] B. O'Sullivan *et al.*, APL (2020); [35] K. Toprasertpong *et al.*, Appl. Phys. A (2022); [36] M. Waltl *et al.*, Adv. Mat. (2022); [37] J. Diaz-Fortuny *et al.*, IRPS (2023); [38] K.-H. Chuang *et al.*, JSSC (2019); [39] Y. Guo *et al.*, IRPS (2024); [40] E. Simoen *et al.*, ECS Trans. (2011); [41] B. Kaczer *et al.*, IRPS (2015); [42] Y. Xiang *et al.*, TED (2021); [43] A. Gaur, PhD Thesis (2020); [44] H. Ravichandran *et al.*, ACS nano (2023);

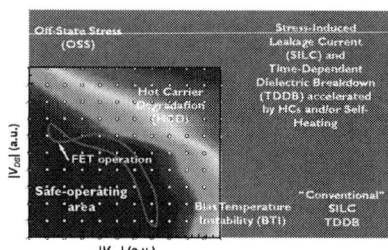

Fig. 1: Degradation mechanisms in FETs taking place in the $\{V_G, V_D\}$ space are caused by trapping in existing and by generation of new defects. For reliable operation, device biases, determined by the encompassing circuit, need to be kept within the Safe Operating Area (dark blue).

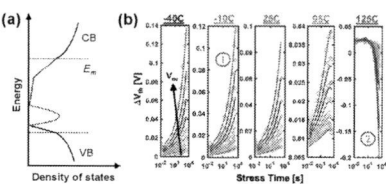

Fig. 4: (a) Complex density of states in InGaZnO used in novel DRAM and BEOL FETs results in carrier transport properties dissimilar to standard semiconductors. (b) "Standard" PBTI trapping in gate oxide (positive ΔV_{th}) (1) is surpassed at elevated Ts by release of H from oxide and its embedding in the IGZO channel, where it acts as a donor (negative ΔV_{th}) (2) [16].

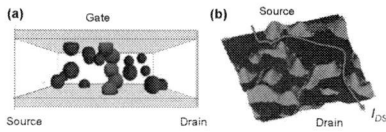

Fig. 7: Percolation in deeply downscaled devices in (a) gate-oxide due to defect generation, contributing to TDDB variability [23], and (b) in channel transport due to random channel potential variations enhanced by charged defects [4], contributing to time-zero, RTN [40], BTI [4], HCD [41] variability.

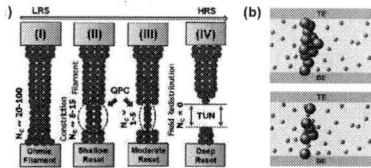

Fig. 10: Two-terminal memory elements, including (a) RRAM [29] and (b) OTS [30], rely on defects forming conduction paths between electrodes. (b) Defects switching between insulating (localized, ~ small conduction radius) states and conductive (delocalized, ~ large conduction radius) states contribute to OTS.

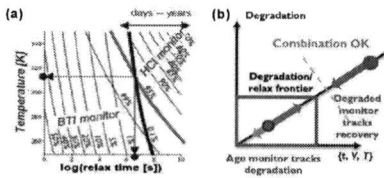

Fig. 13: A silicon odometer to monitor the cumulative use of an IC, e.g. to prevent its failure in mission-critical applications, employs (a) "triangulation" of use time based on different temporal and thermal properties of BTI and HCD mechanisms. (b) Possible tampering, such as annealing of the IC, is detected with a second, pre-stressed monitor [37]

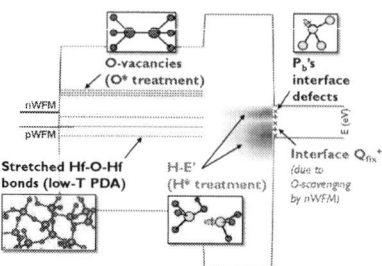

Fig. 2: Defects in FET high-k stack contributing to degradation. Both defect generation and trapping can be modeled as (multiple) state transitions with (sizable) internal barriers, stimulated by electric field, with a contribution of thermalized ("cold") or energetic ("hot") carriers [5].

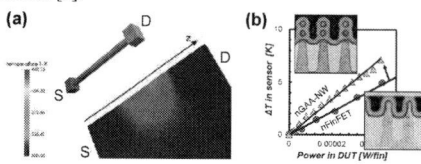

Fig. 5: (a) TCAD simulation of lattice temperature distribution in GAA-NW and FinFET due to Self-Heating in the HCD regime [18]. (b) Increased self-heating in GAA-NW, measured in an adjacent sensor, is caused by a reduced heat conduction path into the substrate [19].

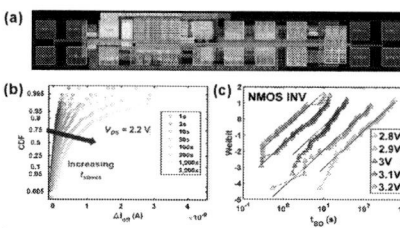

Fig. 8: (a) Complex, in-house designed on-chip "SmartArray" circuits allow extracting time-zero and degradation *distributions* from 1000s of FETs at an arbitrary $\{V_G, V_D\}$ bias point, including (b) OSS [25] and (c) TDDB [24] (cf. Fig. 1).

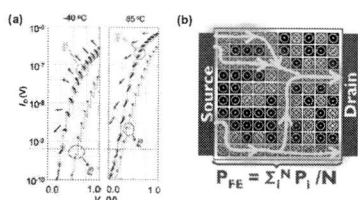

Fig. 9: (a) "Moving bits" distributions in Flash cells due to anomalous charge loss through (b) percolation paths formed by 2 aligned gate oxide defects [28]. Loss of ~10^3 electrons can be easily detected [27].

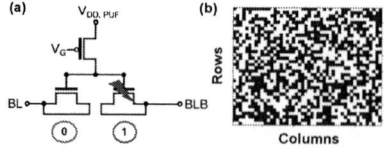

Fig. 11: (a) Interplay between (de)trapping and ferroelectric gate oxide (de)polarization during I_D-V_G Forward and Reverse sweeps (FS and RS) of a FeFET can result in regions of steep sub-threshold (during RS) [33]. (b) Modeling of this effect includes percolative channel conduction caused by ferroelectric oxide crystallinity [42].

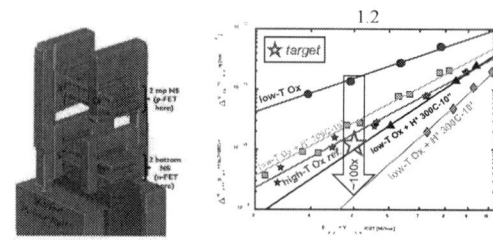

Fig. 3: Tight vertical pitches and complex integration flows of (a) Complementary FET (CFET) will require simple, sacrificial gate-free gate stacks fabricated with low thermal budgets, such as those annealed with remote H or O plasma. Such oxides offer equal or superior BTI (b), HCD, and TDDB reliability with respect to the currently used high-thermal budget oxides [5].

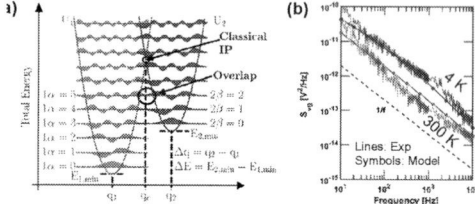

Fig. 6: At cryogenic temperatures (a) trapping does not necessarily freeze out due to appreciable nuclear tunneling between (neutral and charged; red and blue) states [21], while (b) at 4 K, a considerable contribution to low-frequency noise comes from band-tail states formed by random variations in channel potential (cf. Fig. 7b), while at 300 K, the source of low-f noise is gate oxide defects [22].

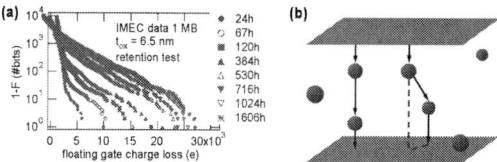

Fig. 12: In TMD 2D FETs, (a) fixed charges (1), border traps (2), interface and channel residues (3), non-passivated bonds (4) and point defects in the channel (5) contribute to mobility degradation and significant reliability issues [43]. (b) Giant Random Telegraph Noise (RTN) manifestation in our deeply scaled devices [44] due to non-uniform, percolative carrier transport (cf. Fig. 7b) in channel modulated by defects and TMD film corrugation.

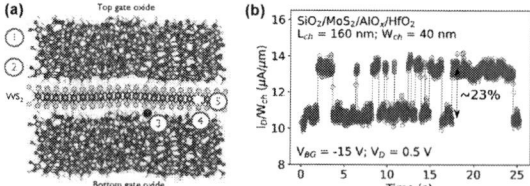

Fig. 14: (a) A 1-bit cell of a designed and fabricated PUF array. Entropy is generated by the competition of TDDB between two n-channel FETs. The p-channel FET driver then prevents a second breakdown. (b) An array of cells provides chip fingerprint with outstanding reliability, uniqueness, and unpredictability [38].

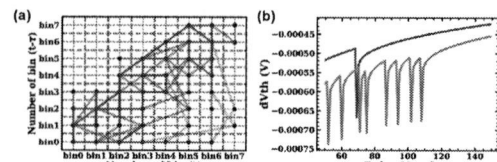

Fig. 15: Implementation of a reservoir computer employing gate-oxide trapping (specifically, NBTI) for one-shot learning. (a) Example of a subject-specific signal converted to phase space. (b) Comphy simulation of V_{th}-shift output from a single phase-space bin (= 1 pFET) resulting from voice data from two subjects, each lasting 10-15 s [39].

979-8-3503-9164-0/24 $31.00 © 2024 IEEE

Heterogeneous CFET: A Pathway to CMOS Performance Balance Engineering

Xin-Ren Yu[1], Wen-Hsiang Lu[1], Shu-Wei Chang[2], Wen-Hsin Chang[3], Darsen D. Lu[1], Yao-Jen Lee[4,5*], Tatsuro Maeda[3], Wen-Fa Wu[5], Yi-Ming Li[6], Yeong-Her Wang[1]

[1]Inst. of Microelectronics, Natl. Cheng Kung University, Tainan, Taiwan, [2]Dept. of Electrical Engineering, Natl. Cheng Kung University, Tainan, Taiwan, [3]National Institute of Advanced Industrial Science and Technology (AIST), Tsukuba, Japan, [4]Inst. of Pioneer Semiconductor Innovation, National Yang Ming Chiao Tung University, Hsinchu, Taiwan, [5]Taiwan Semiconductor Research Inst. (TSRI), Hsinchu, Taiwan, [6]College of Electrical and Computer Engineering, National Yang Ming Chiao Tung University, Hsinchu, Taiwan

Email: *yjlee1976@nycu.edu.tw;

Abstract

The CFETs can provide high-performance characteristics with significant area reduction for Å technology nodes. However, mobilities between n/p FETs need to be considered for the performance balance of CMOS. This paper introduces heterogeneous and hetero-orientational CFETs using wafer bonding and layer-transfer techniques.

Introduction

To adapt to low-energy-consuming units such as the Internet of Things (IoT) and high-performance units such as High-Performance Computing (HPC), the design needs to pay more attention to the distribution of circuit characteristics, such as smaller threshold voltage, better output characteristics, and electrostatic characteristics.

Therefore, Nano-sheets have been proposed as an alternative transistor structure [1]. By vertically stacking, the number of channels can be doubled or more. Overall, this has opened interest in stacking transistor structures into 3-D structures. One of the highest-density stacked structures is the complementary transistor (CFET), which vertically stacks different types of channels through the epitaxial or wafer bonding process. The schematic is shown in Fig.1 [2-6].

However, many technical difficulties must be overcome to stack transistors in three dimensions: mobility and threshold voltage balance between n/p FETs, device isolation, thermal budget control, etc. Therefore, this paper discusses the manufacturing method that uses wafer bonding technology and bonds two different materials together through the front-end process. This can maximize the fabrication's flexibility.

Low-temperature Heterogeneous Bonding Technique

To monolithically fabricate CFET devices, we will perform wafer bonding directly on the un-patterning wafer. Taking the GeOI as an example, the process will start with the epitaxy of the donor wafer and perform CMP on it. Planarization followed by deposition of bonding oxide. Then, to achieve a sufficiently high bonding force under low-temperature annealing (<300 °C), plasma activation of the surface is performed, thereby increasing the number of dangling bonds on the surface. After the wafers are bonded, the backside of the donor wafer will be removed. During this process, we used SOI wafers to obtain the crucial etching stop layer for sufficiently high uniformity. At the end of the process, CMP will be used again to flatten the surface. The process flow is shown in Fig 2.

In addition to bonding heterogeneous materials, wafer bonding also removes the defective epitaxial buffer layer. Fig. 3 (a) shows Rocking curves; (b) & (c) shows the RSM analysis before and after Ge thinning down. The reduced radius and FWHM in RSM and rocking curves indicate the complete removal of the defective Ge layer. Therefore, our previous studies verified that layer-transferred GeOI has superior characteristics beyond those direct epitaxy on Si; the TEM image and I_D-V_G curve are shown in Figs.4 & 5 [7].

Heterogeneous Bonding CFET

Compared to other CFET processes, LT-HBT greatly simplifies the process of CFETs. At the gate region, both the top and bottom channels are commonly controlled after channel release; at the S/D regions, the interlayer of bonding oxide remains; thus, the S/D regions are fully isolated.

The first demonstration of bonding CFET is Ge/Si CFET [8]. To minimize the complexity of CFET fabrication, inversion mode (IM) p-Ge and junctionless (JL) n-Si are adopted. The cross-sectional TEM image is shown in Fig.6 (a); the top-view SEM image is shown in Fig.6 (b), and the inset shows the cross-sectional SEM image of the Ge side S/D contact region, which the Ge and Si layer is isolated by the bonding oxide. The I_D-V_G curve of Ge/Si CFET is shown in Fig.7.

To solve the mobility balance issue in CFET structures, the bonding technique can integrate channels in different orientations with superior mobility into a single CFET structure. Fig.8 (a) shows the cross-sectional TEM image of (111) n-Ge / (100) p-Ge CFET [9]. Both traditional Selected Area Diffraction (SAD) and Converged Beam Electron Diffraction (CBED) have been adopted to distinguish orientation at a nano-meter scale, as shown in Fig.8 (b)-(d). Fig. 9 shows the ID-VG curve of homo-oriented & hetero-oriented Ge/Ge CFET, indicating that the changing channel direction is dramatic.

Finally, we also tried to expand the existing CFET architecture into higher-density 3D CFET SRAM. In [10], we stack IGZO on CFET and apply its extremely high switching characteristics to Pass Gate (PG), effectively reducing static power consumption, as shown in Figs.10 & 11.

Table. I show the difference between monolithic integration, sequential integration, and bonding CFET. This highlights FEOL wafer bonding as a worthy technology for next-generation devices.

Conclusion

This work discusses the applications of bonding technology to heterogeneous/hetero-orientational CFETs, which have the advantage of removing defects due to the epitaxy process. It introduces a variety of previously published and latest bonding CFET structures. In conclusion, through proper design, bonding CFETs will have many development possibilities for the Å technology nodes.

ACKNOWLEDGMENT

This work was supported by the National Science and Technology Council, Taiwan, under Grant No. NSTC 112-2221-E-A49 -168 -MY3, 112-2218-E-006-009-MBK, and 113-2634-F-A49-008 -.

Reference

[1] N. Loubet et al., 2017 VLSI [2] Ryckaert J. et al., 2018 VLSI [3] W. Rachmady et al., 2019 IEDM [4] C. -Y. Huang et al., 2020 IEDM [5] S.-W.Chang et al., 2021 IEDM [6] S.-K. Kim et al., 2022 IEDM [7] X.-R. Yu et al., 2023 IEDM [8] T.-Z. Hong et al., 2020 IEDM [9] X.-R. Yu et al., 2022 VLSI [10] X.-R. Yu et al., 2022 IEDM

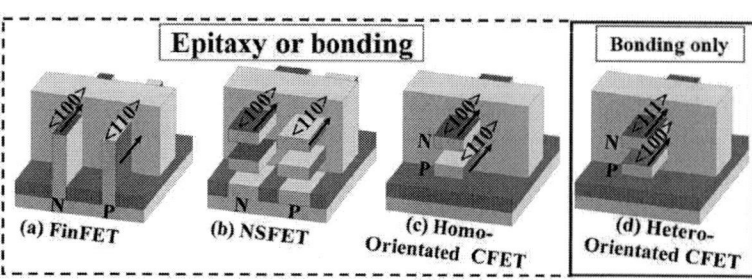

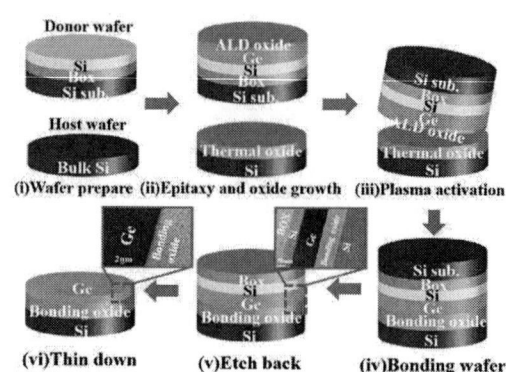

Fig. 1 The schematic diagram of four type of transistors: Homogeneous channel structure of (a) FinFET (b) NSFET and (c) CFET can be fabricated through epitaxial growth or bonding technology. Heterogeneous channel structure of (d) n-Ge (111)/p-Ge (100) CFET can be realized through bonding technology only.

Fig. 2 The schematic diagram of wafer bonding process for GeOI. (i) Wafer preparation; (ii) Epitaxy and oxide growth; (iii)plasma treatment; (iv) Bonded wafer; (v) Etch back; (vi) Thin down.

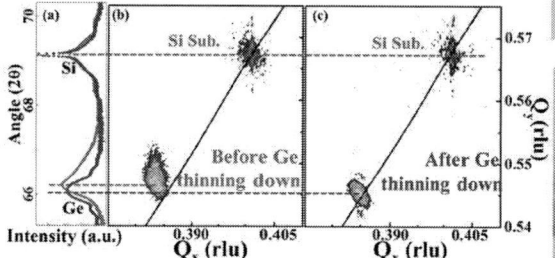

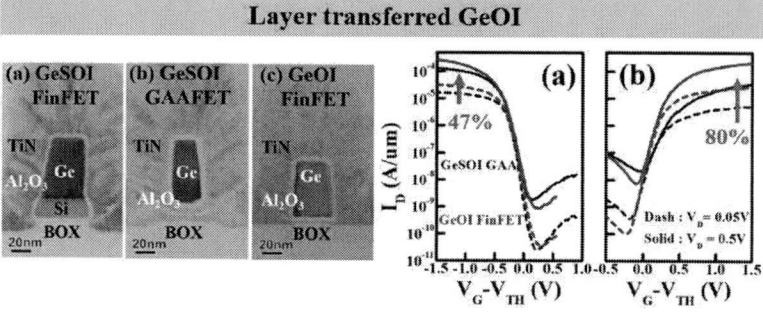

Fig. 3 (a) Rocking curves. (b) & (c) RSM analysis before and after Ge thinning down.

Fig. 4 Cross-sectional TEM image of (a) GeSOI FinFET; (b) GeSOI GAAFET; (c) GeOI FinFET.

Fig. 5 I_D-V_G curve of GeOI FinFET and GeSOI GAAFET (a) for pFET; (b) for nFET.

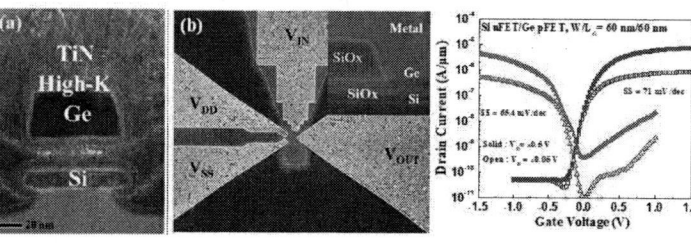

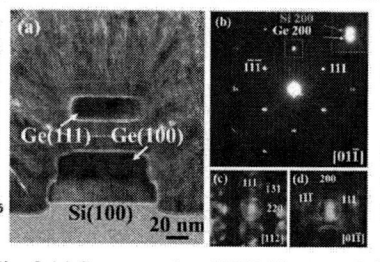

Fig. 6 (a) Cross-sectional TEM image of Ge-Si CFET after HKMG. (b) Top-view SEM image of Ge/Si CFET device. The inset is the cross-sectional SEM image of Ge side S/D contact region.

Fig. 7 I_D-V_G curve of Ge/Si CFET.

Fig. 8 (a)Cross-sectional TEM image of the (111)Ge/(100)Ge CFET after HKMG. (b) SAD pattern; CBED pattern of (c)Ge (111) and (d) Ge (100).

Fig. 9 I_D-V_G curve of homo-oriented/hetero-oriented Ge/Ge CFET.

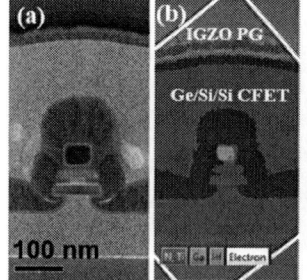

Fig. 10 (a) Cross-sectional TEM image & (b) EDS mapping for heterogeneous 3D CFET SRAM.

Fig. 11 Top-view SEM image of single unit heterogeneous 3D CFET SRAM.

	Monolithic integration	Sequential integration	Monolithic bonding tech.
Technology	Epitaxy	Deposit, Bonding	Epitaxy, bonding
Pros	High density	High mobility	High density, high mobility
Cons	Low flexibility, High leakage	Thermal budget issue	High cost
Ref.	[2,4]	[3,5,6]	[7,8,9]

Table. I Comparison table for monolithic integration; sequential integration and monolithic bonding tech.

Enhanced Electrical Performance of Ultrathin Body Nanosheets

Bo-Wei Huang[1], Yu-Rui Chen[1], Tao Chou[2], Hsin-Cheng Lin[1], Chien-Te Tu[1], Yi-Chun Liu[1], Wan-Hsuan Hsieh[1],
Wei-Jen Chen[1], Min-Kuan Lin[2], Ying-Qi Liu[4], Li-Kai Wang[3], Hung-Chun Chou[2], Yi Huang[3], Ding-Wei Lin[1], and C. W. Liu[1,2,3*]

[1]Graduate Institute of Electronics Engineering, [2]Graduate School of Advanced Technology,

[3]Graduate Institute of Photonics and Optoelectronics, and [4]Department of Materials Science and Engineering

National Taiwan University, Taipei, Taiwan. *E-mail: cliu@ntu.edu.tw

Abstract — Ultrathin body has the immunity to subthreshold swing (SS) degradation by doping and D_{it}, approaching the ideal SS of 60mV/dec. The large effective bandgap (E_g) by the quantum confinement results in high I_{ON}/I_{OFF}. The large E_g weakens the impact ionization to improve the high breakdown voltage. The $Ge_{0.9}Sn_{0.1}$ ultrathin bodies are realized by co-optimization between CVD epitaxy and etching. The nearly ideal SS of 64mV/dec is achieved. The compressive strain of GeSn channel after channel release increases with decreasing body thickness. The ~2nm ultrathin body has a compressive strain of 3.3%. The high breakdown voltage of 11.7V is also achieved, as compared to 6V breakdown voltage of 10nm nanosheets.

Keywords: GeSn, ultrathin body, nanosheets, selective isotropic dry etching

Introduction

The large D_{it} of non-Si channels is an important issue, resulting in a large subthreshold swing (SS) [1, 2]. Ultrathin bodies have the immunity to SS degradation caused doping and D_{it} [3]. Radical-based highly selective isotropic dry etching (HiSIDE) was used to form the GeSn channels without ion damage [3-5]. Ge-based high mobility channels also suffer the large I_{OFF} because of the low bandgap [6-17]. GeSn GAA ultrathin bodies were demonstrated recently to reduce the I_{OFF} due to widened bandgap (E_g) by quantum confinement [4, 5]. Strained GeSn has higher mobility than Ge [1-17] to conquer the mobility degradation by surface roughness. The high output power is important for edge applications in 6G wireless communication. Ultrathin bodies can achieve high breakdown voltage (V_{BD}) by reducing the body thickness (T_{body}).

In this work, the $Ge_{0.9}Sn_{0.1}$ ultrathin bodies with T_{body} down to ~2.4nm and ~2nm are demonstrated to reach low SS, high I_{ON}/I_{OFF}, and high V_{BD}. The strain after the channel release is also simulated by ANSYS. Based on TCAD simulation considering impact ionization, the V_{BD} increases with the decreasing of T_{body} consistent with our experimental results. Note that the S/D and ultrathin body channels are fabricated by the same epilayers without S/D regrowth (**Fig. 1**).

Epilayer Design and Device Fabrication

After the SOI was thinned down, the undoped Ge buffer was grown and then followed by 800°C *in-situ* annealing to improve the epi quality. The $Ge_{0.9}Sn_{0.1}$ channel sandwiched by $Ge_{0.97}Sn_{0.03}$ caps, Ge caps, and Ge:B sacrificial layers (SLs) were grown repeatedly on Ge buffer. For 8 stacked $Ge_{0.9}Sn_{0.1}$ ultrathin bodies, the 11nm $Ge_{0.9}Sn_{0.1}$ channel layers were grown (**Fig. 2**). The 4nm $Ge_{0.9}Sn_{0.1}$ epilayers (**Fig. 3**) were designed to achieve even smaller ultrathin bodies. After channel release, the ~2nm $Ge_{0.9}Sn_{0.1}$ channel has the strain of 3.3% by simulation (**Fig. 4**) due to micro bridge effect. The [B] in Ge:B SLs is as high as $2E21 cm^{-3}$ to reduce the S/D resistance, and the doping in the $Ge_{0.9}Sn_{0.1}$ channel layers is $<1E18 cm^{-3}$ to suppress the impurity scattering for high hole mobility (**Fig. 5**).

After CVD epitaxy, SiO_2 was deposited as hard mask, and e-beam lithography was used to define channels and S/D regions. Cl_2-based RIE was then used to form the fin structure, followed by the FOX definition. HiSIDE was used in the channel release process. The etching selectivity of Ge over $Ge_{0.9}Sn_{0.1}$ is attributed to the 10% [Sn] in the channels. The Sn was reacted with the neutral F radicals from NF_3 etchant to form non-volatile SnF_x to passivate the channels, resulting in a lower etching rate than Ge SLs [3-5]. The double $Ge_{0.97}Sn_{0.03}$ and Ge caps are used to prevent channels from bending and buckling. The 10-cycle TMA passivation was performed to reduce the D_{it} [18, 19]. The *in-situ* $TiN/ZrO_2/Al_2O_3$ was deposited by PEALD for 8 stacked GeSn ultrathin bodies. For 2 stacked ultrathin bodies, *in-situ* $TiN/Hf_{0.2}Zr_{0.8}O_2$ was conformally deposited around channels by PEALD. 400°C FGA was then performed. A thick TiN was deposited by PVD as gate metal pad. Finally, Pt was deposited by PVD on Ge:B SLs to form S/D contact. 400°C PMA was performed for low S/D resistance.

Device Performance and experiment result

The unintentional doping in the channel could result in an electric field and voltage drops on the gate dielectric, making SS nonideal. However, nearly ideal SS can be obtained by ultrathin body with Dir and channel doping (**Fig. 6**). The TCAD simulation also shows the SS decreases with the scaled T_{body} (**Fig. 7**). The ~2.4nm ultrathin bodies have the low SS of 64mV/dec and I_{ON}/I_{OFF} of $1.6x10^7$ at V_{OV}= -0.5V (**Fig. 8 (a)**) with channel width (W_{CH}) =30nm and average T_{body}=2.4nm (**Fig, 8 (b)**). The $Ge_{0.9}Sn_{0.1}$ ultrathin bodies of two stacked channels have I_{ON}/I_{OFF} $\geq 1.7x10^7$ at V_{DS}= -0.5V (**Fig. 9**). The TEM images and HR-TEM is a similar device with T_{body}~2nm (**Fig. 9(b, c, d)**). A ~10nm thick nanosheet, a ~2.4nm ultrathin body, and a ~2nm ultrathin body at V_{OV}= -1.8V have V_{BD}= -6V, -7.7V, and -11.7V, respectively (**Fig. 10**). The V_{BD} is the V_{DS} when I_D per stack achieved 150μA before dielectric breakdown. The V_{BD} increases with the decreasing T_{body}. The TCAD simulation of breakdown is shown in **Fig. 11** using Impact ionization. With the thinner T_{body}, the effective E_g is larger due to quantum confinement effect [3-5]. The impact ionization rate also consequently reduced [20, 21], resulting in high V_{BD}.

Conclusion

The GeSn ultrathin bodies are demonstrated with low SS of 64mV/dec, high I_{ON}/I_{OFF} $\geq 1.7x10^7$, and high V_{BD}. Ultrathin bodies nanosheets enhance the electrical properties. Body thickness is another knob to tune the bandgap for future technology applications.

Acknowledgment

NSTC (112-2218-E-002-024-MBK, 113-2634-F-A49-008 (T-Star center project)) MOE (NTU-CC-113L893401), and TSRI, Taiwan.

References

[1] Y.-S. Huang *et al.*, *VLSI*, 2019, pp. T180. [2] Y.-S. Huang *et al.*, *IEDM*, 2019, pp. 689. [3] C.-E. Tsai *et al.*, *VLSI*, 2022, pp. 401. [4] C.-E. Tsai *et al.*, *IEDM*, 2021, pp. 569. [5] B.-W. Huang *et al.*, *TED*, 2022, 69, 4. [6] S. Gupta *et al.*, *IEDM*, 2011, pp. 398. [7] G. Han *et al.*, *IEDM*, 2011, pp. 402. [8] X. Gong *et al.*, *VLSI*, 2012, pp. 99. [9] M. Liu *et al.*, *VLSI*, 2014, pp. 80. [10] D. Lei *et al.*, *VLSI*, 2018, pp. T197. [11] K. Han *et al.*, *VLSI*, 2019, pp. T182. [12] Y.-S. Huang *et al.*, *IEDM*, 2017, pp. 832. [13] Y.-S. Huang *et al.*, *EDL*, 39, 9, 2018. [14] Y.-S. Huang *et al.*, *IEDM*, 2020, pp. 23. [15] C.-T. Tu *et al.*, *TED*, 68, 4, 2021. [16] Y.-S. Huang *et al.*, *IEDM*, 2016, pp. 822. [17] Y.-S. Huang *et al.*, *TED*, 64, 6, 2017. [18] T.-E. Lee *et al.*, *VLSI*, 2019, T100. [19] T.-E. Lee *et al.*, *TED*, 67, 10, 2020. [20] J. -W. Han *et al.*, *EDL* 2007. [21] B. Kumar *et al.*, *Silicon* 2021.

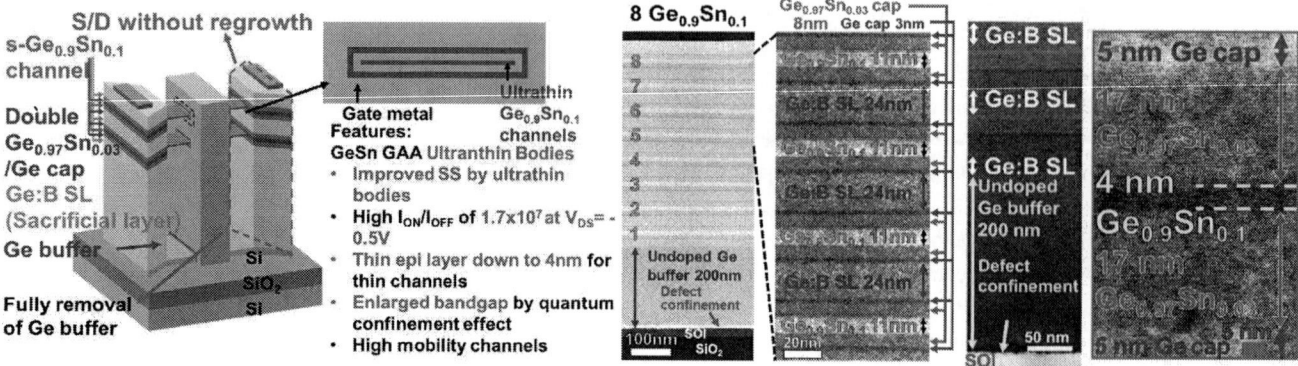

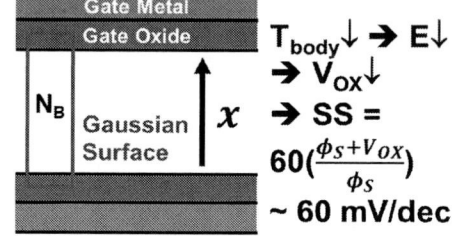

Fig. 1. 3D schematics of Ge$_{0.9}$Sn$_{0.1}$ GAA ultrathin bodies with the key points highlighted.

Fig. 2. TEM of the 8 Ge$_{0.9}$Sn$_{0.1}$ channel layers with Ge$_{0.97}$Sn$_{0.03}$/Ge caps for ultrathin

Fig. 3. TEM of the 2 Ge$_{0.9}$Sn$_{0.1}$ channel layers down to 4nm for ultrathin bodies.

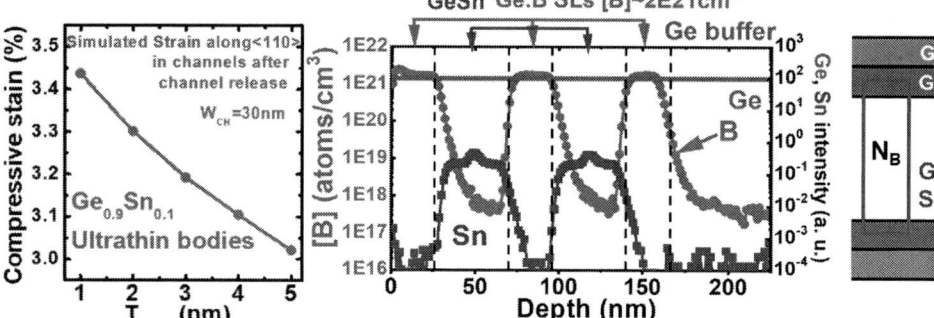

Fig. 4. Simulated strain of Ge$_{0.9}$Sn$_{0.1}$ ultrathin body by ANSYS.

Fig. 5. SIMS of the 2 stacked Ge$_{0.9}$Sn$_{0.1}$ epilayers. [B] in the Ge SLs is as high as 2E21cm^{-3} for low S/D resistance.

Fig. 6. Theoretical concepts of SS. Nearly ideal SS can be achieved by ultrathin bodies with less dopant in the channel

$$T_{body}\downarrow \rightarrow E\downarrow \rightarrow V_{ox}\downarrow \rightarrow SS = 60\left(\frac{\phi_s+V_{ox}}{\phi_s}\right) \sim 60 \text{ mV/dec}$$

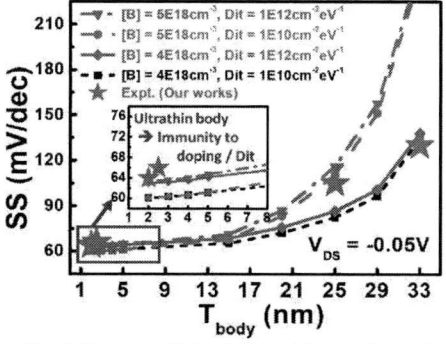

Fig. 7. Simulated SS vs T$_{body}$ with experimental data. The ultrathin bodies have high immunity to SS degradation caused by doping and D$_{it}$.

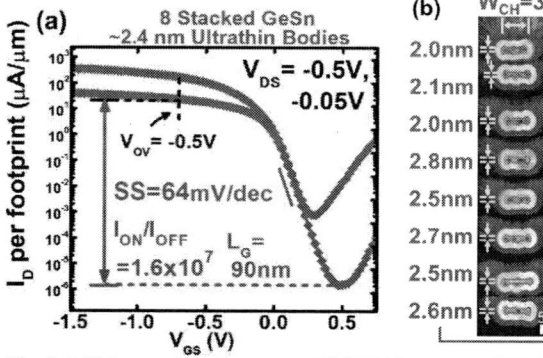

Fig. 8. (a) Measured I$_D$-V$_{GS}$ curve, (b) TEM images, and EDS mapping of 8 stacked GeSn ultrathin bodies. The average T$_{body}$ is 2.4nm. Nearly ideal SS of 64mV/dec is achieved.

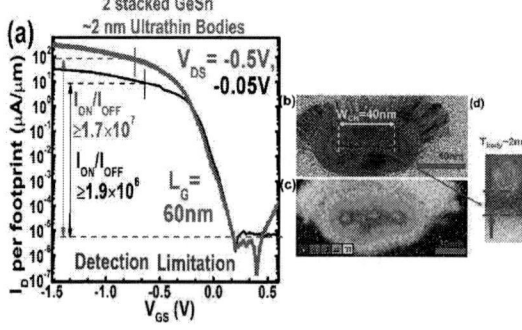

Fig. 9. (a) Measured I$_D$-V$_{GS}$ curve, (b) TEM images, (c) EDS mapping, and (d) enlarged TEM images of 2 stacked GeSn ultrathin bodies.

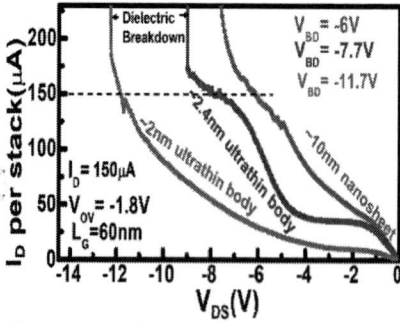

Fig. 10. Breakdown voltage enhancement by ultrathin bodies. The V$_{BD}$ is defined as I$_D$=150µA.

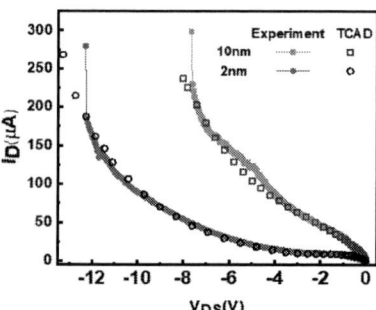

Fig. 11. Simulated breakdown characteristics of the 10nm GeSn nanosheets and the 2nm GeSn ultrathin bodies.

979-8-3503-9164-0/24 $31.00 © 2024 IEEE

Integration Technology for Superconducting Quantum Circuits

Shiro Kawabata[1,2]

[1]Global Research and Development Center for Business by Quantum-AI Technology (G-QuAT), National Institute of
Advanced Industrial Science and Technology (AIST), Japan
[2]Faculty of Computer and Information Sciences, Hosei University, Japan
Email: kawabata@hosei.ac.jp

Abstract — **Superconducting quantum circuits are one of the promising platforms for fault-tolerant universal quantum computers (FTQCs) and quantum annealing machines. We will report our recent activities on integration technology for large-scale superconducting quantum circuits, including 3D integration, fabrication, SPICE simulation, and superconducting control circuits. In addition, theoretical studies on superconducting bosonic qubits and quantum algorithm for FTQC will be presented.**

Keywords: Superconducting quantum circuit, quantum computer, quantum annealing, 3D integration, bosonic qubit, superconducting control circuits

I. INTRODUCTION

Superconducting qubits are the most promising candidates for fault-tolerant quantum computers (FTQCs) and quantum annealing machines. Here, we would like to review our recent activities on integration technology for superconducting quantum circuits. In addition, we will overview theoretical studies on superconducting bosonic qubits and quantum algorithms for FTQCs.

II. SUPERCONDUCTING QUANTUM CIRCUITS

Since the experimental realization of a superconducting charge qubit in 1999 by NEC in Japan, fundamental technologies of superconducting quantum circuits have been developed. In 2011, D-wave systems announced a world-first commercial superconducting quantum annealing machine "D-wave one" which is based on based on Nb superconducting flux qubits. After that, UCSB John Martinis's group has realized highly coherent transmons (a single qubit fidelity of 99.92% and a two-qubit gate fidelity of 99.4%) and scalable FTQC architecture based on transmons.

Nowadays, Noisy Intermediate-Scale Quantum (NISQ) devices using superconducting qubits have been realized by many companies and institutes, e. g., IBM (1,121 qubits), CAS (504 qubits), Beijing University (136 qubits), Rigetti computing (84 qubits), Google (72 qubits), and RIKEN /FUJITSU (64 qubits). In addition, internet access to real superconducting NISQ processors is available through quantum cloud service, e. g., AWS Amazon Braket, IBM-Q, Google Quantum Cloud, Microsoft Azure Quantum, and Alibaba Quantum Cloud. In Japan, RIKEN will be constructed quantum-classical hybrid cloud system with FUGAKU (classical high-performance computer) and IBM

superconducting quantum computer "Heron" (133 qubits) soon.

III. INTEGRATION TECHNOLOGY

To realize practical FTQCs with a million qubits or more, many technical issues need to be resolved. One of the main challenges is integration technology for superconducting large-scale quantum circuit. Here we will overview for our recent activities of superconducting 3D integration (**Fig. 1**) [1-3], qubit fabrication [4], SPICE simulation for quantum circuit design [5], low temperature operation of superconducting quantum annealing machine (**Fig. 2**) [6-9], and cryogenic superconducting control circuit [10, 11]. In addition, we will show our recent theoretical studies on superconducting bosonic qubits [12-18] and FTQC quantum algorithm for radiative transfer equation (**Fig. 3**) [19].

ACKNOWLEDGMENTS

This paper is partly based on the results obtained from a project, JPNP16007, commissioned by the New Energy and Industrial Technology Development Organization (NEDO), Japan.

REFERENCES

[1] W. Feng, K. Kikuchi, M. Hidaka, H. Yamamori, Y. Araga, K. Makise, and S. Kawabata, "*Thermal management of a 3D packaging structure for superconducting quantum annealing machines*", Appl. Phys. Lett. vol. 118, pp. 174004 (2021).

[2] Y. Araga, H. Nakagawa, M. Hashino, and K. Kikuchi, "*Demonstration of 90000 superconductive bump connections for massive quantum computing*", Jpn. J. Appl. Phys. vol. 62, pp. SC1094 (2023).

[3] M. Fujino, Y. Araga, H. Nakagawa, Y. Takahashi, K. Nanba, A. Yamaguchi, A. Miyata, T. Nishi, and K. Kikuchi, "*Nb–Nb direct bonding at room temperature for superconducting interconnects*", J. Appl. Phys. vol. 133, pp. 015301 (2023).

[4] Y. Urade, K.Yakushiji, M. Tsujimoto, T. Yamada, K. Makise, W. Mizubayashi, and K. Inomata, "*Microwave characterization of tantalum superconducting resonators on silicon substrate with niobium buffer layer*", APL Mater., vol. 12, pp. 021132 (2024).

[5] T. Tanamoto, T. Ishikawa, K. Inomata, S. Masuda, T. Onuma, and S. Kawabata, "*Classical SPICE simulation of superconducting quantum circuits*", Appl. Phys. Express, vol. 16, pp. 034501 (2023).

[6] D. Saida, Y. Yamanashi, M. Hidaka, F. Hirayama, K. Imafuku, S. Nagasawa, and S. Kawabata, "*Experimental Demonstrations of Native Implementation of Boolean Logic Hamiltonian in a Superconducting Quantum Annealer*", IEEE Transactions on Quantum Engineering, vol. 2, pp. 3103508 (2021).

D. Saida, M. Hidaka, K. Imafuku, and Y. Yamanashi, FIG.1 Superconducting quantum annealing machine (6 Nb flux qubit) based on Nb flux qubit and Nb multilayer process.

[7] *"Factorization by quantum annealing using superconducting flux qubits implementing a multiplier Hamiltonian"*, Sci. Reports, vol. 12, pp. 13669 (2022).

[8] D. Saida, M. Hidaka, K. Miyake, K. Imafuku, and Y. Yamanashi, *"Superconducting quantum circuit of NOR in quantum annealing"*, Sci. Reports, vol. 12, pp. 15894 (2022).

[9] D. Saida, M. Hidaka, and Y. Yamanashi, *"4-bit Factorization Circuit Composed of Multiplier Units with Superconducting Flux Qubits toward Quantum Annealing"*, arXiv:2308.06566 (2023).

[10] N. Takeuchi, T. Yamae, W. Luo, F. Hirayama, T. Yamamoto, and N. Yoshikawa, *"Scalable flux controllers using adiabatic superconductor logic for quantum processors"*, Phys. Rev. Research, vol. 5, pp. 013145 (2023).

[11] N. Takeuchi, T. Yamae, T. Yamamoto, N. Yoshikawa, *"Scalable quantum-bit controller using adiabatic superconductor logic"*, arXiv:2310.06544 (2023).

[12] T. Kanao, S. Masuda, S. Kawabata, and H. Goto, *"Quantum Gate for Kerr-Nonlinear Parametric Oscillator Using Effective Excited States"*, Phys. Rev. Applied, vol. 18, pp. 014019 (2022).

[13] Y. Suzuki, S. Watabe, S. Kawabata, S. Masuda, Y. Suzuki, S. Watabe, S. Kawabata, and S. Masuda, *"Measurement-based state preparation of Kerr parametric oscillators"*, Sci. Reports, vol. 13, pp. 1606 (2023).

[14] Y. Suzuki, S. Kawabata, T. Yamamoto, and S. Masuda, *"Quantum state tomography for Kerr parametric oscillators"* Phys. Rev. Applied, vol. 20, pp. 034031 (2023).

[15] T. Aoki, T. Kanao, H. Goto, S. Kawabata, and S. Masuda, *"Control of the ZZ coupling between Kerr-cat qubits via transmon couplers"*, Phys. Rev. Applied, vol. 21, pp. 014030 (2024).

[16] T. Yamaji, S. Masuda, A. Yamaguchi, T. Satoh, A. Morioka, Y. Igarashi, M. Shirane, and T. Yamamoto, *"Correlated Oscillations in Kerr Parametric Oscillators with Tunable Effective Coupling"*, Phys. Rev. Applied, vol. 20, pp. 014057 (2023)

[17] Y. Mori, K. Nakaji, Y. Matsuzaki, S. Kawabata, *"Expressive Quantum Supervised Machine Learning using Kerr-nonlinear Parametric Oscillators"*, Quantum Machine Intelligence vol. 6, pp.14 (2024).

[18] Y. Mori, Y. Matsuzaki, S. Endo, S. Kawabata, *"Hardware-Efficient Bosonic Quantum Computing with Photon-loss Detection Capability"*, arXiv:2403.00291 (2024).

[19] A. Igarashi, T. Kadowaki, S. Kawabata, *"Quantum Algorithm for Radiative Transfer Equation"*, Phys. Rev. Applied, vol. 21, pp. 034010 (2024).

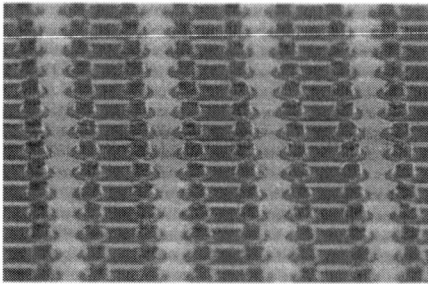

Fig. 1 Superconducting flip-chip bonding through In/Pb bumps for superconducting quantum circuits.

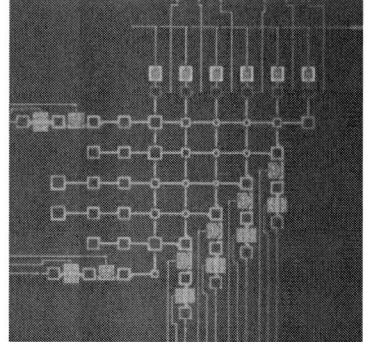

Fig. 2 Superconducting quantum annealing machine (Nb flux qubit) based on the Nb multilayer fabrication process.

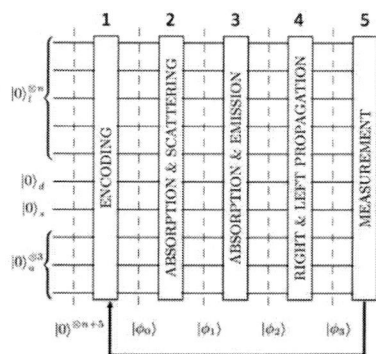

Fig. 3 Quantum algorithm for solving radiative transfer equation.

Demonstration of 99.9% single qubit control fidelity of a silicon quantum dot spin qubit made in a 300 mm foundry process

N. Dumoulin Stuyck[1,2], M.K. Feng[1,2], W. H. Lim[1,2], S. Serrano Ramirez[1,2], C. C. Escott[1,2], T. Botzem[1,2], T. Tanttu[1,2], C. H. Yang[1,2], A. Saraiva[1,2], A. Laucht[1,2], S. Kubicek[3], J. Jussot[3], S. Beyne[3], B. Raes[3], R. Li[3,*], C. Godfrin[3], D. Wan[3], K. De Greve[3,4], and A. S. Dzurak[1,2]

[1]Diraq, Sydney, NSW, Australia
[2]School of Electrical Engineering and Telecommunications, UNSW, Sydney, NSW 2052, Australia
[3]IMEC, Leuven, Belgium
[4]KU Leuven, Leuven, Belgium, *Now at Applied Materials, Inc.
Email: nard@diraq.com, andrew@diraq.com, kristiaan.degreve@imec.be

Abstract – **Commercially viable quantum computing will require large numbers of high-quality qubits. Therefore, leveraging the extensive capabilities of conventional CMOS technology is expected to accelerate the development of a fault-tolerant quantum computer. Here we report on the highest single-qubit control fidelity of 99.9% for any Si/SiO_2 spin qubit fabricated in a 300 mm wafer process. This result represents an important milestone towards a large-scale silicon quantum processor on a single chip.**

INTRODUCTION Spin qubits formed at the Si/SiO_2 interface by charge accumulation with patterned gate electrodes are promising candidates for large-scale quantum computing due to their extensive overlap with existing CMOS technologies. Prototype spin qubit devices fabricated at academic cleanroom facilities with customized fabrication flows have shown to exhibit long coherence times and high-fidelity qubit control [1], even at elevated temperatures above 1 K [2]. However, to fully leverage the existing CMOS industry for scaling up spin qubit quantum processors, an important milestone is to demonstrate that spin qubits fabricated in a 300mm foundry environment can exhibit control fidelities compatible with quantum error correction thresholds. Here, we report the control fidelity of a single Si-MOS spin qubit surpassing 99.9% as analysed using two different and widely used benchmarking techniques. To achieve this high level of control, we employ real-time feedback protocols with state-of-the-art FPGA-based controllers and control pulse shaping [3,4].

RESULTS AND DISCUSSIONS - The device is fabricated in a planar 300 mm process using an epitaxially grown 800 ppm ^{28}Si substrate and a flexible hybrid DUV/electron-beam lithography fabrication flow which were extensively discussed before [5]. Key to the qubit performance is a 20 nm high-quality thermally grown Si/SiO_2 interface. The device layout, as sketched in **Fig. 1,** consists of a charge sensor, or single electron transistor (SET), which is patterned in proximity to a double quantum dot (QD). All measurements are performed in a ^{3}He/^{4}He cryostat with a base temperature of 180 mK [1,6]. SET transport measurements, **Fig. 2,** reveal Coulomb blockade and a lever arm of 0.04. The charge noise spectrum with a low level of 0.4 μeV at 1 Hz, shown in **Fig. 3**, indicates a high quality Si/SiO_2 interface as explored previously in devices with SiO_2 thicknesses of 8 nm and 12 nm [3,4]. Qubit initialization and readout, using Pauli Spin Blockade (PSB), is performed at the (3,1)-(4,0) electron number transition with (P_1,P_2) indicating charge numbers under gates P1/P2, see **Fig. 4** [2]. **Fig. 5** shows coherent spin manipulation by applying and ac current to a patterned ESR-antenna on resonance with the Larmor frequency of 18.89 GHz, set by the global magnetic field of 0.7 T. Coherent driving is observed with increasing applied microwave power showing Rabi frequencies (f_{Rabi}) up to 2 MHz and Q-factors, defined as the product between Rabi coherence time and f_{Rabi}, up to 100 as shown in **Fig 6,** and selected qubit metrics are given in **Table 1**. We employ a Gaussian single-sideband modulation control voltage pulse shape to minimize crosstalk, which would impact the second electron spin and PSB readout and analyse the single qubit control fidelity (F_{SQ}) using two community standard benchmarking techniques. With randomized benchmarking (RBM), **Fig. 7.**, we extract a Clifford gate fidelity of 99.91±0.01%. Using Gate Set Tomography (GST) we extract single qubit fidelities of $99.97^{+0.03}_{-0.04}$ %and 99.97±0.03% for $X_{\pi/2}$ and $Y_{\pi/2}$ gates, respectively, summarized in **Table 2**. The low Hamiltonian and high stochastic contributions in the GST analysis suggest that further improvements can be made by increasing qubit coherence, for example by using a ^{28}Si substrate with increased purification levels - above the level currently used.

CONCLUSION - In this work, we demonstrated 99.9% single qubit gate fidelity measured by the two most used benchmarking techniques. This is the highest single qubit fidelity measured on a 300 mm Si/SiO_2 spin qubit to date and an important milestone towards fault-tolerant quantum computing by integrating large-scale quantum processors using industrial semiconductor facilities.

ACKNOWLEDGEMENTS

We acknowledge support from the Australian Research Council (FL190100167 and CE170100012).

REFERENCES

[1]   M. Veldhorst *et al.*, "A two-qubit logic gate in silicon," *Nature*, vol. 526, no. 7573, pp. 410–414, 2015, doi: 10.1038/nature15263.

[2]   J. Y. Huang *et al.*, "High-fidelity spin qubit operation and algorithmic initialization above 1 K," *Nature*, vol. 627, no. 8005, pp. 772–777, Mar. 2024, doi: 10.1038/s41586-024-07160-2.

[3]   N. Dumoulin Stuyck *et al.*, "Silicon spin qubit noise characterization using real-time feedback protocols and wavelet analysis," *Appl. Phys. Lett.*, vol. 124, no. 11, p. 114003, Mar. 2024, doi: 10.1063/5.0179958.

[4]   L. M. K. Vandersypen and I. L. Chuang, "NMR techniques for quantum control and computation," *Rev. Mod. Phys.*, vol. 76, no. 4, pp. 1037–1069, Jan. 2005, doi: 10.1103/RevModPhys.76.1037.

[5]   A. Elsayed et al., "Low charge noise quantum dots with industrial CMOS manufacturing." arXiv, 2022. doi: 10.48550/ARXIV.2212.06464.

[6]   A. Elsayed *et al.*, "Comprehensive 300 mm process for Silicon spin qubits with modular integration," in *2023 IEEE Symposium on VLSI Technology and Circuits (VLSI Technology and Circuits)*, Jun. 2023, pp. 1–2. doi: 10.23919/VLSITechnologyandCir57934.2023.10185272.

[7]   T. N. Camenzind et al., "High mobility SiMOSFETs fabricated in a full 300 mm CMOS process," Materials for Quantum Technology, vol. 1, no. 4. IOP Publishing, p. 041001, Dec. 01, 2021. doi: 10.1088/2633-4356/ac40f4.

979-8-3503-9164-0/24 $31.00 © 2024 IEEE

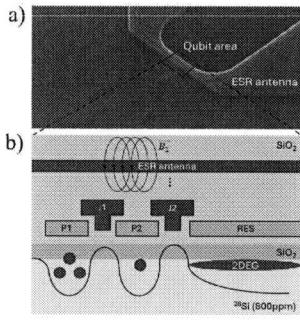

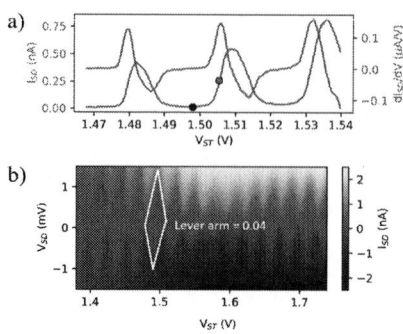

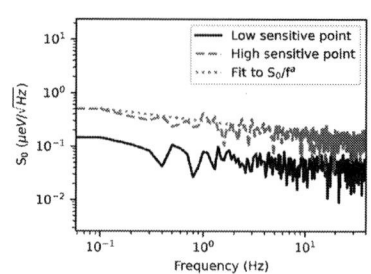

Figure 1. a) CDSEM of a 300 mm spin qubit device. **b)** Schematic cross-section of the device under test depicting the double quantum dot potentials (not to scale). We operate the qubit device by first loading 3 (1) electrons under the P1 (P2) gate and then depleting the two-dimensional electron gas (2DEG) under the J2 and RES gate [2]. Fabrication and device details can be found in [5,6].

Figure 2. a) Charge sensor dc and conductance measurements. **b)** Coulomb diamonds with an extracted lever arm of 0.04, in line with results from 12 nm SiO$_2$ devices [5]. Details on the methods can be found in in [5].

Figure 3. Noise spectral density of the charge sensor for two operation points indicated in **Fig.2a**. SET current fluctuations are converted to energy level fluctuations using dI$_{SD}/d$V$_{ST}$ and extracted lever arm, **Fig. 2** [5]. Dotted line is a fit to S_0/f^{α} with $\alpha = 0.23$.

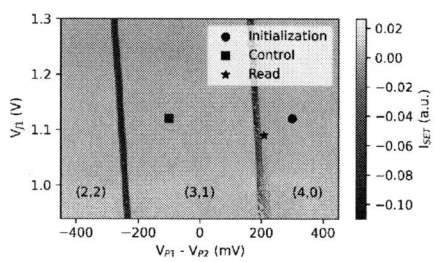

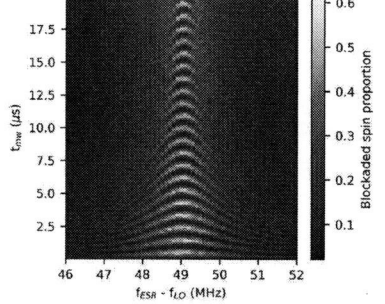

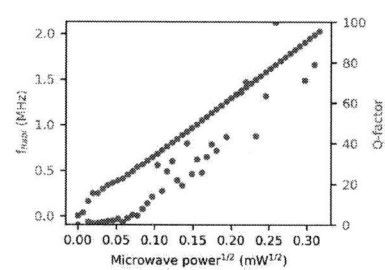

Figure 4. Charge stability map across the (P$_1$,P$_2$) = (2,2) to (4,0) charge occupation regimes [2]. The circle, square and star symbols indicate the initialization, control, and read point, respectively.

Figure 5. Single qubit Rabi chevron demonstrating coherence spin oscillations using the on-chip ESR antenna. f_{LO} is the microwave source local oscillation frequency of 18.44 GHz set by the global magnetic field of 0.7T.

Figure 6. Rabi frequency and Q-factor vs applied microwave power to the electron spin resonance antenna on-chip. Q-factors are calculated from the fitted Rabi frequency and coherence as $f_{Rabi} \cdot T_2^{Rabi}$.

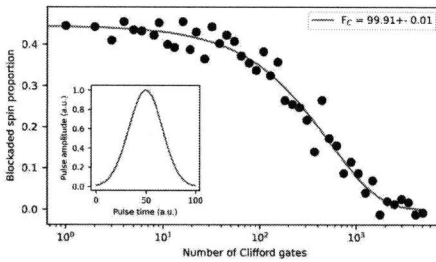

Figure 7. Randomized benchmarking data for 400 randomly generated sequences up to 5000 Clifford gates length with each Clifford comprise of $X_{\frac{\pi}{2}}$ and $Z_{\frac{\pi}{2}}$ single qubit gates.

Inset: Gaussian pulse shape used for qubit gate control voltage pulses.

Table 1. Selected single qubit metrics; Ramsey (T_2^*), Hahn echo (T_2^{Hahn}) and Rabi (T_2^{Rabi}) coherence times [4].

Metric	Value (µs)
T_2^*	4.8 ± 0.7
T_2^{Hahn}	105 ± 11
T_2^{Rabi}	146 ± 8

Table 2. Single qubit gates fidelity and error contribution (EC) estimates using Gate Set Tomography (GST) analysis.

Qubit Gate	$X_{\frac{\pi}{2}}$	$Y_{\frac{\pi}{2}}$
Fidelity estimate	$99.98 \pm 0.04\%$	$99.95 \pm 0.07\%$
Total Hamiltonian EC	$2 \times 10^{-6} \pm 0.0002$	$(1 \pm 6) \times 10^{-5}$
Total stochastic EC	0.0002 ± 0.0007	0.0005 ± 0.002

979-8-3503-9164-0/24 $31.00 © 2024 IEEE

Gap in pagination due to unavailable paper.

Pages 13-14

IGZO Channel VCT(Vertical Channel Transistor) Technology for sub-10nm DRAM : Challenges and Opportunites

Yongjin Lee, Daewon Ha, W. Lee, S. Yoo, J.H. Bae, M.H. Cho, K. Yoo, S.M. Lee, S. Lee, M. Terai, T.H. Lee, K.J. Moon, C. Sung, M. Hong, D.G. Cho, H. Kim, J. H. Seo, K. Park, B.J. Kuh, S. Hyun, S.J. Ahn, and J.H. Song
Semiconductor R&D Center, Samsung Electronics Co. Ltd., Gyeonggi-do, Korea, Email: daewon.ha@samsung.com

Abstract — **In DRAM technology at nodes below 10 nm, scaling down conventional architectures presents significant challenges due to uprising physical limit on integration and electrical properties. As alternatives, innovative structures incorporating deposition channels, such as VCT with IGZO channels, are being explored. This paper presents the IGZO channel VCT as a promising candidate for next-generation DRAM cell architecture with several technical challenges.**
Keywords: DRAM, Oxide semiconductor, VCT, Contact Resistance, Thermal stability, Reliability

I. INTRODUCTION

The landscape of DRAM development has undergone a significant transition from an $8F^2$ planar configuration to a more advanced $6F^2$ cell structure, largely driven by the adoption of the buried cell array transistor (BCAT) architecture as depicted in **fig. 1**. The industry has witnessed a leap in patterning precision with the development of Extreme Ultraviolet lithography, enabling continuous device miniaturization. However, scaling down below the 10 nm threshold have faced substantial barriers due to the intrinsic limitations of physical patterning, leading to the detrimental short channel effect (SCE). In response, the exploration of a $4F^2$ architecture is emerging as a strategic intermediary step, anticipating the eventual transition towards a fully realized 3D stack architecture, marking a critical evolution in DRAM development history.

II. IGZO VERTICAL CHANNEL TRANSISTOR

Fabricating VCT structure requires a channel material suitable for deposition. IGZO stands out as an optimal candidate for VCT DRAM cells due to its extremely low leakage current without floating body effects (FBE) and gate-induced drain leakage current (GIDL) [1]. Furthermore, utilizing traditional silicon wafers as Front-End-Of-Line (FEOL) devices, in conjunction with the application of IGZO as Back-End-Of-Line (BEOL) devices, facilitates the implementation of a Cell-over-Periphery architecture (**fig. 2**). This approach enables simultaneous integration of silicon and IGZO while capitalizing on their distinct material properties in semiconductor device fabrication.

III. EXPERIMETNAL RESULT AND DISCUSSION

A. Electrical characteristics and Contact resistance
Fig. 3a illustrates the transfer characteristics of IGZO VCT measured at 85°C for V_{DS}=1.0V and 0.05V, displaying a large on-off ratio of 12 orders of magnitude, a SS of 104 mV/dec, and DIBL of 230 mV/V. **Fig. 3b** indicates that at elevated gate voltages, the device does not present an ideal linear relationship, suggesting Schottky characteristics. This phenomenon becomes more pronounced with decreasing temperature, likely due to thermionic emission transport within the contact region, as shown in **fig. 4**.

In emerging devices, device performance below the 100nm scale is increasingly determined by a contact resistance rather than a channel mobility [2]. The formation of metal oxide layers during IGZO channel deposition can significantly impact contact resistance. Therefore, it is crucial to minimize oxidation during the integration process. Contact resistance values, extracted using a $1/V_{OV}$ method [3], highlight the sensitivity of bit-line contact resistance to various IGZO channel deposition processes (fig. 6), underscoring the importance of controlling metal oxide formation for optimal device performance.

B. Thermal Stability and Reliability
Unlike the utilization of IGZO material in the display industry, memory device fabrication requires higher process temperatures and higher operation voltages which pose significant challenges in terms of thermal stability and reliability. **Fig. 7a** demonstrates that an enhanced hydrogen anneal process induces a negative shift in V_T without any SS deterioration, suggesting that hydrogen either acts as a shallow donor or passivates acceptor-like defects [4]. Thus, the incorporation of hydrogen during the integration process is critical for controlling V_T in the design and operation of IGZO-based DRAM devices. Additionally, **fig. 7b** shows that annealing temperatures beyond the BEOL process temperature result in V_T and SS deterioration. In the evaluation of negative bias temperature instability (NBTI), the V_T shift exhibits a pronounced power-law relationship with respect to time, as quantitatively indicated by the slope evident in **fig. 8**. This slope for IGZO significantly exceeds that observed in Si DRAM, underscoring the importance of controlling time-related degradation factors to ensure long-term reliability in DRAM technologies.

IV. SUMMARY

High performance IGZO VCT has been experimentally demonstrated for sub-10nm DRAM, addressing technical challenges as an intermediary step beyond Moore's scaling.

REFERENCES

[1] D. Ha *et. al.*, IEDM 2023, pp. 6.3.1-6.3.4
[2] S. Das *et. al.*, Nature Electronics 2021, 4 pp. 786-799
[3] S. Yoo *et. al.*, VLSI 2024 accepted
[4] K. Nomura *et. al.*, Solid State Sci. Technol 2013, pp. 5-8

979-8-3503-9164-0/24 $31.00 © 2024 IEEE

4.1

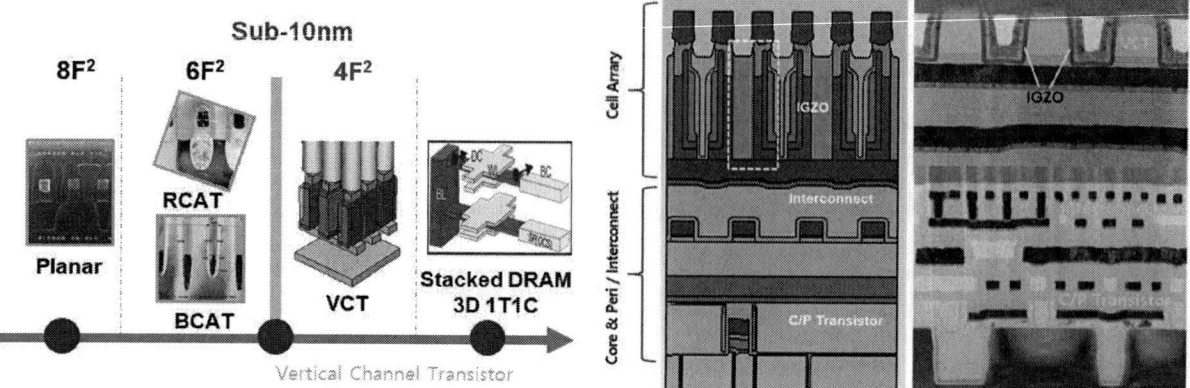

Fig. 1. DRAM Cell Transistor Evolution: $8F^2$ Planar, conventional $6F^2$ BCAT, $4F^2$ VCT and 3D stacked DRAM.

Fig. 2. Schematic and TEM image of cell-over-peripheral (COP) DRAM architecture.

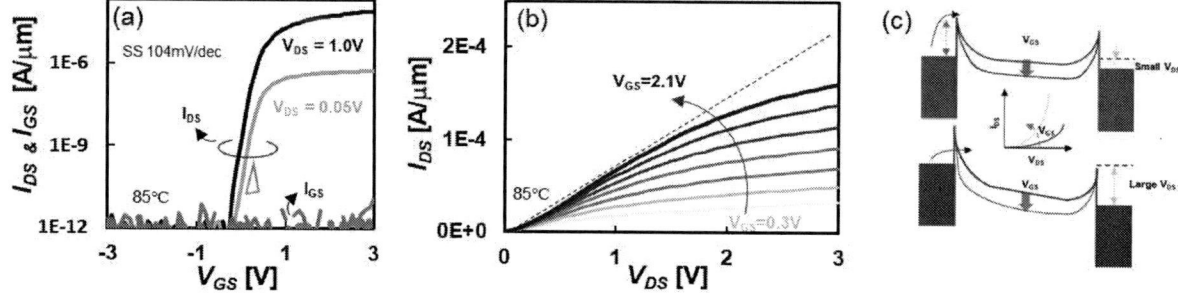

Fig. 3. Electrical characteristics of VCT. (a) Transfer curve (I_{DS}-V_{GS}) at high and low V_{DS} and (b) Output (I_{DS}-V_{DS}) curves at V_{GS} = 0.3 to 2.1V in steps of 0.3V. (c) Enery band diagram of Schottky contact.

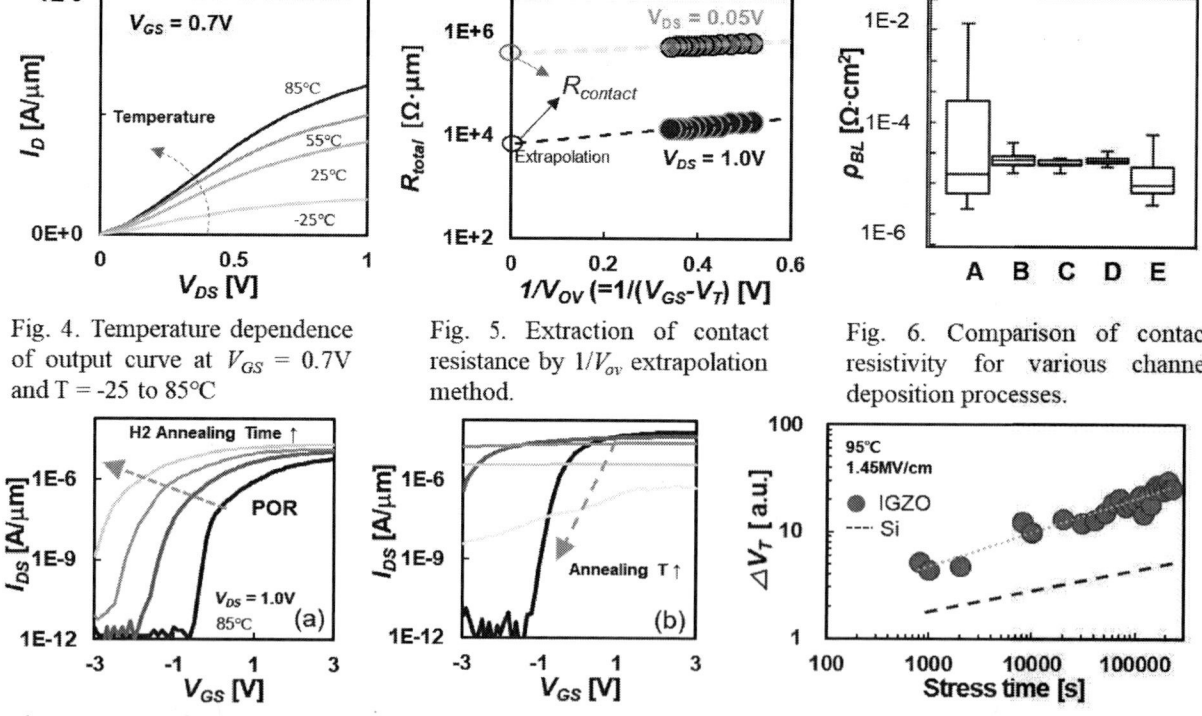

Fig. 4. Temperature dependence of output curve at V_{GS} = 0.7V and T = -25 to 85°C

Fig. 5. Extraction of contact resistance by $1/V_{ov}$ extrapolation method.

Fig. 6. Comparison of contact resistivity for various channel deposition processes.

Fig. 7. Comparison of Transfer curve (I_{DS}-V_{GS}) (a) as a function of annealing time in H_2 ambient and (b) annealing temperature in N_2 ambient.

Fig. 8. Comparison of V_T shift as a function of NBTI stress time for IGZO and Si. Note that $\triangle V_T$ has a power-law dependence with respect to time.

979-8-3503-9164-0/24 $31.00 © 2024 IEEE   16

A Novel 2-Tier Stacked Vertical Channel GAA Transistor Architecture with Area-Saving 3-Dimensional Local Interconnect to Achieve Gb-Density 6T SRAM with Acceptable Low Standby Chip Leakage (~1mA)

Hang-Ting Lue, Weichen Chen, Teng-Hao Yeh, Keh-Chung Wang, and Chih-Yuan Lu

Macronix International Co., Ltd., 16 Li-Hsin Road, Hsinchu Science Park, Hsinchu, Taiwan.
E-mail: htlue@mxic.com.tw

Abstract

Among memory devices, 6T SRAM stands out as the premier classic memory, offering the best switching speed and bandwidth performances. Although CMOS will continue to scale to ~1nm node, the scaled CMOS devices have quite high leakage currents that make Gb-density SRAM a very high-power consuming memory at standby. In this paper, we introduce a groundbreaking 2-tier stacked vertical-channel Gate-All-Around (GAA) device, achieving high-density SRAM equivalent to N1 node (simulation unit cell area ~0.006um²). A novel 3D local interconnect efficiently connects contacts and routing between 3 devices (PG, PU, PD) without footprint wastage. The vertical-channel device, with a long channel length (Lg=30nm) and suitable drain-gate offset (Ls=20nm), significantly reduces Gate-Induced Drain Leakage (GIDL) to ~0.2pA per transistor at Vcc. This enables designing a 1Gb standalone SRAM memory chip with affordable standby leakage at only ~1mA level, which is already smaller than DRAM refresh power. Very high-speed switching time (~ns) together with very low switching energy per bit (~31fJ/bit) can be achieved to meet ultra high-bandwidth memory.

I. Introduction

The classic 6T SRAM is the best-performance computing memory. Compared with DRAM, SRAM does not need refresh and write back, and the read/write speed is only in the nano second range. There is no retention issue when the Vcc is sustained. However, the drawback of SRAM is the large cell size and high standby leakage. In **Fig. 1(a)**, the 6T SRAM has many contacts and complex metal routing that restrict the cell size scaling. Furthermore, to shrink CMOS aggressively for Lg ~10nm often produces very high leakage current at nearly nA range per transistor. This produces nearly ~1A standby leakage when producing Gb-density SRAM. The junction contact and spacer process are all very critical and withdraw sizable foot print. The issues can be resolved by adopting vertical-channel CMOS. This paper propose a novel approach to satisfy scaling and low leakage.

II. Structure Explanations

Figure 1(b) illustrates the proposed structure. There are a total of 4 holes in a unit SRAM cell, where there are stacked 2-tier CMOS connected by a junction arranged in the vertical way. There are a total of 4 NMOS and 4 PMOS, where we shunt 2 PMOS together to provide double currents to match NMOS to better design 6T SRAM.

The proposed manufacturing process can be modified from 3D NAND manufacturing concept. We can deposit multi-layer ONOPONO (PS: "N" stands for the sacrificial nitride, "P" stands for the sacrificial poly), and use selective wet etching method to remove SiN to become a metal gate for CMOS, while to remove poly to make it a contact metal. The beauty of this structure is that the contact in the middle layer between two-tier CMOS automatically connect the junctions between PG, PD, PU devices in the SRAM circuit, without sacrificing additional footprint. Meanwhile, the gates of M2/M3 and M4/M5 are connected in a straightforward way. This largely saves the metal routing complexity. The "all-around contact" for this vertical-channel structure is naturally lower-resistance than the conventional 2D planar contact.

A three-layer staircase process is needed to complete the SRAM routing. There are several critical dimensions, listed in the **Fig. 1(c)** for various technology nodes. The lateral dimension includes L1~L5 for isolation rules, L6 for GAA diameter, and GOX thickness. Vertical dimension includes Lg for channel length, Lsd for contact height, Ls for drain-to-gate offset. For cell footprint scaling, L1 ~L5 consumes the most area, while they are only about isolation and process accuracy, not related to device. For device characteristics, the most important parameters Lg and Ls are vertical dimension without foot print concern.

We can enlarge the Lg and Ls for smaller CMOS leakage.

In **Fig. 1(c)**, we list the proposed dimension to fit the same SRAM cell size for CMOS technology roadmap from N7 to N1. Scaling are mostly relevant to the isolation rule (L1 ~L5) shrinkage, while the key device parameters Lg and Ls remain constant. GAA dimeter is scaled to 12nm for N1 node.

Figure 1(d) illustrates the plane-view of unit cell. For N1 node simulation, the X/Y size are 248nm and 24nm respectively, producing equivalent cell size =0.006um². **Figure 1(e)** illustrates the 3D view of the 6T SRAM, including metal routing to complete the staircase and nodes a/b. **Figure 1(f)** illustrate the array with many WL's and BL/BLB. **Figure 1(g)** illustrate the top view to connect Vcc and BL/BLB.

We do not yet complete the whole processing to form the 6T SRAM, but we have previously developed a 3-WL stacked GAA transistor for GCT DRAM [1] for a process reference in this paper. **Figure 2(a)** shows the 3-layer staircase process. **Figure 2(b)** shows the plane-view of the GAA transistor. **Figure 2(c)** shows the cross-sectional view of selective epi growth channel for the stacked GAA device.

III. Device characteristics and TCAD Simulation:

We first based on our previous measured N/P-channel vertical-GAA devices in [1,2] to calibrate the basic parameters including mobility, Vt tuning, and GIDL parameters to ensure the correctness of TCAD simulation for various dimension. 3D TCAD automatically includes the structure capacitance thus memory switching timing and displacement current can be used to calculate the energy consumption.

In **Figs. 3(a) and (b)**, we can adjust the drain-to-gate offset length from 10nm to 20nm. In **Fig. 3(c)**, at Lg=30nm, the GIDL current can be greatly suppressed to ~0.2pA for PMOS. In **Fig. 4(d)**, the leakage current is reduced to < 0.2pA for NMOS. This means that to produce Gb-density SRAM, the total standby leakage current is in the order of ~1mA. It's already much smaller than Gb-density DRAM refresh power to sustain data.

Very low-standby leakage of SRAM to sustain all the Gb-density data is very useful. In modern AI network, the model weights are quite often re-used thus keeping the data in SRAM with low leakage power is very useful. SRAM has huge bandwidth to support AI computing.

Figure 4(a) shows the SRAM butterfly curves at Vcc from 1.2V to 0.8V for N1 node device. Higher Vdd is more preferred for better margin and speed, while lower Vdd is low-power. **Figure 4(b)** shows the SRAM write inverter check. It can be easily switched in the sub-ns range. The proposed SRAM has pretty small total parasitic capacitance in the range of several 10fF and the CMOS Ion is in the few 10uA range that is enough to support high-speed switching. **Figure 4(c)** shows the SRAM butterfly curves with various GAA diameter variations (range from 5.5nm to 8.5nm from Monte Carlo simulation). The GAA device is quite robust to the finite GAA diameter variations because Vt is not dominant by GAA size. We don't even introduce body doping (undoped) for the extremely scaled GAA device.

To cover various operation requirements including read stability, write margin, speed and power, we can allow some flexibility of different channel diameters of N- and P-channel devices for DTCO.

IV. Conclusion:

Table 1 summarize the characteristics of our proposed vertical-channel 3D SRAM. It can be operated at Vcc ranging from 0.8V to 1.2V, with ~ns write/read speed. The small switching energy per bit ~31fJ/bit can be designed for ultra high bandwidth memory at low chip active power. Estimated 1Gb chip SRAM die size is only ~9.3mm² (at 65% array efficiency). Such 3D SRAM is potential to serve as the L3 3D V-Cache [3] connected to CMOS chip by Cu hybrid bonding 3D IC integration to enhance the AI computing performance.

References: [1] W. C. Chen, H. T. Lue, et al, IEDM 2023, S6-5. [2] C. L. Sung, H. T. Lue, et al, IMW 2021. [3] John Wu, VLSI Short Course 2, 2022.

979-8-3503-9164-0/24 $31.00 © 2024 IEEE

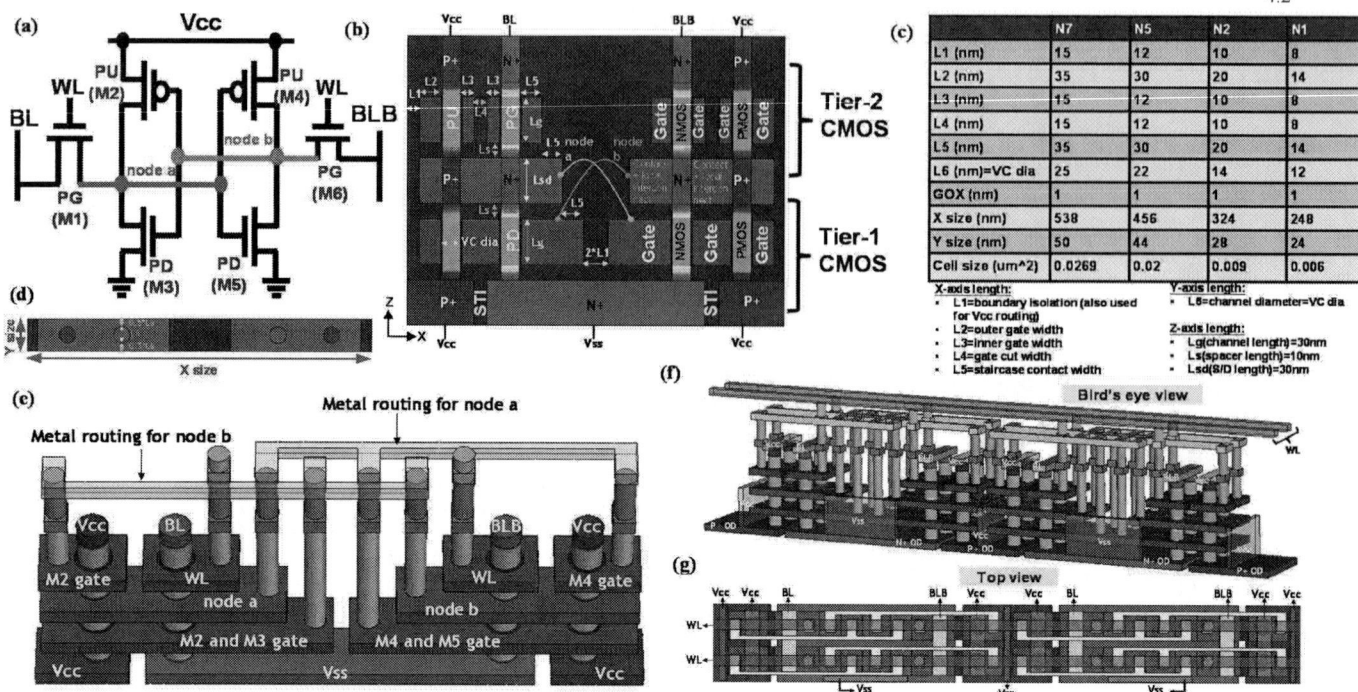

	N7	N5	N2	N1
L1 (nm)	15	12	10	8
L2 (nm)	35	30	20	14
L3 (nm)	15	12	10	8
L4 (nm)	15	12	10	8
L5 (nm)	35	30	20	14
L6 (nm)=VC dia	25	22	14	12
GOX (nm)	1	1	1	1
X size (nm)	538	456	324	248
Y size (nm)	50	44	28	24
Cell size (um^2)	0.0269	0.02	0.009	0.006

X-axis length:
- L1=boundary isolation (also used for Vcc routing)
- L2=outer gate width
- L3=inner gate width
- L4=gate cut width
- L5=staircase contact width

Y-axis length:
- L6=channel diameter=VC dia

Z-axis length:
- Lg(channel length)=30nm
- Ls(spacer length)=10nm
- Lsd(S/D length)=30nm

Fig. 1 Overall summary of the proposed 2-tier stacked vertical-channel 6T SRAM. **(a)** The classic 6T SRAM circuit diagram involves a total of 4 NMOS (2 PG, 2 PD) and 2 PMOS (PU) devices, with 12 contacts and complex metal local interconnects that consume significant area. The ultra-scaled CMOS with shorter channel length (Lg < 10nm) often exhibits high leakage. **(b)** The proposed 2-tier stacked CMOS employs vertical-channel GAA devices, utilizing 4 holes to create 4 NMOS + 4 PMOS. By shunting 2 PMOS together (2 for 1 PU), current is doubled to match the NMOS. Adopting a 3D NAND-like structure with a 3-layer metal configuration, the 2nd layer serves as a local interconnect with sidewall contact. A key advantage is the natural connection between PU, PG, and PD without any footprint penalty. **(c)** The proposed dimension parameters from N7 to N1, revealing an equivalent SRAM cell size comparable to CMOS from 7nm to 1nm. Importantly, the scaling of this structure primarily involves shrinking isolation rules without impacting CMOS devices, while keeping the device channel length constant (Lg=30nm). The channel diameter of the GAA cell is scaled to 12nm for N1. **(d)** The plane view of the structure depicts 4 holes with 4 NMOS + 4 PMOS. **(e)** The 3D schematic of the 6T SRAM demonstrates simpler metal routing compared to conventional planar 6T SRAM. **(f)** The 3D bird's eye view of the novel SRAM array with multiple WL's and BL/BLB. **(g)** The top view illustrates the novel 6T SRAM, with 3D TCAD used to evaluate the structure and devices.

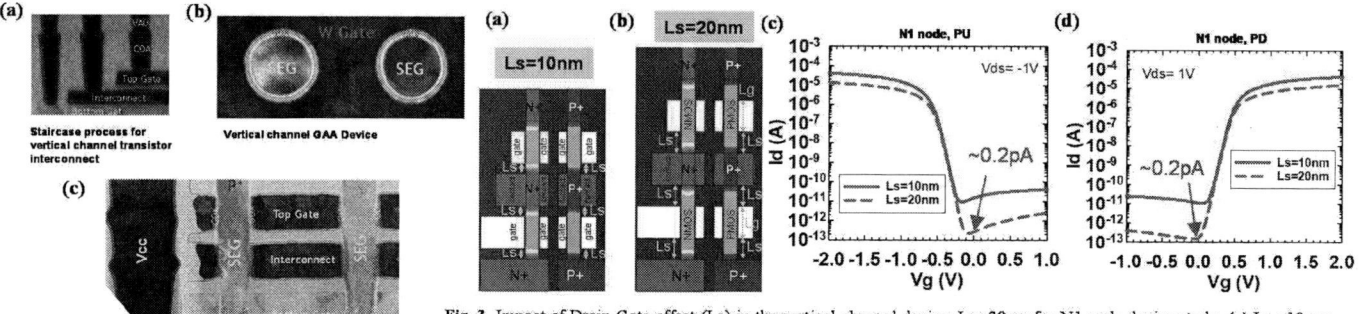

Fig. 2 Structure DEMO of (a) staircase contacts; (b) GAA transistor device; (c) Cross-sectional view.

Fig. 3 Impact of Drain-Gate offset (Ls) in the vertical-channel device. Lg=30nm for N1 node device study. **(a)** Ls =10nm; **(b)** Ls=20nm. **(c)** PMOS (PU) device Id-Vg curves with Ls=10 or 20nm. At Ls=20nm, the GIDL leakage current can be reduced to nearly 0.2pA per cell. **(d)** NMOS (PD) device Id-Vg curves with Ls=10 or 20nm. At Ls=20nm, the GIDL leakage current can be reduced to nearly 0.2pA per cell. These results imply that suitable drain-gate extension space (Ls>10nm) can greatly reduce the GIDL leakage current; thus it's possible to produce Gb-density SRAM chip with leakage current at ~1mA range which is much smaller than DRAM chip standby refresh current consumptions.

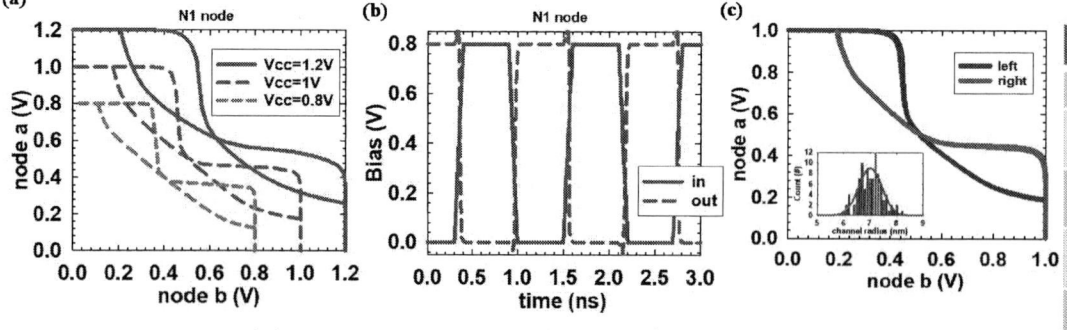

Items	Results
Vcc	0.8 ~ 1.2V
Read/Write Speed	~ns range
Cell size (um^2)	0.026 (N7) 0.006 (N1)
Estimated 1Gb die Size	40 mm2 (N7) 9.3 mm2 (N1)
Energy per switching (cell only)	~31 fJ/bit

Fig. 4 Summary of N1 node simulation results with device dimensions as listed in Fig. 1(c) (Lg=30nm, channel diameter=12nm). **(a)** SRAM read butterfly check at various Vcc (ranging from 1.2V to 0.8V) shows a smaller window for smaller Vdd, but successful memory switching is observed. **(b)** SRAM write inverter check exhibits high-speed switching with a period of 1ns, where each input/output switching time is 0.5ns. Instant memory switching indicates fast intrinsic operation speed due to small loading capacitance and robust operation mechanism. **(c)** Monte Carlo simulation for GAA diameter variations, with channel radius (L6/2) ranging from 5.5 to 8.5nm, demonstrates stable SRAM butterfly curves. The device proves to be insensitive to hole CD variations.

Table I Summary table of the merit of this novel stacked vertical-channel 6T SRAM.

979-8-3503-9164-0/24 $31.00 © 2024 IEEE

A Comparative Study of HBL and VBL 3D DRAM:
Signal Margin, Bit-cell Density, and Scalability

Xiangjin Wu[1], Luke R. Upton[1], Po-Kai Hsu[2], Jian Chen[1], Shimeng Yu[2], and H.-S. Philip Wong[1]

[1]Stanford University; [2]Georgia Institute of Technology; Email: xiangjin@stanford.edu , hspwong@stanford.edu

Abstract—Ongoing evolution of DRAM scaling can be enabled by monolithically stacking DRAM cells in the vertical direction (3D DRAM). However, there is no consensus as to whether 3D DRAM with horizontal bitline (HBL) or vertical bitline (VBL) is more scalable. Here we evaluate the signal margin and bitcell density of horizontal bitline (HBL) vs. vertical bitline (VBL) 3D DRAM using Coventor process model and circuit simulation. We show that VBL has 37% smaller BL-BL coupling noise and 2x higher iso-density signal margin than HBL at 64 layers. To surpass the density of 12 nm $4F^2$ DRAM while maintaining robust signal margin, VBL 3D DRAM requires ~ 50 layers, which is 35% fewer than HBL.

I. INTRODUCTION

DRAM scaling to sub-10 nm range faces barriers such as increasing WL resistance and decreasing storage capacitance. Vertically stacking monolithic DRAM cells can overcome these limitations, enabling increased density by freeing line width and space constraints [1, 2]. In 1T1C 3D DRAM, WLs and BLs can be arranged in different orientations, leading to two different architectures: the horizontal bitline (HBL, **Fig. 1(a)**) structure with BL staircases or contact vias, and the vertical bitline (VBL, **Fig. 1(b)**) structure with WL staircases. CMOS circuits [sense amplifier (SA), sub-WL driver (SWD), decoders, etc.] can be fabricated on a separate wafer and bonded to the memory cell wafer.

To evaluate the bitcell density of 3D DRAM, it is essential to understand the required storage capacitor length, L_{cap}, for providing sufficient signal margin during sensing. There is no consensus in literature for minimum L_{cap} in HBL and VBL structures, respectively. Unlike 2D DRAM, the required L_{cap} of 3D DRAM depends on the number of memory layers, staircase geometry, and SA structure. To investigate the L_{cap} requirements, we developed Coventor [3] process flow models for HBL (final structure in **Fig. 1c**) and VBL (**Fig. 1d**) 3D DRAM based on ref [1-5], and used the extracted RC for circuit simulation and signal margin analysis. Simulation results show that VBL has up to 2x higher signal margin than HBL at equal bitcell density. Hence, VBL requires 35% fewer layers than HBL to surpass the density of 12 nm $4F^2$ DRAM while maintaining robust signal margin.

II. BL/WL STAIRCASE STRUCTURE AND CONTACT VIAS

Fig. 2(a) shows two BL contact structures for HBL 3D DRAM. Conventional BL staircase structure (right panel) is limited by the staircase pitch due to patterning process (here we use 500 nm pitch). Alternatively, BL contact vias[6] that can vertically pass through the layers of BL (through-BL vias) could reduce the pitch and diameter, allowing more compact BL contact. We build process models for BL contact (**Fig. 2(c)**) using Minimal Incremental Layer Cost (MiLC) with parameters in ref [6]. SiN is used for sidewall isolation. For VBL, we build WL staircase shown in **Fig. 2b**.

Fig. 2d shows the extracted components of BL capacitance for the three structures. The model accounts for contribution from the staircase, through-BL via, and hybrid bonding pads. VBL shows lower total BL capacitance and BL-BL component due to the absence of BL staircase or contact via.

III. SIGNAL MARGIN & BITCELL DENSITY

Compared with traditional 2D DRAM, 3D DRAM shows larger BL-BL capacitance. This is due to BLs directly facing each other without shielding from metal contact, which is also an issue in $4F^2$ DRAM[7]. To mitigate BL coupling noise, we adopt a twisted-BL-in-SA structure in our simulation[8], which eliminates the post-SA coupling noise (i.e., noise during the latching phase in read). Pre-SA BL-BL coupling noise remains. We extract BL-BL coupling noise using methods in ref [9, 10] and calculate signal margin in **Fig. 3b**. Sufficient signal margin is needed to combat SA mismatch and tail-bit charge leakage. VBL shows 2x and 4x higher signal margin than contact-via HBL and staircase HBL, respectively, at 64 layers. The signal margin in HBL can be improved by > 2x when contact via pitch is aggressively scaled to < 100 nm (projection in **Fig. 3c**), while VBL signal margin is less affected by staircase pitch.

Fig. 4a shows the impact of # of memory layers (N), capacitor length (L_{cap}) and contact via pitch (P_{via}), or [N, L, P], on bitcell density and signal margin. Bitcell density and signal margin can be traded off with each other by tuning L_{cap}, while decreasing P_{via} improves both metrics (60 nm P_{via} is required for HBL to match VBL). **Fig. 4b** displays the achievable bitcell density vs. # of layers. We adjust L_{cap} to ensure that signal margin > 80 mV in all cases[11]. VBL shows the best scalability, followed by contact via HBL. HBL with staircase at current pitch (~ 400 nm) struggles to surpass $4F^2$ DRAM with robust signal margin, even at 128 layers. **Fig. 5** summarizes WL loading time of the two structures as well as their pros and cons.

IV. CONCLUSIONS

Signal margin and coupling noise should be considered when evaluating pitch scalability and bitcell density of 3D DRAM. HBL requires tighter staircase pitch than VBL in order to maintain robust signal margin and density gain. Through-bitline vias in HBL allow more compact bitline contacts than conventional staircase HBL, improving signal margin & scalability. Other considerations (e.g., process variability and material growth quality) could be included for future study.

Acknowledgement: This work is supported by PRISM and CHIMES, two of the SRC JUMP 2.0 centers, and by the Stanford NMTRI. We are thankful to Dr. David Fried (Lam Research) for technical support and the use of Coventor. **References:** [1] J.W. Han, *VLSI*, 2023, TFS1-1 [2] M. Huang, *IMW*, 2023, pp. 1-4 [3] https://www.coventor.com/products/semulator3d/ [4] S. Varghese, US20220344339 [5] Y. He, US20220335982 [6] S.-H. Chen, *IEDM* 2012 [7] S. Park, *IEDM*, 2023, pp. 6-1 [8] S. Watanabe *IEEE JSSC*, 1995, 30, 9, pp. 960-971 [9] J. S. Yuan, *IJE*, 1993, 10.1080/00207219308925833 [10] Y. Nakagome, *IEEE JSSC*, 1988, 23, 5, pp. 1120-1127 [11] Rambus DRAM model, http://www.rambus.com/energy [12] X. Peng, IEDM 2019.

979-8-3503-9164-0/24 $31.00 © 2024 IEEE

4.3

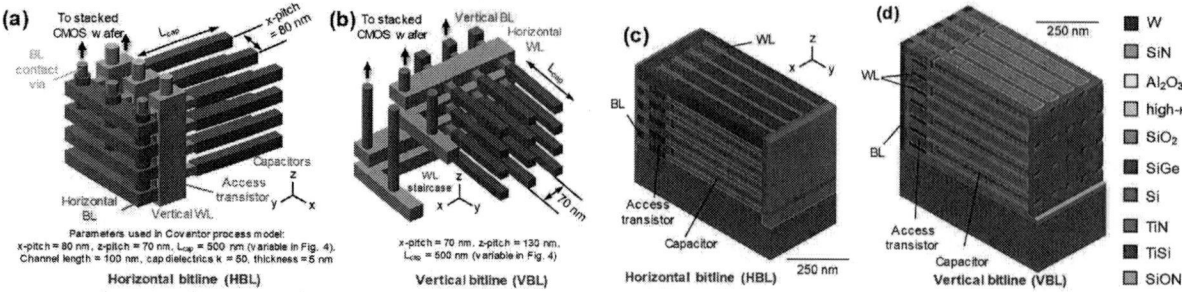

Fig. 1: Schematics of HBL **(a)** and VBL **(b)** 3D DRAM. **(c, d)** Coventor process model. Only the final structures are shown due to space limit. Si/SiGe superlattice structure is grown by epitaxial growth where SiGe serves as sacrificial layer. TiN is used as capacitor electrode, and W as WL.

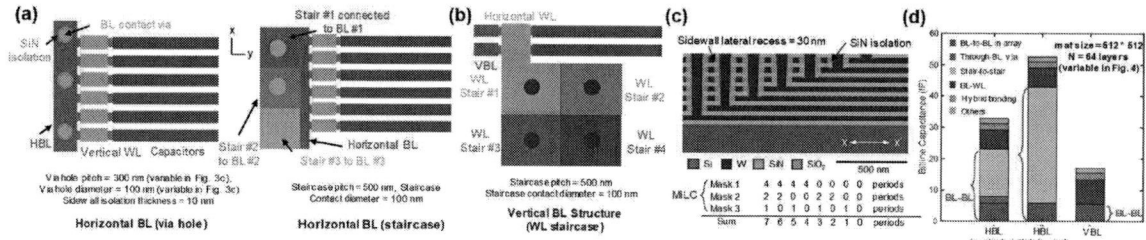

Fig. 2: (a) Top-view schematics of two bitline contact structures for HBL (not to-scale). **(b)** WL staircase for VBL. Two staircases per group are used. **(c)** Coventor process model for MiLC HBL contact hole array. Cross-section along x-x' direction is shown. Sidewall lateral recess enhances electrical isolation. **(d)** Components of BL capacitance extracted for three architectures. HBL has greater BL to BL capacitance.

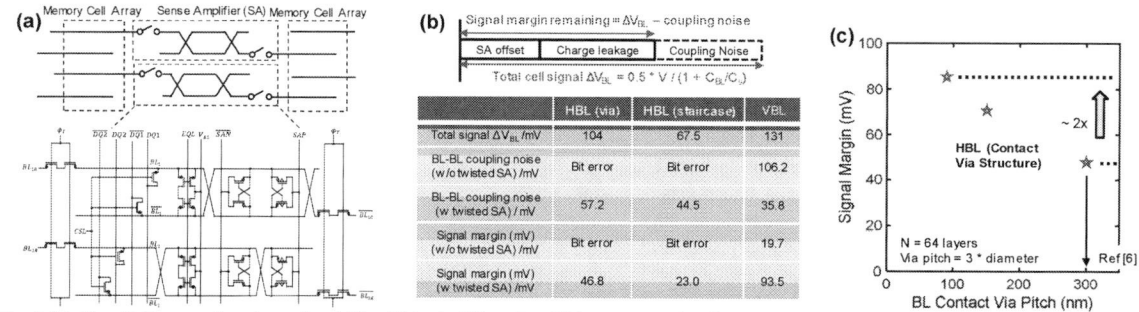

Fig. 3 (a) Circuit diagram showing twisted BL within the SA region. This unconventional structure eliminates post-SA noise in open BL arrays. **(b)** Cell signal, BL-BL coupling noise, and signal margin with and without twisted BL in SA. Coupling noise is simulated with SPICE and cross-checked with analytical solutions in ref [9, 10] **(c)** Singal margin vs. BL contact via pitch, showing ~2x increase in signal margin with denser vias.

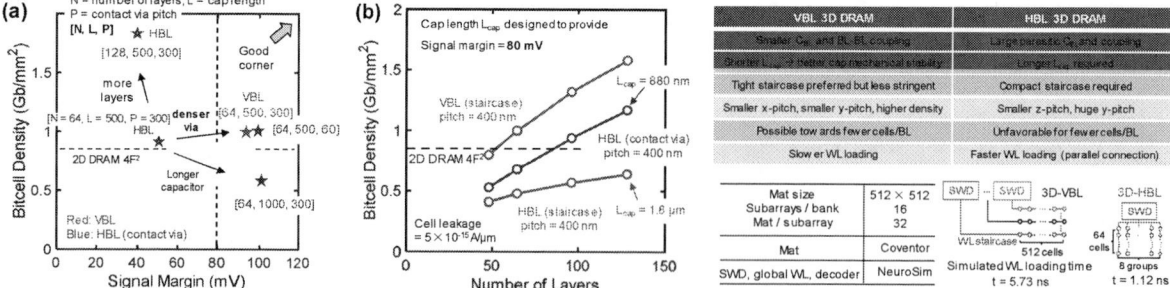

Fig. 4 (a) Effect of # of layers (N), capacitor length (L) and contact via pitch (P) on 3D DRAM signal margin and bitcell density. Tradeoff between bitcell density and signal margin can be achieved. For 64-layers HBL 3D DRAM, scaling down the via to sub-100 nm is required to simultaneously achieve sufficient signal margin and density. **(b)** Bitcell density vs. number of layers. To obtain sufficient signal margin, HBL needs larger (and thus longer) capacitors (as labeled). Even with compact in-BL contact vias, HBL is less scalable than VBL.

Fig. 5 Summary of pros and cons of VBL vs. HBL 3D DRAM. WL loading time for both structures are simulated with NeuroSim[12] with parameters extracted from Coventor process model.

979-8-3503-9164-0/24 $31.00 © 2024 IEEE

Gap in pagination due to unavailable paper.

Pages 21-22

A Novel RRAM-based Ternary Strong-PUF with High Security Intensity Feasible for Low-earth Orbit Satellites in the 6G Era

P. H. Huang[1], K. Y. Lee[1], S. H. Lin[1], E Ray Hsieh[1,*], Yu-Cheng Lin[2], Ching-Ju Lin[2], Jen-Chou Tseng[2], Chien-Fan Wang[2], Wen-Ting Chu[2], Yu-Der Chih[2], Jonathan Chang[2]

[1]Dept. of Electrical Engineering, National Central University, Taoyuan, Taiwan, [2]TSMC, Hsinchu, Taiwan,
*E-mail: linshih910103@gmail.com

Abstract—We propose a novel RRAM-based ternary-PUF chip, which utilizes 3 modes of RRAM, i.e., the FORMing, SET, and RESET, to increase entropy of electrical variations so as to strengthen security-intensity. This ternary-PUF can be operated as not only a weak PUF but also a strong PUF, feasible for the applications of the low-earth orbit satellites in the 6G era. The experiments demonstrate the SET shows the most significant advantage, yielding an Inter-Hamming distance (Inter-HD) of 49.49%, an intra-HD of 0%; near 50% HW. The randomness meets the NIST tests at temperatures of 25°C and 75°C.

I. INTRODUCTION

Security protection are escalating in the era of rapid information growth, especially with the widespread application of the low-earth orbit satellites in the 6G era in a near future. The demand for protecting products against manipulation and reverse engineering is increasing. While most software protections are not only vulnerable to hacking but also highly energy thirsting. The Physical Unclonable Function (PUF) offers a 100% encrypted hardware solution. The PUF utilizes the process or electrical variations to generate unique identifiers (keys), thwarting chip replication and bolstering security. The non-volatile memories, such as Resistive Random-Access Memory (RRAM), exhibits greater development potential than traditional volatile memory (the SRAM or DRAM). The RRAM benefits gains of design for its simple structure, rapid speeds, low cost, and retention, well-suited for the PUF implementation. Moreover, the RRAM can swiftly regenerate new keys without any compromise. This advantage negates the need to consume the entire array to store the keys, thus effectively preserving functionality and ensuring key randomness.

In this paper, we implement a 1 M-bit RRAM PUF MACRO with a 1T1R unit cell design on the 40nm logic CMOS technology. We will operate the PUF functionalities of the 1T1R cell through three operational modes (FORMing, SET, and RESET), demonstrating the novel "Ternary-PUF" Then, we will investigate whether it can achieve high randomness, high uniqueness, and high reliability under three operational modes and different temperatures (25°C and 75°C) to realize optimal Ternary-PUF functionalities.

II. DESIGN AND PREPARATION OF THE RRAM PUFs

Fig. 1 illustrates the unit-cell structure comprising a control transistor and a storage RRAM MIM. Two cells share a source line to minimize the layout overhead. Fig 2 displays a schematic of a 6.5-kb array for RRAM PUF, consisting of 256×256 cells. Fig. 3 presents the memory block diagram, incorporating the main memory array, control lines for the X and Y directions of the array, word lines and bit lines decoder, a level shifter for supplying the program voltage, and a sense amplifier for reading. Table 1 provides the operational conditions for the RRAM-based Ternary-PUF.

III. RESULTS AND DISCUSSIONS

A. Schematic of the RRAM PUFs

Fig. 4(a), (b), and (c) display the Shmoo plots of programming voltage versus time. We identify the optimal '1' and '0' distributions in the shmoo plots of FORMing, SET, and RESET, under the program voltage and time of the operating conditions: (a) 3.48 V and 250 ns ; (b) 1.2V and 70 ns; (c) 1.2V and 60 ns, respectively. Fig. 5 and 6 show the resistance distribution during FORMing, SET, and RESET under the aforementioned operating conditions at 25°C and 75°C. We concluded that the SET mode offers a balanced distribution of "1" and "0" bits with minimal temperature effect, resulting in more stable outputs.

B. Cryptographic Parameters of the RRAM PUFs

The Inter-HD evaluates the uniqueness of the PUFs. Fig. 7, 8, and 9 compare the Inter-HD of the PUF with the FORMing, SET, and RESET operations at 25°C and 75°C. At 25°C, the SET operation exhibits superior mean values, rendering it suitable for the PUF operations again. At 75°C, mean values are similar across all operations, but the SET operation shows a smaller standard deviation, indicating its superiority over the RESET or FORMing operations even at 75°C. The intra-hamming distance proves the stability of the PUFs. Fig. 10 shows the Intra-Hamming distance of PUFs with the FORMing, SET, and RESET operations. All of these results indicate that the PUF maintains near zero even at 25°C or 75°C. The Hamming weight (HW) refers to the total number of non-zero bits in a string f bits. Fig. 11 and 12 show the HW of PUFs with FORMing, SET, and RESET operations at 25°C and 75°C. It can be observed that whether at normal or high temperatures, under SET operation, the mean value of the Hamming Weight is close to 50%, indicating better randomness in the distribution of "1" and "0". Fig. 13 and 14 depict the mean and standard deviation values of the Inter-Hamming Distance (HD) of the PUF after XOR processing. Results suggest that after several XOR processing iterations, the mean value and standard deviation of the PUF string are much closer to 50% and 0%, respectively.

C. Reliability and Randomness of the RRAM PUFs

Fig. 15 shows the data retention of the RRAM PUF at 75°C for one month. Table 2 compares the performance of the proposed RRAM PUF with previous designs [1-11]. Table 3 shows that the proposed PUF passes all NIST randomness tests in both 25°C and 75°C.

We have demonstrated that our RRAM Ternary-PUF can perform PUF functionality in all three operational modes, exhibiting excellent reliability (with data retention exceeding 1 month at 75 ℃). Additionally, it demonstrates high randomness (50% of hardware) and high uniqueness (49.49% of Inter-HD and 0% of Intra-HD), specifically in the SET operation mode, thus showcasing its potential in hardware security.

Acknowledgments This work was supported by the National Science and Technology Council, Taiwan, under NSTC 112-2628-E-008-003.

Reference:
[1] B. Lin et al., *JEDS*, pp. 1257-1265, 2020. [2] Y. Pang et al., *ISSCC*, pp. 402-404, 2019. [3] B. Lin et al., *JSSC*, pp. 1641-1650, 2021. [4] B. Gao et al., *IEEE TED*, pp. 536-542, 2022. [5] X. Xue et al., *ASSCC*, pp. 29-32, 2019 [6] S. K. Mathew et al., *ISSCC*, pp. 278-279, 2014. [7] Y. Shifman, et al., *LSSC*, pp. 138-141, 2018. [8] W. C. Wang et al., *IEDM*, pp. 31.6.1-31.6.4, 2020. [9] K.-H. Chuang, et al., *IRPS*, pp. P-CR.2-1-P-CR.2-5, 2018. [10] M. -Y. Wu et al., *ISSCC*, pp. 130-132, 2018. [11] Jiahao Song et al., *LSSC*, pp. 58-61, 2022

979-8-3503-9164-0/24 $31.00 © 2024 IEEE

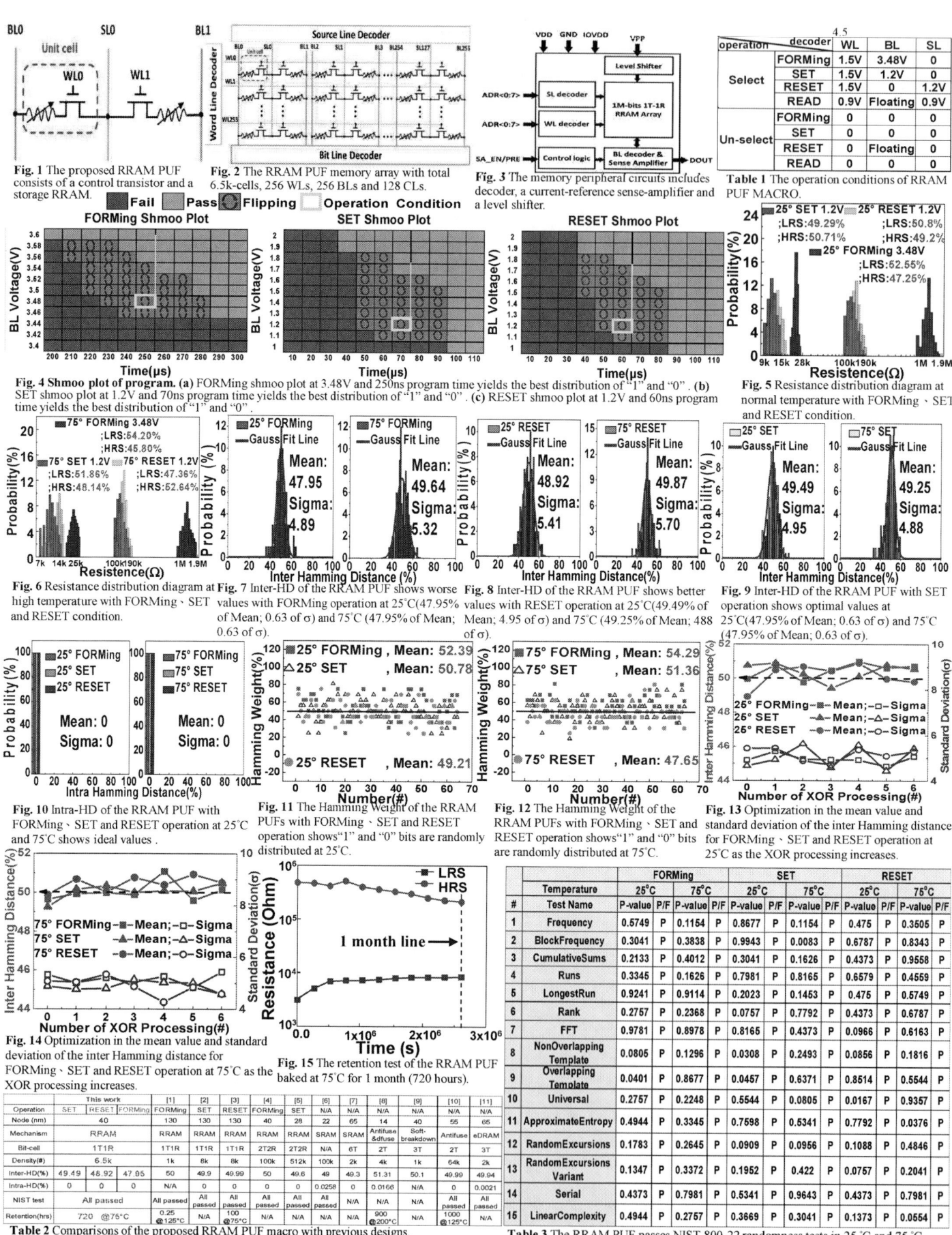

Fig. 1 The proposed RRAM PUF consists of a control transistor and a storage RRAM.

Fig. 2 The RRAM PUF memory array with total 6.5k-cells, 256 WLs, 256 BLs and 128 CLs.

Fig. 3 The memory peripheral circuits includes decoder, a current-reference sense-amplifier and a level shifter.

Table 1 The operation conditions of RRAM PUF MACRO.

operation	decoder	WL	BL	SL
Select	FORMing	1.5V	3.48V	0
	SET	1.5V	1.2V	0
	RESET	1.5V	0	1.2V
	READ	0.9V	Floating	0.9V
Un-select	FORMing	0	0	0
	SET	0	0	0
	RESET	0	Floating	0
	READ	0	0	0

Fig. 4 Shmoo plot of program. (a) FORMing shmoo plot at 3.48V and 250ns program time yields the best distribution of "1" and "0". **(b)** SET shmoo plot at 1.2V and 70ns program time yields the best distribution of "1" and "0". **(c)** RESET shmoo plot at 1.2V and 60ns program time yields the best distribution of "1" and "0".

Fig. 5 Resistance distribution diagram at normal temperature with FORMing、SET and RESET condition.

Fig. 6 Resistance distribution diagram at high temperature with FORMing、SET and RESET condition.

Fig. 7 Inter-HD of the RRAM PUF shows worse values with FORMing operation at 25℃(47.95% of Mean; 0.63 of σ) and 75℃ (47.95% of Mean; 0.63 of σ).

Fig. 8 Inter-HD of the RRAM PUF shows better values with RESET operation at 25℃(49.49% of Mean; 4.95 of σ) and 75℃ (49.25% of Mean; 488 of σ).

Fig. 9 Inter-HD of the RRAM PUF with SET operation shows optimal values at 25℃(47.95% of Mean; 0.63 of σ) and 75℃ (47.95% of Mean; 0.63 of σ).

Fig. 10 Intra-HD of the RRAM PUF with FORMing、SET and RESET operation at 25℃ and 75℃ shows ideal values.

Fig. 11 The Hamming Weight of the RRAM PUFs with FORMing、SET and RESET operation shows "1" and "0" bits are randomly distributed at 25℃.

Fig. 12 The Hamming Weight of the RRAM PUFs with FORMing、SET and RESET operation shows "1" and "0" bits are randomly distributed at 75℃.

Fig. 13 Optimization in the mean value and standard deviation of the inter Hamming distance for FORMing、SET and RESET operation at 25℃ as the XOR processing increases.

Fig. 14 Optimization in the mean value and standard deviation of the inter Hamming distance for FORMing、SET and RESET operation at 75℃ as the XOR processing increases.

Fig. 15 The retention test of the RRAM PUF baked at 75℃ for 1 month (720 hours).

Table 2 Comparisons of the proposed RRAM PUF macro with previous designs.

Operation	This work			[1]	[2]	[3]	[4]	[5]	[6]	[7]	[8]	[9]	[10]	[11]
	SET	RESET	FORMing	FORMing	SET	RESET	FORMing	SET	RESET	N/A	N/A	N/A	N/A	N/A
Node (nm)	40			130	130	130	40	28	22	65	14	40	55	N/A
Mechanism	RRAM			RRAM	RRAM	RRAM	RRAM	RRAM	SRAM	SRAM	Antifuse &dfuse	Soft-breakdown	Antifuse	eDRAM
Bit-cell	1T1R			1T1R	1T1R	1T1R	2T2R	2T2R	N/A	6T	2T	3T	2T	3T
Density(#)	6.5k			1k	8k	8k	100k	512k	100k	2k	1k	64k	2k	3T
Inter-HD(%)	49.49	48.92	47.95	50	49.9	49.99	50	49.6	49	49.3	51.31	50.1	49.99	49.94
Intra-HD(%)	0	0	0	N/A	0	0	0	0.0258	0	0.0166	N/A	0	N/A	0.0021
NIST test	All passed			All passed	All passed	All passed	All passed	All passed	N/A	N/A	N/A	N/A	All passed	All passed
Retention(hrs)	720 @75℃			0.25 @75℃	N/A	100 @75℃	N/A	N/A	N/A	900 @125℃	N/A	1000 @125℃	N/A	N/A

Table 3 The RRAM PUF passes NIST 800-22 randomness tests in 25 ℃ and 75 ℃.

		FORMing		SET		RESET	
Temperature		25℃	75℃	25℃	75℃	25℃	75℃
#	**Test Name**	P-value P/F	P-value P/F	P-value P/F	P-value P/F	P-value P/F	P-value P/F
1	Frequency	0.5749 P	0.1154 P	0.8677 P	0.1154 P	0.475 P	0.3505 P
2	BlockFrequency	0.3041 P	0.3838 P	0.9943 P	0.0083 P	0.6787 P	0.8343 P
3	CumulativeSums	0.2133 P	0.4012 P	0.3041 P	0.1626 P	0.4373 P	0.9558 P
4	Runs	0.3345 P	0.1626 P	0.7981 P	0.8165 P	0.6579 P	0.4559 P
5	LongestRun	0.9241 P	0.9114 P	0.2023 P	0.1453 P	0.475 P	0.5749 P
6	Rank	0.2757 P	0.2368 P	0.0757 P	0.7792 P	0.4373 P	0.6787 P
7	FFT	0.9781 P	0.8978 P	0.8165 P	0.4373 P	0.0966 P	0.6163 P
8	NonOverlapping Template	0.0805 P	0.1296 P	0.0308 P	0.2493 P	0.0856 P	0.1816 P
9	Overlapping Template	0.0401 P	0.8677 P	0.0457 P	0.6371 P	0.8514 P	0.5544 P
10	Universal	0.2757 P	0.2248 P	0.5544 P	0.0805 P	0.0167 P	0.9357 P
11	ApproximateEntropy	0.4944 P	0.3345 P	0.7598 P	0.5341 P	0.7792 P	0.0376 P
12	RandomExcursions	0.1783 P	0.2645 P	0.0909 P	0.0956 P	0.1088 P	0.4846 P
13	RandomExcursions Variant	0.1347 P	0.3372 P	0.1952 P	0.422 P	0.0757 P	0.2041 P
14	Serial	0.4373 P	0.7981 P	0.5341 P	0.9643 P	0.4373 P	0.7981 P
15	LinearComplexity	0.4944 P	0.2757 P	0.3669 P	0.3041 P	0.1373 P	0.0554 P

Gap in pagination due to unavailable papers.

Pages 25-28

Ferroelectricity Engineered AlScN Thin Films Prepared by Hydrogen Included Reactive Sputtering for Analog Applications

Si-Meng Chen[1,2], Hirofumi Nishida[1], Takuya Hoshii[1],
Kazuo Tsutsui[3], Hitoshi Wakabayashi[3], Edward Yi Chang[2], and Kuniyuki Kakushima[1]

[1]Department of Electrical and Electronic Engineering, Tokyo Institute of Technology, Yokohama, Kanagawa, Japan
[2]International College of Semiconductor Technology, National Yang Ming Chiao Tung University, Hsinchu, Taiwan
[3]Institute of Innovative Research, Tokyo Institute of Technology, Yokohama, Kanagawa, Japan
Email: chen.s.ap@m.titech.ac.jp

Abstract — Recently, various studies have been conducted to understand the material properties, ferroelectricity, and device applications of nitride ferroelectrics. In this research, we propose breakdown strength and ferroelectricity engineered AlScN thin films by utilizing reactive sputtering with H_2 gas inclusion. Higher breakdown field/coercive field (E_{BD}/E_c) ratio was demonstrated, indicating high endurance potential. Moreover, via changing the H_2 flux percentage, AlScN film exhibited linear-like dependence between remanent polarization (P_r) and electric field (E), showing the possibility of multi-state operation on future analog devices.

Keywords: Ferroelectric AlScN, ferroelectricity tuning, multi-state operation, endurance

I. INTRODUCTION

Since the first experimental demonstration of nitride ferroelectrics, researchers and engineers have committed massive endeavours to explore their potential for emerging functional devices [1], [2]. However, limited by large unwanted leakage and low breakdown field/coercive field (E_{BD}/E_c) ratio, AlScN films typically exhibit endurance around 10^5 cycles [3]. Fortunately, our previous work successfully reported an endurance over 10^7 cycles by utilizing reactive sputtering with 1% H_2 inclusion, showing the feasibility of reducing ambient for enhanced reliability [4]. In this research, more H_2 flux was blended into the sputtering environment to form AlScN thin films with potential of multi-state operation and further endurance improvement.

II. CAPACITOR FABRICATION

Figs. 1(a) and **1(b)** show the schematic and process flow of $Al_{0.74}Sc_{0.26}N$ metal-ferroelectric-metal (MFM) capacitors. The fabrication was initiated by degreasing n^+-Si wafers with chemical solutions. 50-nm-thick ferroelectric $Al_{0.74}Sc_{0.26}N$ films, sandwiched by TiN electrodes, were deposited through *in-situ* reactive sputtering at 400°C. Herein, an extra H_2 gas inlet was utilized to blend H_2 (0 – 16.67 vol %) into the Ar/N_2 ambient for ferroelectric deposition. **Table I** summarizes the detailed deposition conditions. Note that the deposition rate showed neglectable discrepancy for all sputtered ferroelectric films. The fabrication process was eventually completed by evaporating Al metal on the backside.

III. RESULTS AND DISCUSSIONS

Butterfly loops in capacitance-voltage (C-V) measurements confirm the ferroelectricity of sputtered $Al_{0.74}Sc_{0.26}N$ films up to 11.77% of H_2 inclusion, as illustrated in **Fig. 2(a)**. The disappearance of ferroelectricity in 16.67% H_2 flux film could be attributed to the distinction of c-axis orientation grains [5]. **Fig. 2(b)** shows that static dielectric constant (ε_i) decreased monotonically with higher H_2 flux. Moreover, the imprint field (E_{imp}), which could be derived from the peak positions of C-V curves [6], shifts toward zero with more H_2 gas flow.

Results of time-zero dielectric breakdown (TZDB) tests are demonstrated in **Fig. 3**. The 0% H_2 flux film shows high leakage current due to high Sc content of the ferroelectric film with low bandgap value [4]. Fortunately, as H_2 flux increased to 11.77%, suppressed leakage and boosted E_{BD} are prominent. However, the 16.67% H_2 flux film not only loses its ferroelectric properties, but also exhibits drastic decline in E_{BD}.

Positive-up negative-down (PUND) measurements are especially convenient to extract the remanent polarization (P_r) without the influence of leakage current. **Figs. 4(a)** and **4(b)** illustrate the switching and non-switching response in 0% and 6.25% H_2 flux films at PUND measurements. Here, different voltages were applied to meet a target P_r of 70 $\mu C/cm^2$. **Fig. 4(c)** displays the overall P_r-electric field (E) relationship of PUND measurements. Consistent with our previous work, sputter deposition with higher H_2 flux results in lower P_r at all E region, along with more ambiguous semi-saturation kinks [4]. The former phenomenon could be explained by deteriorated c-axis orientation growth. E_c values, which were determined by linear extrapolation of the uprising stage of P_r-E relationship, also increased with H_2 flux [7]. In addition, the linear-like uprising slope of 11.77% H_2 flux film indicates the possibility of multi-state operation for analog accelerators. **Fig. 4(d)** depicts the tendency of E_c and E_{BD}/E_c ratio obtained from TZDB and PUND tests. The increased E_{BD}/E_c ratio suggests better reliability potential of AlScN thin films prepared by H_2 gas involved sputtering.

IV. CONCLUSION

In summary, we successfully fabricated ferroelectric AlScN thin films by reactive sputtering with H_2 flux up to 11.77%. The C-V results show that ε_i and E_{imp} are tunable characteristics via varying the percentage of H_2 flux. PUND results also suggest that sputtering deposition in reducing ambient lead to higher E_c values. Linear P_r gain over a broad range of E is observed in 11.77% H_2 flux film, indicating the possibility of multi-state operation for future analog NVM technology. Additionally, enhanced E_{BD}/E_c ratio and suppressed leakage are demonstrated by H_2 flux films. Therefore, we expect further improved endurance strength in AlScN thin films deposited by H_2 involving reactive sputtering, as field cycling tests results are yet to be shown in this research.

979-8-3503-9164-0/24 $31.00 © 2024 IEEE

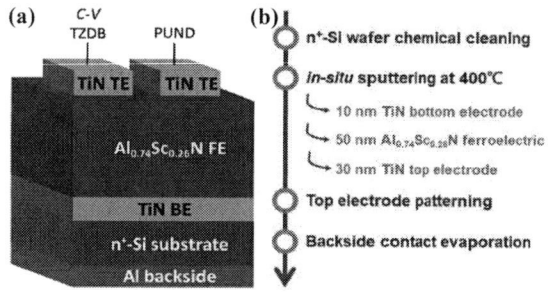

Fig. 1. (a) Schematic illustration and (b) process flow of $Al_{0.74}Sc_{0.26}N$ MFM capacitors.

TABLE I
FERROELECTRIC DEPOSITION CONDITION

Sputter target	$Al_{0.53}Sc_{0.47}$
DC power supply (W)	300
Substrate temperature (°C)	400
Target-substrate distance (cm)	10.0
Chamber pressure (Pa)	0.7
Ar/N_2 gas flow rate (sccm)	5/10
Amount of H_2 gas addition (%)	0, 6.25, 11.77, 16.67

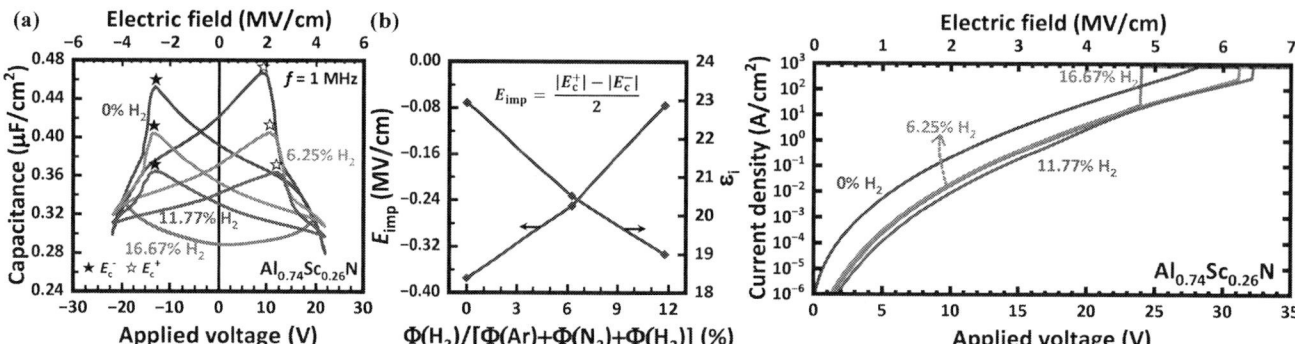

Fig. 2. (a) *C-V* curves of fabricated $Al_{0.74}Sc_{0.26}N$ MFM capacitors. Ferroelectricity sustained up to 11.77% of H_2 flux during sputter deposition. (b) E_{imp} and ε_i in dependence with the H_2 flux percentage. E_{imp} was calculated based on the positions of *C-V* curve peaks in (a). ε_i was calculated from the cross-point of the *C-V* curves in (a).

Fig. 3. TZDB results of all deposited films. Leakage suppression observed in films deposited below 11.77% H_2 flux.

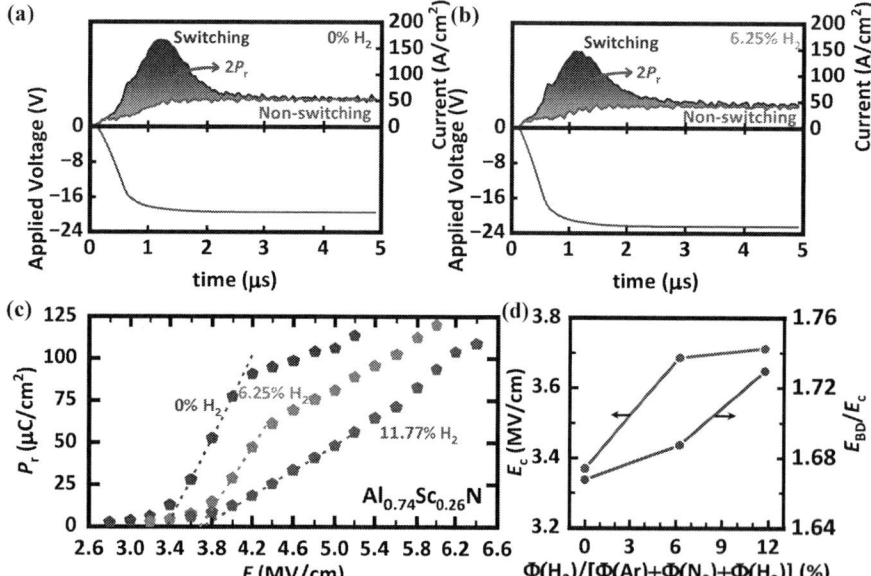

ACKNOWLEDGMENTS

This work was supported by MEXT Initiative to Establish Next-generation Novel ICs Centers (X-NICS), under Grant No. JPJ011438.

REFERENCES

[1] S. Fichtner *et al.*, *J. Appl. Phys.*, vol. 125, pp. 114203, March 2019. [2] D. Wang *et al.*, *Appl. Phys. Lett.*, vol. 124, pp. 150501, April 2024. [3] T. Mikolajick *et al.*, *J. Appl. Phys.*, vol. 129, pp. 100901, March 2021. [4] S.-M. Chen *et al.*, *Jpn. J. Appl. Phys.*, vol. 63, no. 03SP45, February 2024. [5] Y.-J. Yong *et al.*, *J. Vac. Sci. Technol., A*, vol. 15, pp. 390, March 1997. [6] Y. Hiranaga *et al.*, *J. Appl. Phys.*, vol. 128, pp. 244105, December 2020. [7] S. Yasuoka *et al.*, *J. Appl. Phys.*, vol. 128, pp. 114103, September 2020.

Fig. 4. Demonstration of PUND measurements for (a) 0% of H_2 flux and (b) 11.77% of H_2 flux films. (c) P_r-E relationship of fabricated $Al_{0.74}Sc_{0.26}N$ films in PUND measurements. (d) Extracted E_c and E_{BD}/E_c ratio with respect to H_2 flux percentage.

979-8-3503-9164-0/24 $31.00 © 2024 IEEE

Investigation of HfO₂ and ZrO₂ Nanolamination Thickness on Superlattice HZO FeRAMs Exhibiting Enhanced Remnant Polarization and Improved Endurance

Dong-Ru Hsieh[1, **], Zi-Yang Hong[1], Wei-Ju Yeh[1], Jia-Chian Ni[1], Huai-En Luo[1],
Yan-Kui Liang[2], Shang-Lin Hsieh[1], Kuan-Lin Chen[1], and Tien-Sheng Chao[1, *]

[1]Dept. of Electrophysics, National Yang Ming Chiao Tung University (NYCU), Hsinchu, Taiwan
[2]International College of Semiconductor Technology, NYCU, Hsinchu, Taiwan
Tel: +886-3-5131367 Fax: +886-3-5725230 *E-mail: tschao@nycu.edu.tw, **E-mail: drhsieh0425@gmail.com

Abstract — In this work, we reported that HfO₂-ZrO₂ superlattice (SL) HfZrO₂ (HZO) FeRAMs with a HfO₂ and ZrO₂ nanolamination (NL) thickness of 1 nm and a ZrO₂ seed layer of 1.5 nm achieve an excellent two remnant polarization (2P$_r$) of 43.3 μC/cm². Compared to the conventional HZO, the SL HZO can exhibit an improved HZO quality and suppress the formation of thin TiO$_x$N$_{1-x}$ layers. Therefore, the SL HZO FeRAMs with the HfO₂ and ZrO₂ NLs of 1 nm exhibit a wake-up free behavior, enhanced fatigue effect immunity, and large residual 2P$_r$ of 20.5 μC/cm² after the endurance test of 10^9 cycles. From the ferroelectricity and reliability viewpoints, the proposed SL HZO FeRAMs are a promising candidate for embedded nonvolatile memories (eNVMs).

I. Introduction

HfO₂-based ferroelectrics (FEs) that contain HZO [1-23], HfAlO₂ (HAO) [24], [25], HfSiO₂ (HSO) [26-28] have attracted considerable attention for eNVMs [1-4]. They have many attractive features including a desirable scaling capability, fast switching speed, low-power operation, and great nonvolatile data storage [1-3]. Among the HfO₂-based FEs, the HZO exhibits highly competitive benefits including a high CMOS compatibility, large 2P$_r$, and low post-metal annealing (PMA) temperature. Recently, several studies have reported that the SL HZO exhibit a better ferroelectricity and endurance performance than the conventional HZO [16-19]. However, the effect of the HfO₂ and ZrO₂ NL thickness on the SL HZO has not yet been investigated and discussed in detail. Therefore, in this work, we optimized the HfO₂ and ZrO₂ NL thickness of the SL HZO to further design a CMOS-compatible HfO₂-ZrO₂ SL HZO FeRAM for eNVMs.

II. Device Fabrication

Fig. 1 shows the schematic structures of HfO₂-ZrO₂ SL and conventional HZO FeRAMs. First, a 200 nm TiN as the bottom electrode was deposited by PVD onto a 500 nm thermal SiO₂ grown on a 6-in Si wafer. Next, a 1.5 nm ZrO₂ seed layer was deposited by ALD before the FE layer to induce more FE o-phase [12]. Next, the HZO with a total thickness of 10 nm on the ZrO₂ seed layer was deposited by ALD. TDMAHf, TDMAZr, and H₂O precursors were used as the Hf, Zr, and O sources. The SL HZO with the periodicity of HfO₂ NL/ZrO₂ NL are denoted as S(HMZM)$_N$, where M is the number of ALD deposition cycles for the HfO₂ and ZrO₂ NL formation and N is the number of HfO₂ NL/ZrO₂ NL periods for the SL HZO formation. The three types of HfO₂ and ZrO₂ NL thicknesses fabricated are 0.5 nm (when M=5 and N=10), 1 nm (when M=10 and N=5), and 2.5 nm (when M=25 and N=2), respectively. Next, a 100 nm TiN as the top electrode was deposited by PVD and patterned by dry etching to complete the HfO₂-ZrO₂ SL and conventional HZO FeRAMs. The latter is denoted as S(H1Z1)$_{50}$. PMA was performed by RTA at 650 °C for 30 s in N₂ ambient.

III. Results and Discussion

Fig. 2(a) and **(b)** show that S(H5Z5)$_{10}$ and S(H10Z10)$_5$ exhibit a more desirable P–E characteristic than S(H1Z1)$_{50}$. With an increase in the NL thickness up to 1 nm, S(H10Z10)$_5$ exhibits an obviously higher 2P$_r$ of 37 μC/cm² and lower 2E$_c$ of 2.55 MV/cm than S(H1Z1)$_{50}$ in **Fig. 2(c)** and **(d)**. This mainly results from that the SL HZO can extra supply a lattice/thermal expansion coefficient mismatch-induced tensile stress between the HfO₂ and ZrO₂ NLs to facilitate more FE o-phase formation during PMA. This indicates that the SL HZO at the pristine state possesses a lower amount of V$_o$, resulting in a better HZO quality [29]. **Fig. 2(c)** and **(d)** show that S(H25Z25)$_2$ exhibits a lower 2P$_r$ and higher 2E$_c$ than S(H10Z10)$_5$, which might be attributed to more m-phase formation. For all the HZO conditions, an obvious crystalline peak corresponding to the o-phase (111)/t-phase (011) can be clearly recognized in **Fig. 3(a)** [1], [2]. The m-phase (-111)/m-phase (111) crystalline peaks become significant when the NL thickness is increased up to 2.5 nm in **Fig. 3(a)**. S(H10Z10)$_5$ possesses a higher relative o-phase (111) percentage and lower relative t-phase (011) percentage than S(H1Z1)$_{50}$, confirming S(H10Z10)$_5$ with more FE o-phase in **Fig. 3(b)** and **(c)**. In general, the O binding energy for HfO₂ is relatively higher than ZrO₂ [30]. Therefore, in the periodic SL HZO, the 1 nm HfO₂ NLs as the diffusion barrier layers can effectively suppress the O and Ti diffusions at the TiN/HZO interfaces during PMA and then repress the formation of thin TiO$_x$N$_{1-x}$ layers at the TiN/HZO interfaces and V$_o$-rich HZO [29] than S(H1Z1)$_{50}$ in **Fig. 4**. Therefore, S(H10Z10)$_5$ has a higher O relative atomic percentage within the HZO, confirming the reduction in the V$_o$ concentration. **Fig. 5** shows the endurance measurements performed at a frequency of 1 MHz (or pulse width of 500 ns). Compared to S(H1Z1)$_{50}$, the wake-up effect and FE dielectric breakdown (BD) can be effectively suppressed by using the SL HZO in **Fig. 6**. With an increase in the NL thickness up to 1 nm, S(H10Z10)$_5$ exhibits a stronger fatigue effect immunity than S(H1Z1)$_{50}$ and S(H5Z5)$_{10}$ in **Fig. 6**. After the endurance test up to 10^9 cycles, S(H10Z10)$_5$ still possesses a sufficiently large 2P$_r$ of 20.5 μC/cm². **Fig. 7(a)** and **(b)** show that the SL HZO FeRAMs with the HfO₂ and ZrO₂ NLs of 1 nm exhibit not only an excellent 2P$_r$ up to 43.3 μC/cm² at a relatively low electric field of 2.5 MV/cm but also a superior endurance performance up to 10^9 cycles than other published conventional HZO/HAO/HSO FeRAMs [5-15, 18, 23-28].

IV. Conclusions

The SL HZO FeRAMs with an increased NL thickness up to 1 nm exhibit an excellent 2P$_r$ of 43.3 μC/cm² than the conventional HZO FeRAMs. Furthermore, the HZO quality and the formation of thin TiO$_x$N$_{1-x}$ layers can be improved and suppressed by using the 1-nm-thick HfO₂ and ZrO₂ NLs, respectively. Therefore, the HfO₂-ZrO₂ SL HZO FeRAMs exhibit a wake-up free characteristic, enhanced fatigue effect immunity, and large residual 2P$_r$ of 20.5 μC/cm² after the endurance test of 10^9 cycles. This indicates that they possess a relatively high potential for eNVMs.

Acknowledgement: This work was supported in part by the National Science and Technology Council (NSTC), Taiwan, under Contract MSTC-112-2221-E-A49-167-MY3; by the Taiwan Semiconductor Research Institute (TSRI), Taiwan, under Contract JDP112-Y1-045; by Materials Analysis Technology Inc. (MA-tek), Taiwan.

979-8-3503-9164-0/24 $31.00 © 2024 IEEE

5.4

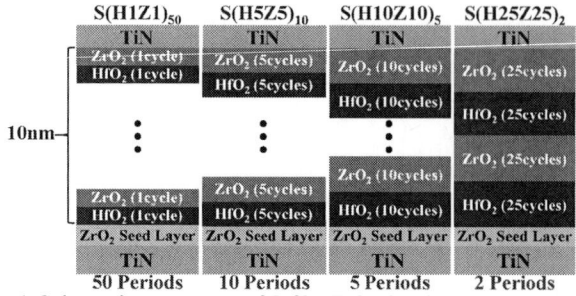

Fig. 1 Schematic structures of HfO_2-ZrO_2 SL (M = 5, 10, 25) and conventional HZO FeRAMs (M = 1).

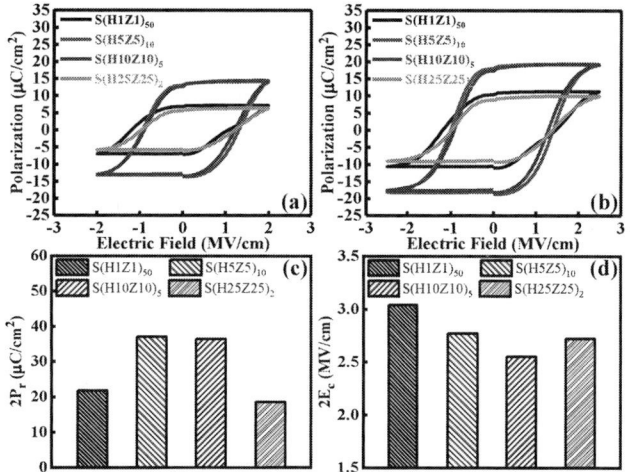

Fig. 2 P–E curves of SL and conventional HZO FeRAMs measured with the PUND at (a) 2.0 and (b) 2.5 MV/cm. (c) $2P_r$ and (d) $2E_c$ of SL and conventional HZO FeRAMs extracted by the P–E curves.

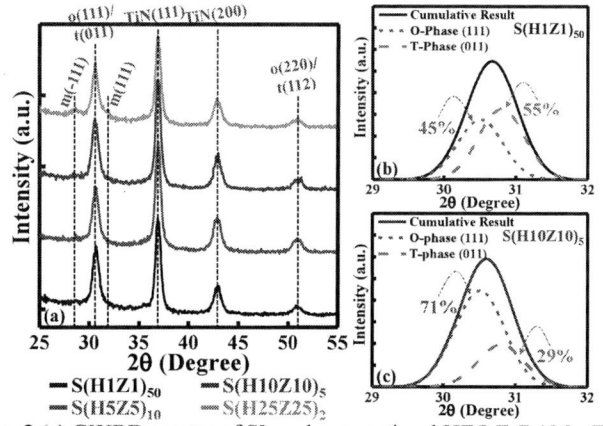

Fig. 3 (a) GIXRD spectra of SL and conventional HZO FeRAMs. The relative o- and t-phase percentages of crystalline peaks extracted by the GIXRD curves of (b) $S(H1Z1)_{50}$ and (c) $S(H10Z10)_5$ in Fig. 3(a).

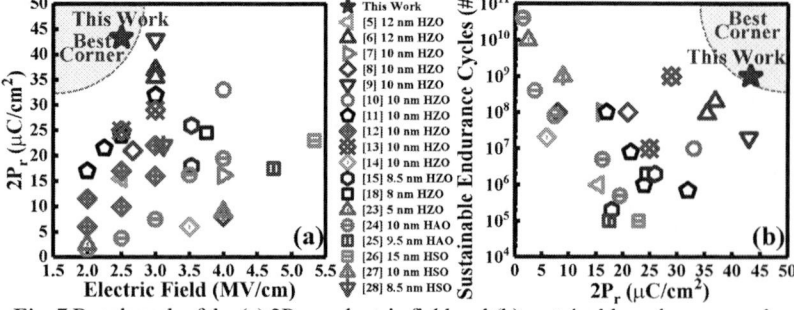

Fig. 7 Benchmark of the (a) $2P_r$ vs. electric field and (b) sustainable endurance cycles before the FE dielectric BD vs. $2P_r$ for this work with other published FeRAMs.

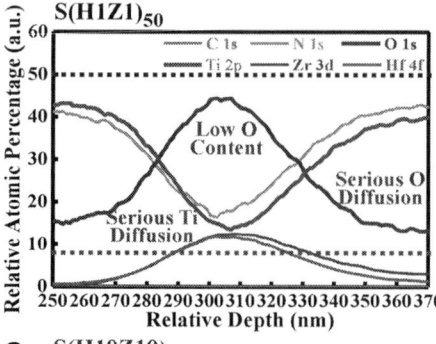

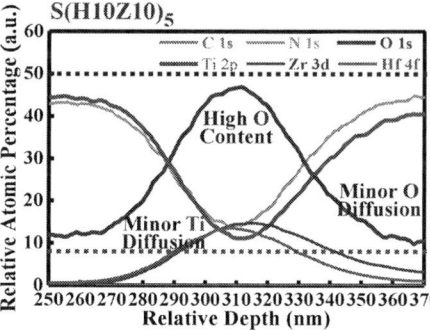

Fig. 4 XPS depth analyses of C, N, O, Ti, Zr, and Hf elements across the MFM configuration for the SL and conventional HZO FeRAMs.

Endurance Measurement Waveform

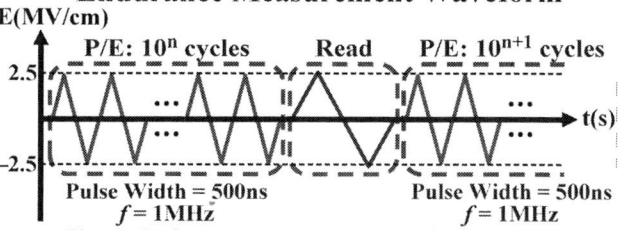

Fig. 5 Endurance measurements performed at a frequency of 1 MHz (i.e., the pulse width of 500 ns).

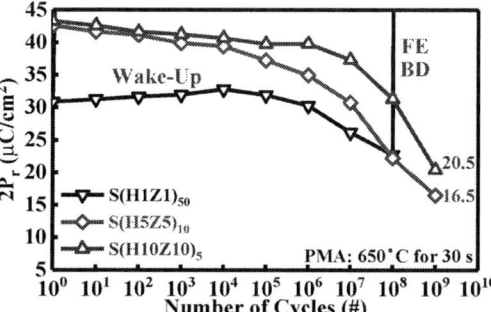

Fig. 6 Endurance characteristics of SL and conventional HZO FeRAMs measured by using the test manner in Fig. 5.

References: [1] J. Müller, *Nano Lett.*, 2012. [2] J. Müller, *IEDM*, 2013. [3] M. H. Park, *Adv. Mater.*, 2015. [4] M. Pešić, *Adv. Funct. Mater.*, 2016. [5] K. Y. Chen, *EDL*, 2018. [6] Y. C. Liu, *EDL*, 2023. [7] H. K. Peng, *EDL*, 2022. [8] Y. F. Chen, *EDL*, 2023. [9] T. Huang, *EDL*, 2023. [10] Y. Xu, *TED*, 2022. [11] J. Hur, *TED*, 2021. [12] T. Onaya, *Appl. Phys. Exp.*, 2017. [13] B. S. Kim, *Nanosc. Res. Lett.*, 2020. [14] V. Gaddam, *EDL*, 2021. [15] Y. K. Liang, *Trans. Nanotechnol.*, 2022. [16] S. L. Weeks, *Appl. Mater. Interfaces*, 2017. [17] M. H. Park, *Appl. Phys. Rev.*, 2019. [18] B. Cui, *EDL*, 2023. [19] Y. Peng, *EDL*, 2022. [20] Y. K. Liang, *EDL*, 2022. [21] Z. Zhao, *EDL*, 2022. [22] X. Li, *EDL*, 2023. [23] S. J. Kim, *Appl. Phys. Lett.*, 2018. [24] J. Zhou, *EDL*, 2020. [25] K. Florent, *VLSI*, 2017. [26] K. Florent, *IEDM*, 2018. [27] L. Grenouillet, *VLSI*, 2020. [28] T. S. Böscke, *IEDM*, 2011. [29] D. R. Hsieh, *TED*, 2022. [30] Z. Gong, *Appl. Phys. Lett.*, 2022.

979-8-3503-9164-0/24 $31.00 © 2024 IEEE

32

Impact of Time Delay Schemes on Reliability Degradation during Program/Erase Cycling in HZO-based FeFETs

Xiaopeng Li[1], Guoqing Zhao[1], Lu Tai[1], Pengpeng Sang[1], Xiaoyu Dou[1], Xuepeng Zhan[1], Xiaolei Wang[2], Jixuan Wu[1, *], and Jiezhi Chen[1]

[1]School of Information Science and Engineering (ISE), Shandong University, 266237 Qingdao, China;
[2]Key Laboratory of Microelectronic Devices and Integrated Technology, Institute of Microelectronics of Chinese Academy of Sciences, 100029 Beijing, China. *Email: jixuanwu@sdu.edu.cn

ABSTRACT — The commercialization of HZO-based FeFETs is hindered by issues such as MW degradation and poor endurance. In this study, we systematically investigate degradation suppression using time delay schemes. Our findings reveal that symmetric time delay during cycling fails to effectively mitigate MW and SS degradation. Interestingly, we observe that the optimization of MW or SS can be achieved through asymmetric delay (i.e., time delay only after program or erase). It's found that delay after programming (P-delay) effectively suppresses MW degradation, while delay after erasure (E-delay) proves effective in mitigating SS degradation. This study offers valuable insights into the reliability optimization of HZO-based FeFETs.

I. INTRODUCTION

HZO-based FeFETs are emerging as highly promising alternatives to traditional charge trap flash memory, boasting rapid operation speed, low operation voltage, and excellent CMOS compatibility [1]. Indeed, despite silicon's historical prominence as a traditional channel material for FeFETs, significant degradation issues persist, including memory window (MW) and subthreshold swing (SS) degradation, in addition to poor endurance [2-3]. These challenges are closely linked to the low-k oxide interface (SiOx IL) and stack defects (Fig. 1) [3]. Recent studies have demonstrated prolonged endurance through interface and cycle operation optimization [4-6]. Consequently, the imperative for MW optimization has become increasingly critical, necessitating a thorough investigation of degradation phenomena.

In this work, degradation suppression using time delay schemes is investigated in depth. It is observed that, 1) Symmetric time delay fails to suppress degradation; 2) Asymmetric delay presents distinct benefits: delay after program (P-delay) effectively suppresses MW degradation, whereas delay after erase (E-delay) is effective in mitigating SS degradation. The underlying mechanisms are elucidated through comprehensive charge defect analysis.

II. DEVICE FABRICATION

The process flow of the main steps and structure diagram are shown in Figs. 2(a-b). The gate stack includes 0.7nm SiO_2 and 10nm $Hf_{0.5}Zr_{0.5}O_2$ (HZO). TiN/W is deposited as gate and source/drain electrodes. 550°C RTA for 60s in N_2 is used to form the orthorhombic phase (Fig. 2(c)), and 30min H_2/N_2 forming gas annealing (FGA) at 450°C is performed.

III. RESULTS AND DISCUSSIONS

Fig. 3 shows the mechanism of degradation induced by charge trapping/de-trapping during cycling, and it would be beneficial to reduce the interfacial damage through relaxation [7]. Different symmetric delay times (0~10ms, time delay after both program and erase) are designed to test the effect,

as shown in Fig. 4(a). Unfortunately, there is no significant improvement in both MW and SS (Figs. 4(b-c)). To further explore the relationship between degradation and charge relaxation, asymmetric time delay schemes (time delay only after program or erase) are employed (Fig. 5(a)). In Figs. 5(b-c), optimization effects are evident, showcasing contrasting trends between the two schemes regarding SS and MW: P-delay scheme slows MW degradation but accelerates SS degradation, while E-delay scheme has the opposite effect.

Considering the interface and FE layer degradation during cycling, the one-time current measurement is utilized to characterize remanent polarization (Pr) [8]. The measurement waveform and associated equations are shown in Fig. 6. Let's focus on a delay of 10ms as an illustrative example. Fig. 7 shows the extracted results, indicating that the P-delay scheme enhances Pr, while the E-delay scheme weakens them compared to the without-delay scheme. This suggests a significant impact of the two schemes on interfacial charge and FE domain pinning. To verify this hypothesis, the relaxation of V_{th} after program is further analyzed. As shown in Fig. 8, V_{th} gradually shifts due to de-trapping as the waiting time increases, saturating at approximately 10s. Here, we perform V_{th} relaxation comparisons for devices after 10^4 cycles (Fig. 9). As expected, E-delay scheme weakens V_{th} and ΔV_{th}, and P-delay scheme enhances them. It is reasonable to conclude that the opposite degradation inhibition trend under P/E asymmetric delay is highly correlated with the charge behavior during the time delay. With P-delay, higher electron enrichment leads to stronger electon-hole recombination and weaker FE domain pinning. Consequently, MW and SS are larger. Conversely, with E-delay, Vo^{2+} leads to FE domain pinning and less defect generation. Therefore, MW and SS are smaller (Fig. 10).

IV. CONCLUSION

A comprehensive discussion regarding the influence of time delay schemes on FeFET degradation is presented. It becomes evident that asymmetric delay schemes yield distinct advantages: P-delay effectively suppresses MW degradation, whereas E-delay proves effective in mitigating SS degradation. This research furnishes valuable insights into understanding the degradation mechanisms of Si-channel HZO-based FeFETs, thus guiding subsequent optimizations.

ACKNOWLEDGMENTS — This work was supported by China Key Research and Development Program under Grant (2022YFB3603900, 2023YFB4402400), National Natural Science Foundation of China (Nos. U23B2040, 92264201, 62034006), Natural Science Foundation of Shandong Province (ZR2023QF054, ZR2023LZH007), and TaiShan Scholars (TSQN202306059).

REFERENCES — [1] S. Beyer, IMW, 2020; [2] M. K. Kim, Science Advances, 2021; [3] F. Mo, VLSI, 2019; [4] J. Min, EDL, 2021; [5] A. J. Tan, EDL, 2021; [6] Y. Zhou, IEDM, 2022; [7] E. Yurchuk, TED, 2016; [8] R. Ichihara, VLSI, 2020.

979-8-3503-9164-0/24 $31.00 © 2024 IEEE

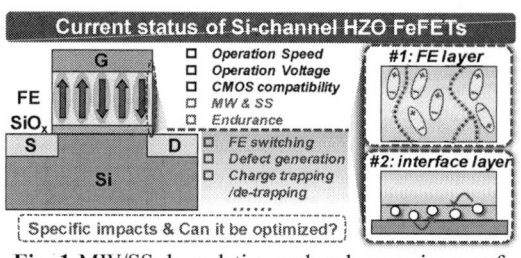

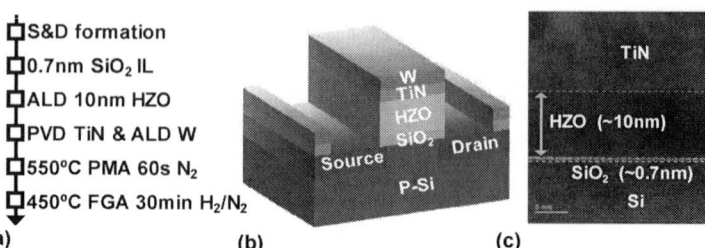

Fig. 1 MW/SS degradation and endurance issues of Si-channel HZO FeFETs.

Fig. 2 (a) The key fabrication process flow, and **(b)** schematic diagram of the FeFET. **(c)** The TEM image of TiN/HZO/SiO$_2$/Si gate structure.

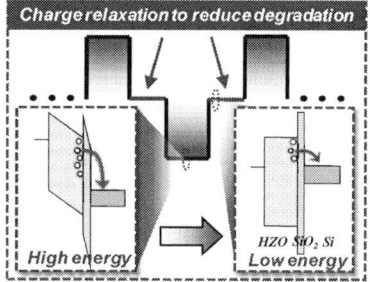

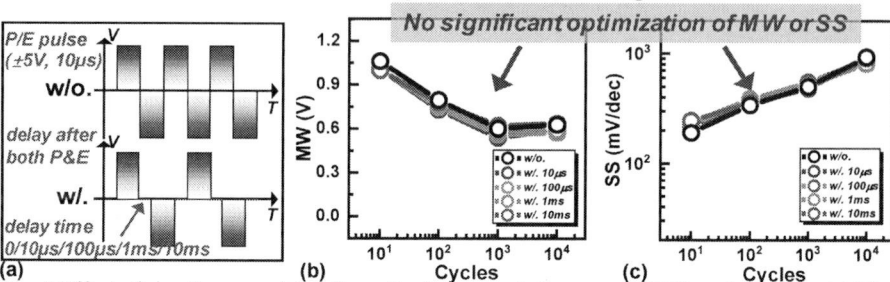

Fig. 3 Delay time induced charge relaxation to reduce IL degradation.

Fig. 4 Effect of simultaneous delay time after Program & Erase on FeFET performance. **(a)** The corresponding cycling sequence. **(b)** MW, and **(c)** SS. This scheme has not significantly decelerated the degradation.

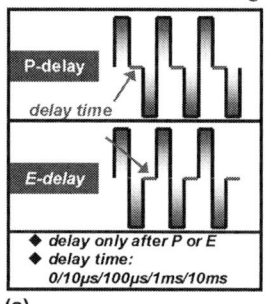

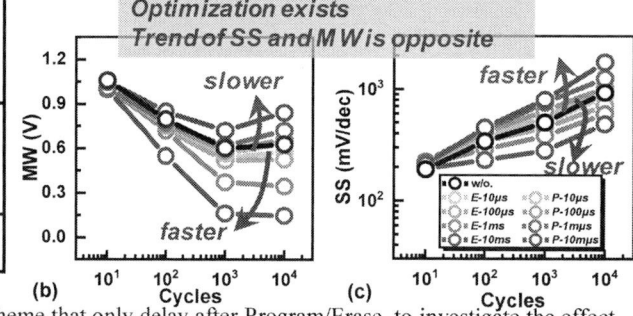

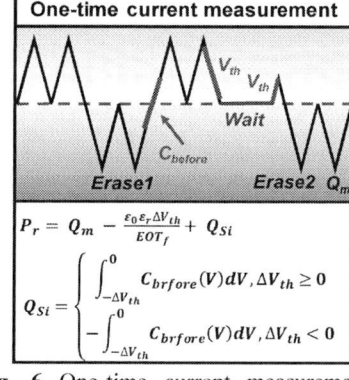

Fig. 5 (a) Pulse cycling scheme that only delay after Program/Erase, to investigate the effect of delay position on degradation. Effect of delay after Program/Erase on FeFETs performance. **(b)** MW, and **(c)** SS. Program-delay slows MW degradation intensifying SS degradation, while Erase-delay has the opposite effect.

Fig. 6 One-time current measurement waveform, and associated equations.

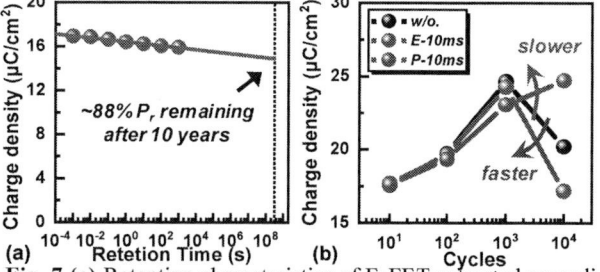

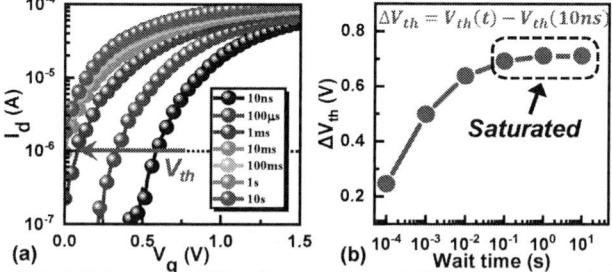

Fig. 7 (a) Retention characteristics of FeFET extracted according to one-time current measurement. **(b)** Variation of Pr with cycles under different delay conditions.

Fig. 8 Relaxation of V$_{th}$ after program. **(a)** Transfer characteristic, and **(b)** ΔVth versus wait time.

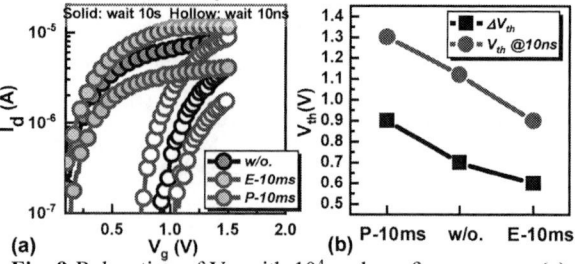

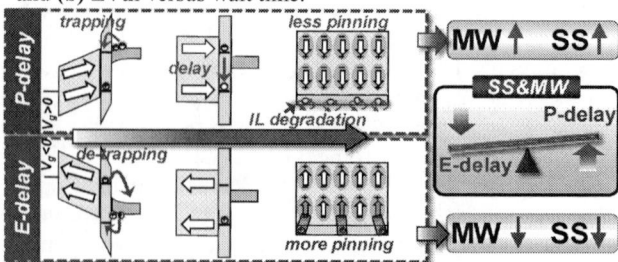

Fig. 9 Relaxation of V$_{th}$ with 10^4 cycles, after program. **(a)** Transfer characteristic, and **(b)** Extracted V$_{th}$ and ΔV$_{th}$.

Fig. 10 Mechanisms of defect behaviors. The opposite degradation inhibition trend under P/E asymmetric delay is highly correlated with the charge behavior during the time delay

979-8-3503-9164-0/24 $31.00 © 2024 IEEE

Study on the Anomalous Characteristics of Random Telegraph Noise in FeFETs

Puyang Cai[1,#], Yishan Wu[2,#], Zixuan Sun[1], Hao Li[1], Xiaolei Wang[3], Zhigang Ji[2,*], Runsheng Wang[1,*], and Ru Huang[1]

[1]School of Integrated Circuits, Peking University, Beijing, China, [2]National Key Laboratory of Advanced Micro and Nano Manufacture Technology, Shanghai Jiaotong University, Shanghai, China, [3]Key Laboratory of Microelectronics Devices and Integrated Technology, Institute of Microelectronics, Chinese Academy of Sciences, Beijing, China

[#]Equal contribution; [*]Email: r.wang@pku.edu.cn, zhigangji@sjtu.edu.cn

Abstract — **In this work, we investigate the behavior of random telegraph noise (RTN) in HfO$_2$-based FeFETs. Anomalously high location factors of RTN in FeFETs are observed for the first time, which can be successfully explained by the lateral domain wall motion, with the help of a dynamic 2-D phase-field simulation. This work reveals the unique characteristics of RTN in HfO$_2$-based FeFETs and provides insight into the ferroelectric switching process.**

I. INTRODUCTION

HfO$_2$-based FeFET has attracted much attention as the candidate for emerging non-volatile memory in the past decade, due to its CMOS compatibility, excellent scalability, and high energy efficiency [1]. However, the trap charge-related reliability issue is still one of the challenges for the application of HfO$_2$-based FeFET. On one hand, the trap charges in the gate oxide screen parts of the ferroelectric (FE) polarization charge, leading to a degraded memory window [2]. On the other hand, the process of charge trapping and de-trapping induces instability in the I_d of FeFET, resulting in the variability of the obtained V_{th} [3]. The signal caused by charge trapping and de-trapping, in turn, can serve as an effective approach to studying the microscopic information of traps [4]. In this work, random telegraph noise (RTN) is detected and used to analyze the trap behavior in HfO$_2$-based FeFETs. Anomalous RTN characteristics in FeFETs are discovered, and a mechanism of the interaction between the charge trapping/de-trapping process and FE polarization switching is revealed.

II. EXPERIMENTS

The fabrication flow and device structure of the n-type FeFETs measured in this work are shown in Fig. 1. The gate stack structure of the FeFETs is TiN/Hf$_{0.5}$Zr$_{0.5}$O$_2$(HZO)/SiO$_2$/Si (MFIS), with an 8.5 nm HZO FE layer and a 0.7 nm interfacial dielectric (DE) layer. Fig. 2 shows the I_d-V_g curves of FeFET after different bipolar pulse cycles. During the first 10^2 cycles, the memory window widens as V_{th} in the program (PRG) state decreases and V_{th} in the erase (ERS) state increases. After 10^4 cycles, due to the trap generation at the interface and within the gate stack, V_{th} in both PRG and ERS states of FeFET is shifted, leading to the degradation of the memory window.

II. RESULTS AND DISCUSSION

To investigate the trap information, a measurement waveform shown in Fig. 3 is adopted. Before the RTN measurement, FeFETs are written to the PRG state. Then, with V_d biased at 50 mV, a series of stepping V_g is applied and the signal of I_d is monitored. Fig. 4 shows the measured RTN raw data and the results extracted by a hidden Markov model (HMM) [5] of Trap I and Trap II. The dependence of capture (τ_c) and emission time constant (τ_e) on V_g of Trap I and Trap II is shown in Fig. 5(a)-(b), respectively. As V_g increases, Trap I exhibits a consistent increase of τ_e and maintains a stable τ_c. For Trap II, its τ_e increases while τ_c decreases with increasing V_g. The values of $\ln(\tau_c/\tau_e)$ of

Trap I and Trap II are further calculated, and they both present a good linearity with V_g, as shown in Fig. 5(c)-(d). This linear relationship indicates the band bending induced by V_g increments, allowing for the obtainment of the trap's physical location within the gate oxide based on Eq. (1)-(4) in Fig. 6 [6]. When the trap is located at the TiN/HZO interface, the location factor (LF) reaches an upper limit of 38.68 V^{-1}. Fig. 7 summarizes the LF of traps extracted from different FeFETs, with nearly half of the LF exceeding the upper limit. To our knowledge, this is the first discovery of such anomalous LF in RTN analysis.

To understand the anomalous RTN characteristics in FeFETs, a 2-D phase-field simulation is conducted based on the time-dependent Landau-Ginzburg (TDLG) theory [7]. A FE/DE stack is set up in the simulation framework, and the simulated polarization-voltage (P-V) loop is shown in Fig. 8, confirming the FE property of the FE layer. Then, a stepping V_g is applied to investigate the polarization switching process. In FE systems, the polarization switching progresses through FE domain nucleation and subsequent domain wall (DW) propagation [8]. As shown in Fig. 9, as V_g increases, FE polarization gradually switches via the lateral motion of DW. Therefore, for the defect located in the vicinity of DW, its corresponding polarization state in the thickness direction is changed. The band diagrams corresponding to different polarization states are shown in Fig. 10(a). After the FE polarization switching, the depolarization field induced by the polarization charge becomes opposite to the direction of the applied electric field, lowering the voltage drop on the FE layer and elevating the voltage drop on the DE layer and the surface potential. Thus, the energy level of the trap in the gate oxide is pulled down significantly. As shown in Fig. 10(b), the LF along the gate oxide is remarkably increased after the regulation of DW motion, which is in agreement with our experiment results. Therefore, the lateral DW motion can induce an anomalously high LF of RTN in FeFETs, distinctly different from MOSFETs with dielectric gate oxide, and RTN, in turn, can serve as a tool to probe the FE switching process in FeFETs.

IV. SUMMARY

In this paper, we investigate the RTN characteristics in HfO$_2$-based FeFETs. We identify the anomalously high LF in FeFETs and perform a 2-D phase-field simulation to reveal that the band bending regulated by lateral DW motion is the cause. This work uncovers the unique RTN characteristics in HfO$_2$-based FeFETs and provides a novel understanding of the impact of FE polarization switching on trap behavior.

REFERENCES: [1] A. I. Khan *et al.*, *Nature Electron.*, 3, 588 (2020). [2] P. Cai *et al.*, *IEDM*, 32.2 (2022). [3] F. Zhang *et al.*, *EDL*, 45, 566 (2024). [4] R. Wang *et al.*, *IEDM*, 17.2 (2018). [5] J. Zhang *et al.*, *IEDM*, 17.3 (2018). [6] T. Gong *et al.*, *TED*, 68, 3716, (2021). [7] A. K. Saha *et al.*, *APL*, 114, 202903 (2019). [8] X. Lyu *et al.*, *VLSI Tech.*, T07-2 (2022).

ACKNOWLEDGEMENTS: This work was supported by NSFC (61927901, 62125401) and the 111 Project (B18001).

5.6

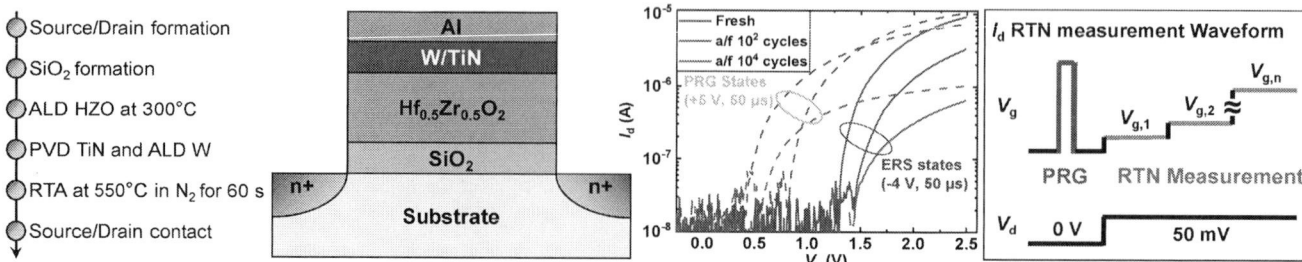

Fig. 1. Fabrication process flow and cross-section structure of the n-type FeFETs measured in this work.

Fig. 2. I_d-V_g curves of the FeFET in PRG and ERS states after different bipolar stress cycles.

Fig. 3. I_d RTN measurement waveform. While V_d is kept at 50 mV, V_g is set to be stepping voltage.

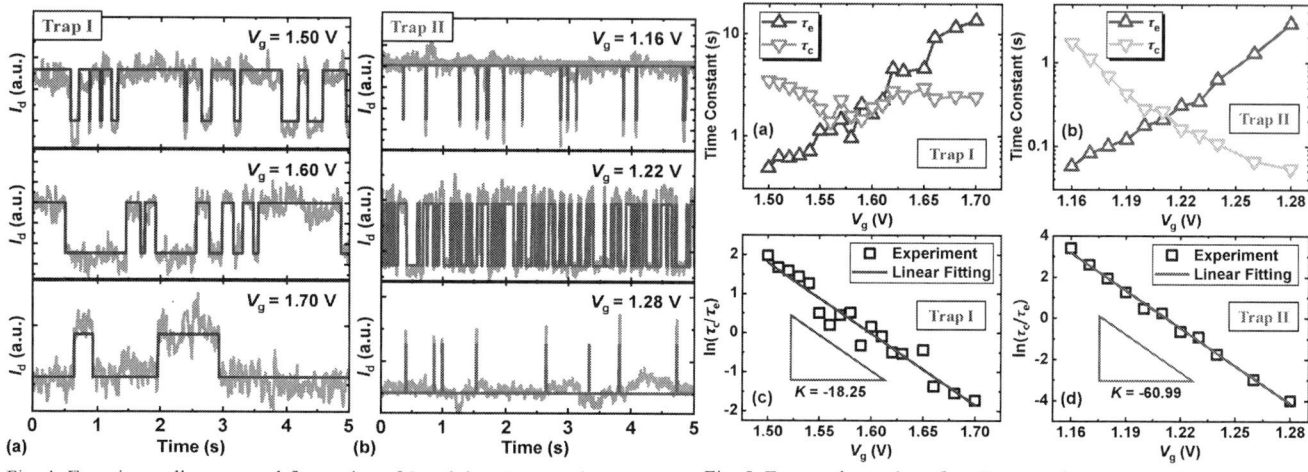

Fig. 4. Experimentally measured fluctuation of I_d and the HMM method extracted RTN of (a) Trap I and (b) Trap II.

Fig. 5. Extracted τ_c and τ_e of (a) Trap I and (b) Trap II. The dependence of $\ln(\tau_c/\tau_e)$ on V_g of (c) Trap I and (d) Trap II.

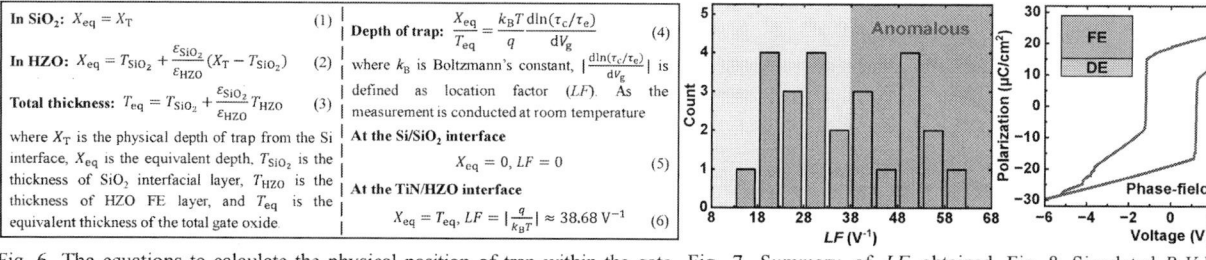

In SiO₂: $X_{eq} = X_T$ (1)

In HZO: $X_{eq} = T_{SiO_2} + \dfrac{\varepsilon_{SiO_2}}{\varepsilon_{HZO}}(X_T - T_{SiO_2})$ (2)

Total thickness: $T_{eq} = T_{SiO_2} + \dfrac{\varepsilon_{SiO_2}}{\varepsilon_{HZO}}T_{HZO}$ (3)

where X_T is the physical depth of trap from the Si interface, X_{eq} is the equivalent depth, T_{SiO_2} is the thickness of SiO₂ interfacial layer, T_{HZO} is the thickness of HZO FE layer, and T_{eq} is the equivalent thickness of the total gate oxide.

Depth of trap: $\dfrac{X_{eq}}{T_{eq}} = \dfrac{k_B T}{q}\dfrac{d\ln(\tau_c/\tau_e)}{dV_g}$ (4)

where k_B is Boltzmann's constant, $|\frac{d\ln(\tau_c/\tau_e)}{dV_g}|$ is defined as location factor (*LF*). As the measurement is conducted at room temperature

At the Si/SiO₂ interface
$$X_{eq} = 0, \; LF = 0 \quad (5)$$

At the TiN/HZO interface
$$X_{eq} = T_{eq}, \; LF = |\frac{q}{k_B T}| \approx 38.68 \text{ V}^{-1} \quad (6)$$

Fig. 6. The equations to calculate the physical position of trap within the gate oxide. The maximum and minimum of *LF* at room temperature are analyzed and given.

Fig. 7. Summary of *LF* obtained from the RTN measured in different FeFETs.

Fig. 8. Simulated *P-V* loop of the FE layer based on the phase-field model [7].

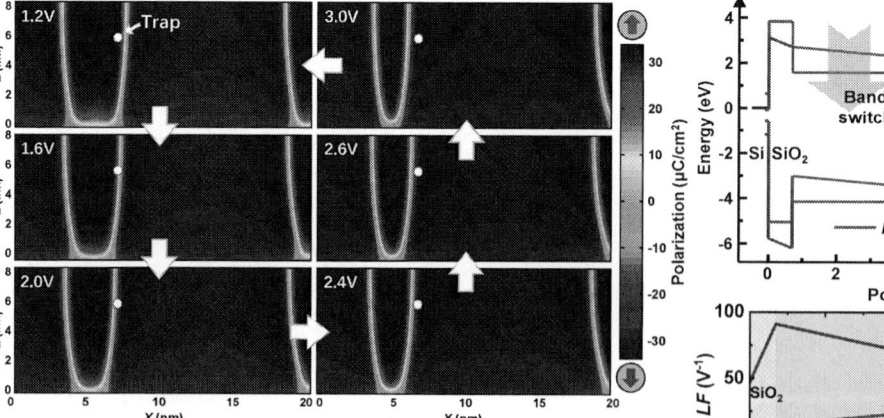

Fig. 9. Simulated polarization profiles in the FE layer at different applied voltages. Z is along the thickness direction. The trap in the vicinity of DW is plotted as the white circle in each profile. It is worth noting that the DW motion is reversible.

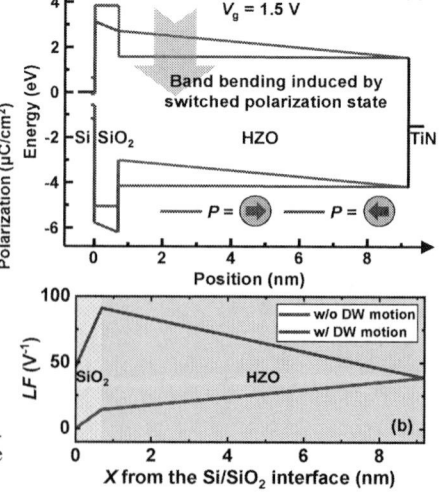

Fig. 10. (a) Energy band diagrams of the MFIS stack with FE layer in different polarization states. The band diagram is given by the FeFET simulation based on kinetic Monte Carlo nucleation-limited-switching model (MCNLS) [2]. (b) Calculated *LF* along the gate oxide with and without the regulation of DW motion. The left side is Si/SiO₂ interface and the right side is TiN/HZO interface.

979-8-3503-9164-0/24 $31.00 © 2024 IEEE

Reservoir computing using nonlinear polarization and charge dynamics of anti-ferroelectric HZO/Si FETs

<u>Shin-Yi Min</u>, Kasidit Toprasertpong, Eishin Nako, Ryosho Nakane,
Mitsuru Takenaka, and Shinichi Takagi

Department of Electrical Engineering and Information Systems, The University of Tokyo, Japan
E-mail: symin@mosfet.t.u-tokyo.ac.jp

Abstract — We demonstrate the operation of anti-ferroelectric (AFE) $Hf_{1-x}Zr_xO_2$ (HZO)/Si FETs and its potential for high-performance reservoir computing. The Zr content is adjusted to modulate the AFE properties of HZO system. The AFE behaviors with double polarization switching in FET with an AFE-HZO gate stack on a Si substrate are demonstrated. High reservoir computing capacities for time-series computational tasks are experimentally achieved thanks to rich polarization switching dynamics and polarization/charge interaction in AFE-FETs. Further enhanced reservoir computing performance is achieved by utilizing multiple current components of AFE-FETs.

I. INTRODUCTION

Reservoir computing (RC), which is a machine learning technique derived from recurrent neural networks, is promising for real-time information processing [1, 2]. In RC, only weights in the output layer are trained with simple methods such as ridge regressions (Fig. 1(a)), resulting in fast and low-power training [3]. Here, the reservoir should generate time-series response signals to realize a history-dependent nonlinear transformation of time-series inputs into higher-dimensional node states. We have demonstrated a CMOS-compatible physical RC system utilizing a $Hf_{1-x}Zr_xO_2$ (HZO)-based FeFET [4, 5]. The challenge of RC using conventional FeFETs is the insufficient complexity of polarization dynamics for nonlinear transformation. In this study, we propose anti-ferroelectric (AFE) FETs as a promising solution to the monotonous polarization dynamics of FeFET reservoirs (Fig. 1(b)). It is well known that Zr-rich HZO films exhibit AFE properties [6, 7]. Unlike a single polarization loop for FE-HZO, AFE-HZO shows double hysteresis loops, which makes response signals have a variety of features. However, HZO/Si FETs with proper AFE characteristics are not easily achieved on Si substrates [8] and experimental evidence is still lacking. In this paper, AFE behaviors with double polarization switching in FET with an AFE-HZO gate stack on a Si substrate are demonstrated. We experimentally present a high potential for effective RC system through rich dynamical characteristics of HZO/Si AFE-FETs and further computing performance enhancement is achieved by utilizing multiple current components of AFE-FETs.

II. EXPERIMENTAL DETAILS

A. Impact of Zr content in HZO system

Fig. 2 summarizes the process flow of TiN/HZO/TiN MFM capacitors and HZO FETs with different Zr contents. Fig. 3(a) shows the polarization switching properties of the MFM capacitors. As Zr content increases, the FE hysteresis loop appears with higher remanent polarization (P_r). Further increasing Zr content leads to a polarization loop thinning at zero bias with decreasing coercive voltage (V_c), resulting in AFE-like double-loop hysteresis behavior. The maximum polarization (P_{max}) and P_r of [Zr] = 75 % are higher than [Zr] = 100 % due to partial o-phase (Fig. 3(b)). Figs. 4(a, b) show the P-V_g loop, $2P_r$, and P_{max} values for HZO/Si FETs with different Zr contents. The AFE properties with four polarization switching current peaks can be successfully observed in Si FETs with the AFE-HZO gate stacks (Fig. 4(c)).

B. RC system utilizing HZO FET reservoir

Fig. 5 shows the operating scheme of an RC system using a HZO FET physical reservoir. The time-series random binary sequence $u(n)$ is encoded into $v(t)$ using a mask function to feed into the physical reservoir. The $I(t)$ response is sampled as virtual nodes (VNs) to create multiple-node response signals and the sum of weighted VNs gives the system output $y(n)$ [9]. The weight values are trained by using ridge regression when a specific task is given [5]. We evaluated Short-term memory (STM), Temporal-exclusive OR (XOR), and Parity check (PC) tasks as the fundamental index of RC capacities. Here, the XOR and PC tasks can evaluate both short-term memory and nonlinearity.

III. RESULTS AND DISCUSSION

A. Visualization of HZO FET reservoir by t-SNE analyses

Fig. 6(a) shows the $I_d(t)$ responses of HZO FETs with different Zr contents under same $v(t)$. The much more complicated $I_d(t)$ waveform is obtained for AFE-FET reservoir with [Zr] = 75 %. The nonlinear mapping capability of a reservoir can be visually evaluated by (t-distributed stochastic neighbor embedding) t-SNE analyses [10]. Fig. 6(b) shows the separation of VN states for 4-bit input. The HZO FET with [Zr] = 75 % achieves the highest number of distinguishable 12~13 classes, indicating that its dynamical polarization switching enhances RC performance.

B. RC performance improvement in AFE-FET

In evaluating the RC performance, an RC system utilizes a combination of VNs for multiple current components (I_d, I_s, $\bar{I}_d$, and $\bar{I}_s$) of AFE-FETs by applying $v(t)$ and $\bar{v}(t)$, inverted data of $v(t)$, as input signals (Fig. 7(a)). The different current components including the complementary responses ($\bar{I}_d$ and $\bar{I}_s$) can provide rich information [5, 11]. Fig. 7(b) shows the squared correlation r^2 between $y(n)$ and target output $d(n)$ as a function of the delay time step, T_{delay}. The combined VNs for I_d, I_s, $\bar{I}_d$, and $\bar{I}_s$ result in better RC performance than using VNs for only I_d. Fig. 8 summarizes the total RC capacities of HZO FETs with different Zr contents. The highest computing performance with $C_{STM} = 2.99$, $C_{XOR} = 2.93$, and $C_{PC} = 2.73$ has been achieved for HZO AFE-FETs with [Zr] = 75 % by combining VNs for I_d, I_s, $\bar{I}_d$, and $\bar{I}_s$. Fig. 9(a) shows the history-dependent $I_d(t)$ responses to different 6-bit input sequences, where the left 5 bits are the past history (leftmost bit is the input at T_{delay} of 5) and the rightmost bit is the present input (Fig. 9(e)). Fig. 9(b) shows the difference in $I_d(t)$ with the same present input data of '1' but different past histories and Fig. 9(c) summarizes the maximum values of $\Delta I_d(t)$. A large difference in history-dependent $I_d(t)$ is advantageous for discerning the input patterns and it is found that large $\Delta I_d(t)$ is obtained for HZO FETs with [Zr] = 75 %. As shown in Fig. 9(d), the response difference is clearer in AFE-FET with [Zr] = 75 % than in FeFET with [Zr] = 50 % under the same $v(t)$ period due to dynamical polarization switching. These results indicate that AFE properties enlarge the difference in the time response after different input histories and, thus, enhance the RC performance.

C. Prediction of nonlinear time-series data

We examine more practical tasks predicting time-series data from an Nth-order nonlinear dynamic system (NARMA-N) (Fig. 10(a)) [12]. Figs. 10(b, c) show the input $u(n)$ and predicted output $y(n)$ of the trained HZO FET-based RC system (solid line is theoretical model output $d(n)$) for NARMA-3, and the correlation between $d(n)$ and $y(n)$. In Fig. 10(d), the prediction accuracy is improved by combining VNs in the HZO FET with [Zr] = 75 %, achieving the lowest normalized mean square error (NMSE) value of 9.56×10^{-3} for NARMA-3 task.

IV. CONCLUSIONS

We have experimentally clarified that the HZO/Si AFE-FETs have a high potential for effective RC. The rich polarization switching dynamics of AFE-FETs demonstrate the enhanced RC performance with complicated time-series response signals.

ACKNOWLEDGMENTS

This work was supported by JST CREST Grant Number JPMJCR20C3 and JSPS KAKENHI Grant Number 21H01359, Japan.

REFERENCES

[1] H. Jaeger, GMD Report 148 (2001). [2] W. Maass et al., Neural Comp. **14**, 2531 (2002). [3] G. Tanaka et al., Neural Networks **115**, 100 (2019). [4] E. Nako et al., VLSI Symp., TN1.6 (2020). [5] K. Toprasertpong et al., Commun. Eng. **1**, 21 (2022). [6] J. Muller et al., Nano Lett. **12**, 8 (2012). [7] Y. C. Chen et al., IEDM, 338 (2021). [8] Z. Liang et al., IEDM, 382 (2021). [9] L. Appeltant et al., Nat. Commun. **2**, 468 (2011). [10] L. van der Maaten et al., J. Mach. Learn. Research **9**, 2579 (2008). [11] S. Takagi et al., IEDM, 22.2 (2023). [12] A. F. Atiya et al., IEEE Trans. Neural Netw. **11**, 697 (2000).

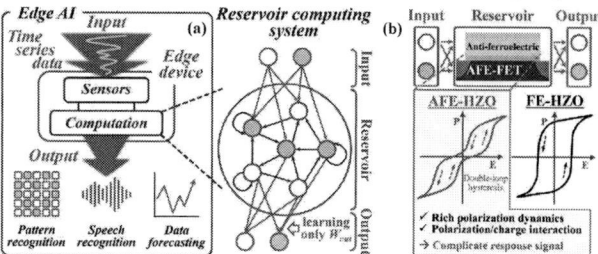

Fig. 1 (a) Concept of an RC system for edge-AI technologies and (b) proposed RC by using AFE-FETs.

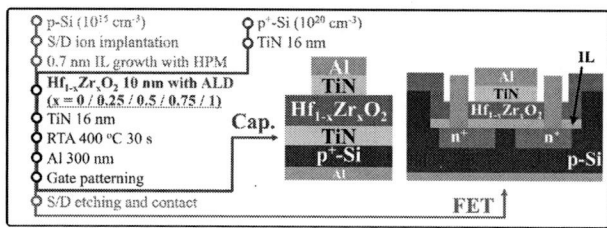

Fig. 2 Process flow of HZO-based capacitors and FETs with different Zr contents.

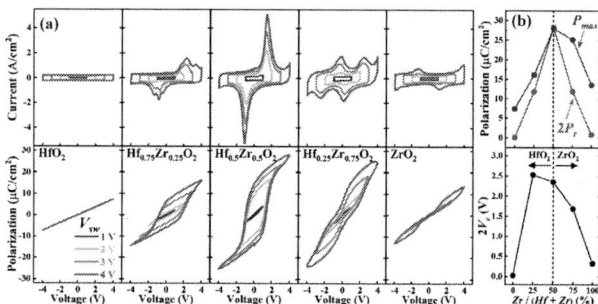

Fig. 3 (a) Polarization switching currents and P-V hysteresis loops of TiN/HZO/TiN MFM capacitors with different Zr contents. (b) $2P_r$, P_{max}, and $2V_c$ values.

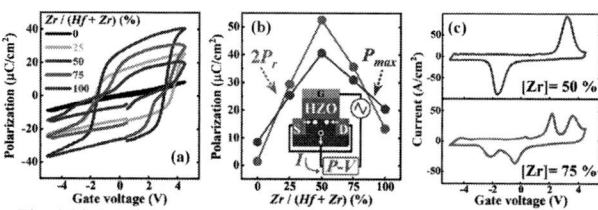

Fig. 4 (a) P-V_g loop, (b) $2P_r$ and P_{max} values for HZO FETs with different Zr contents. (c) I-V_g characteristics for HZO FETs with [Zr]=50 and 75 %.

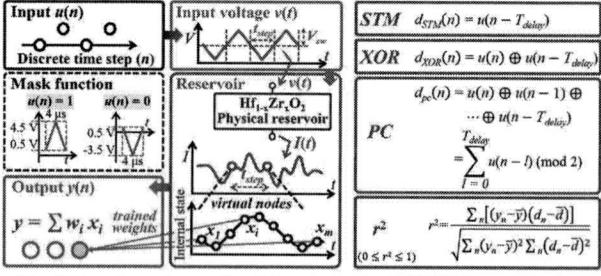

Fig. 5 Operating scheme of an RC system using HZO FET-based physical reservoir and the definition of calculation tasks for the fundamental index of RC capacities.

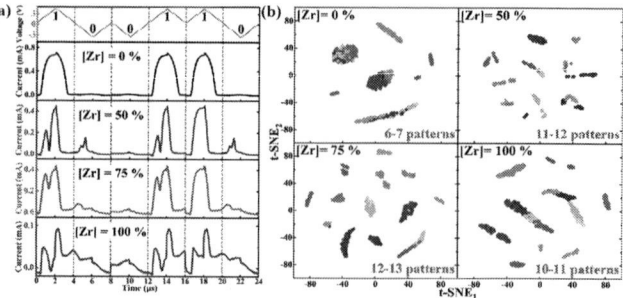

Fig. 6 (a) $I_d(t)$ responses of HZO FETs with different Zr contents. (b) t-SNE analyses of the I_d response of HZO FETs with [Zr]=0, 50, 75, and 100 % for 4-bit digital input.

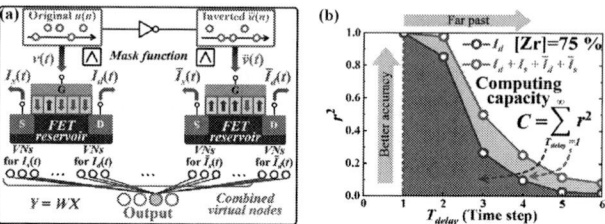

Fig. 7 (a) HZO FETs-based RC system combining VNs of multiple current components. (b) r^2 of the STM task for HZO FETs with [Zr]=75 % and the definition of RC capacity.

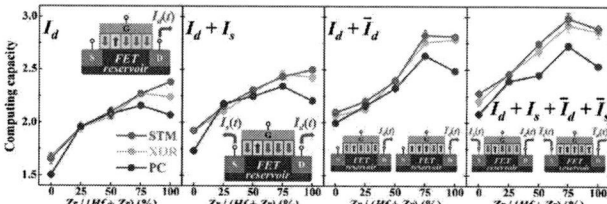

Fig. 8 Comparison of STM, XOR, and PC capacities for RC using combined VNs of HZO FETs with different Zr contents.

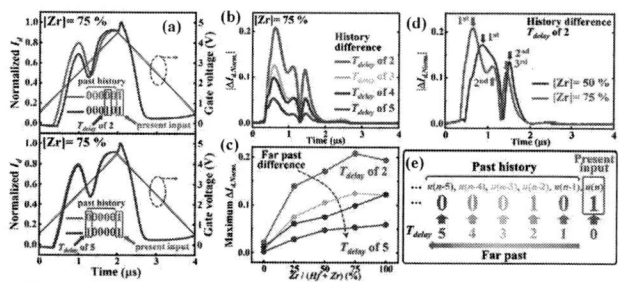

Fig. 9 (a) History-dependent normalized $I_d(t)$ responses and (b) $|\Delta I_{d.Norm}|$ for different input history (T_{delay} difference from 2 to 5) for HZO FETs with [Zr]=75 %. (c) Maximum $\Delta I_{d.Norm}$ for input history of HZO FETs with different Zr contents. (d) $|\Delta I_{d.Norm}|$ for HZO FETs with [Zr]=50 and 75 %. (e) Meaning of input history.

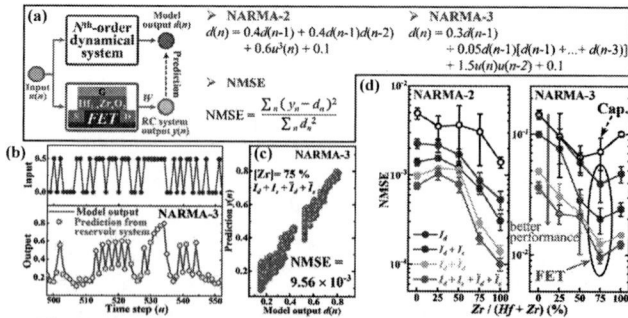

Fig. 10 (a) Schematic of NARMA task. (b) Input signal and prediction results of NARMA-3. (c) Correlation of the model output and the predicted system output. (d) Comparison of NMSE of NARMA-2 and NARMA-3 tasks for HZO FETs with different Zr contents using combined VNs.

Design space trade-offs in networks of coupled injection-locked Ring Oscillators for Ising-based optimum search

Ali Bazzi, Franck Badets, Louis Hutin

CEA-Leti, Univ. Grenoble Alpes, F-38000 Grenoble, France

email: ali.bazzi@cea.fr

Abstract—Based on theoretical considerations and experimental measurements, we provide insight on the operating range for Ising networks constructed with injection-locked CMOS ring oscillators (ROs) for energy-efficient optimum search. We propose and validate a design in which ROs are electrically connected through weighted coupling blocks featuring back-to-back inverters with adjustable supply voltage. Experimental measurements on our reconfigurable 22nm CMOS test chip cover the frequency locking range for network synchronization under various coupling conditions, and the resolution of several weighted MAX-CUT graph partitioning problems, with a focus on the circuit parameters influencing solution optimality.

INTRODUCTION

Current research in energy-efficient computing hardware explores multiple promising avenues, among which the development of specialized accelerators to address NP-hard optimization problems. These problems pose significant challenges due to the trade-offs between solution quality, time to solution, and power consumption when processed on traditional von Neumann architectures [1]. When tackled by local search heuristics, they can be expressed by the Ising model [2], and thus can be mapped to networks of directly interacting binary devices or circuits that spontaneously adjust their states through local interactions, in a way that still minimizes a global objective function. In particular, the elementary "Ising spins" σ_i can be encoded in the phase of injection-locked oscillators, and electrically connected through weighted coupling blocks **Fig. 1** [3-6]. We focus in this work on solving weighted MAX-CUT instances on a test chip featuring CMOS Ring Oscillators, reconfigurable weights and built-in phase readout. We investigate in the following the relationships between graph connectivity, locking range and solution quality.

ARCHITECTURE AND DESIGN SPACE

Our binary spins σ_i are materialized by 7-stage Ring Oscillators of free-running frequency ω_0, each of them receiving a shared synchronization signal at $2\omega_0$ that causes their phase to stabilize to one of two states mutually separated by π **Fig. 1**. The global strength of this synchronization signal is parametrized by a current I_{SYNC} **Fig. 2(a)**. Back-to-back inverter blocks are used for bidirectional coupling of the ROs. Creating a connection between the same stages of two ROs σ_i and σ_j causes them to be out-of-phase, translating into a negative J_{ij}. The supply voltages of back-to-back inverters (V_{B2B}) are tunable separately. A reference oscillator can be chosen to act as both a computation element and a time reference to generate a sampling signal in quadrature with respect to the sensed outputs **Fig. 2(b)**. This sampling signal is then fed to D-latches that convert the RO outputs to two voltage levels (V_{DD} and 0) depending on their phase, thus greatly facilitating state readout across the network. SPICE simulations on **Fig. 3** describe the relationship between V_{B2B} and coupling strength, expressed qualitatively as the time needed by two negatively-coupled oscillators to become out-of-phase. It is shown to be strongly non-linear, and also dependent on I_{SYNC}. Furthermore and crucially, the ratio between V_{B2B} and I_{SYNC} delineates the design space boundaries, due to the necessary balance between $2\omega_0$ synchronization perturbations forcing the system to explore a binary search space, and ω_0 coupling perturbations encoding the problem itself. In other words, a too large ratio could cause instability or failure to binarize in some of the oscillators, whereas a too small ratio would cause fast convergence to an inaccurate result. The optimal regime depends on graph connectivity and weights definition, which is discussed and supported by measurements in the following section.

TEST CHIP AND MEASUREMENTS

A test chip containing a 3×6 array of ROs coupled in King's graph topology was designed in 22nm technology. The adjacency matrix **Fig. 4** reflects the accessible graph connectivity and maximal density within the routing constraints. I_{SYNC} is set externally and the V_{B2B} values are adjustable on the test board using potentiometers. We measured the frequency locking range for an isolated oscillator (RO1) at varying I_{SYNC} values **Fig. 5(a)**. Phase binarization ideally occurs when the injected signal matches twice the oscillator frequency, with tolerance improving as the synchronization strength I_{SYNC} increases [7]. Yet at fixed I_{SYNC}, the locking range can be significantly altered by coupling to other oscillators (RO2). **Fig. 5(b)** illustrates a shift downward in the central frequency and wider boundaries, primarily due to two factors. First, the extra load coming from connections decreases the oscillator's own frequency. Second, shared current coming from connected neighbors, who are also undergoing phase synchronization, exerts a stabilizing effect, thereby extending the locking range. However, it is important to note that the extent of shared injection signal at any given time is contingent upon the relative states of connected oscillators. Specifically, current sharing occurs only when the phase states oppose the intended coupling sign, such as when two negatively-coupled oscillators are in-phase. Consequently, the upper and lower boundaries of the locking range differ significantly for each node in a network, and vary with each new problem instance that has reconfigured weights and connectivity. Despite these fluctuations, we aimed to define a set of coupling and synchronization parameters that would enable solving different graphs with regular (G1) or irregular (G2) connectivity, and support multi-level weights (G3) **Fig. 6(a-c)**. The outputs were scanned sequentially with an oscilloscope, the bar plots on **Fig. 6(a-c)** counting the occurrence of sampled 0 and V_{DD} for each oscillator integrated over $110\mu s$. We derive from these bar plots the node coloring corresponding to steady-state, and define optimality as the cut size, i.e. the weighted sum of edges connecting out-of-phase oscillators, normalized by that of the known optimal solution. The solution quality is shown in **Fig. 7** for two values of I_{SYNC}. While a higher $I_{SYNC}=7\mu A$ did provide an extended locking range and enhance network synchronization, it also led overall to solutions of lower quality. The lower $I_{SYNC}=5.5\mu A$ could lead to optimal solutions for G1 and G3, and 96.15 % optimality on G2.

CONCLUSION

We designed and tested a functional network of Ring Oscillators to solve MAX-CUT problems based on the Ising model. Our investigation into the trade-offs of the design space focused on phase binarization through injection-locking and how these might vary with each problem instance for individual oscillators within the network. Leveraging both simulation and experimental data, we refined our selection of circuit parameters to optimize solution quality across various graphs.

979-8-3503-9164-0/24 $31.00 © 2024 IEEE

6.2

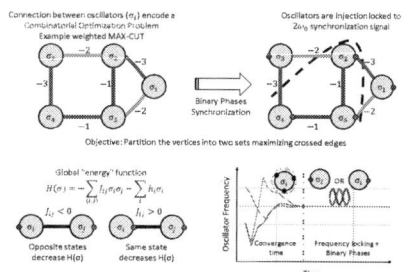

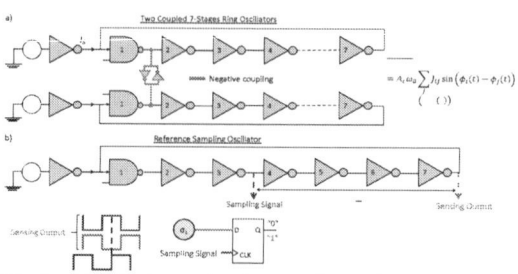

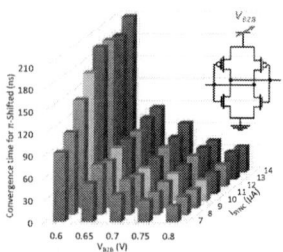

Fig.1: Ising formulation of weighted MAX-CUT problem, an NP-hard optimization problem, using a network of injection-locked oscillators. The coupled oscillators operate as a dynamic system, that converges into a steady state where the MAX-CUT solution is encoded within their phases.

Fig.2: **(a)** Hardware implementation of binary nodes (injection-locked ring oscillators) and bidirectional coupling edges (back-to-back inverter blocks). **(b)** The reference sampling oscillator acts, after injection locking, as a clock for D-latches converting each oscillator phase state into either V_{DD} or 0V.

Fig.3: Time needed for a pair of negatively-coupled Ring Oscillators to converge to opposite phases, as function of the coupling block supply voltage (V_{B2B}) and synchronization currents (I_{SYNC}).

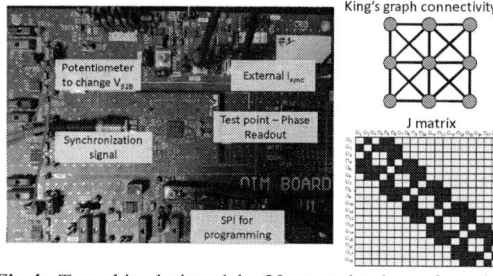

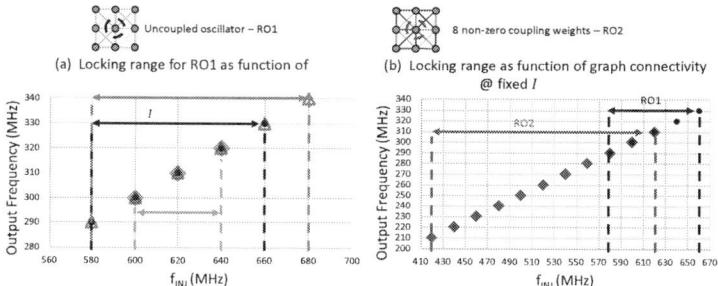

Fig.4: Test chip designed in 22nm technology, featuring 3x6 network of RO connected in King's graph topology. The coupling and synchronization strengths are tuned on the testboard. Phase sampling is included in the design and a single testpoint is used to measure by oscilloscope the phase sampling of the activated network.

Fig.5: **(a)** Locking range measurements for an isolated oscillator (RO1) demonstrating the dependency of the locking range on the synchronization strength represented by (I_{SYNC}). **(b)** Effects of the connectivity on the locking range for a coupled oscillator (RO2). The coupling induces a downward shifting of the central frequency as well as a wider range.

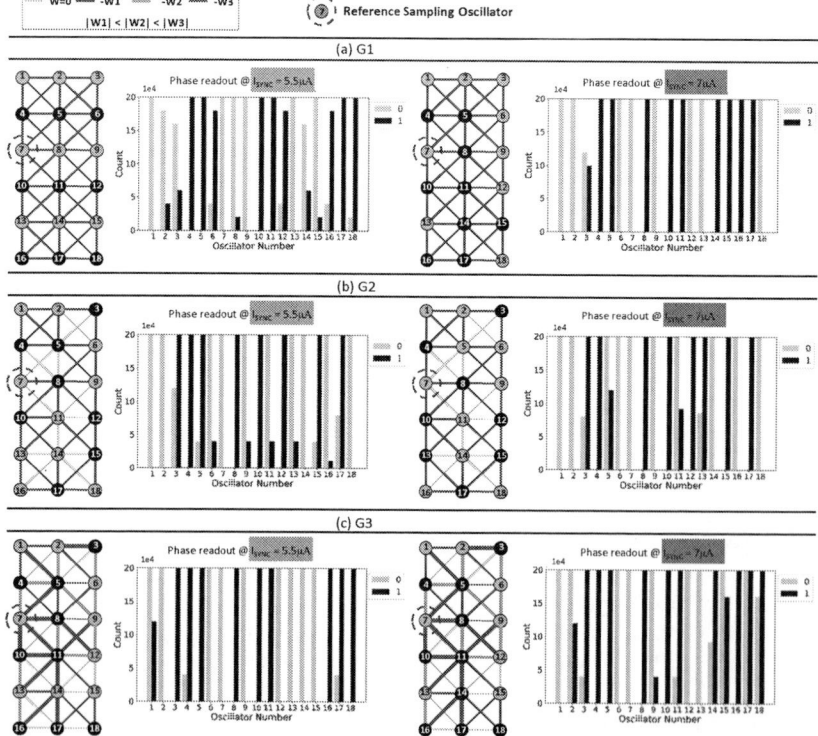

Fig.6: **(a-c)** Measurements for three different MAX-CUT problems having different connectivity and edge weights. Oscillator 7 serves as the sampling clock utilized by the D-latches. The bar plots depict the distribution of oscillator phases in either of the states '0' and '1'. Changing the I_{SYNC} value directly impacts the convergence of the steady state.

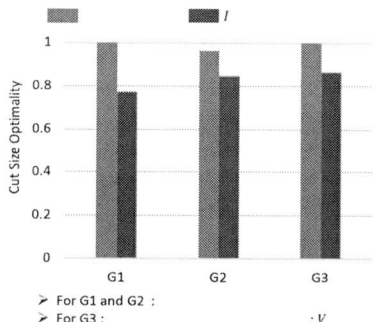

> For G1 and G2 :
> For G3 :   ; V

Fig.7: Normalized cut size for graphs shown in **Fig. 6(a-c)** at different synchronization strengths. Increasing the synchronization strength represented by (I_{SYNC}) tends to degrade the solution quality for all graphs.

ACKNOWLEDGMENT

The authors acknowledge support by the European Commission and the French government through the IPCEI and Nano22 programs.

REFERENCES

[1] I. L. Markov, *Nature*, pp. 147–154, 2014.
[2] A. Lucas, *Frontiers in Physics, 2014.* [3] J. Roychowdhury *et al.*, *Proc.IEEE 103*, 2015.
[4] I. Ahmed *et al.*, *IEEE JSSC 56*, 2021. [5] M. Graber, K. Hofmann, *IEEE SOCC*, 2022.
[6] M. K. Bashar *et al.*, *IEEE JXCDC*, 6, 2020.
[7] B. Hong *et al.*, *IEEE JSSC 54*, 2019.

979-8-3503-9164-0/24 $31.00 © 2024 IEEE   40

Background-Pattern-Dependency-Tolerant Weight Transfer Method for Accurate NAND-Flash-Based Spiking Neural Networks

Bosung Jeon, Jin Ho Chang, and Woo Young Choi[*]

Department of Electrical and Computer Eng. and Inter-university Semiconductor Research Center (ISRC),
Seoul National University, 1 Gwanak-ro, Seoul, Republic of Korea
[*]Email: wooyoung@snu.ac.kr

Abstract —This study focuses on the background pattern dependency (BPD) issue of NAND flash, where threshold voltage (V_{TH}) or bit line current (I_{BL}) change due to the differences between verify and read operations. In NAND-flash-based spiking neural networks (SNNs), which utilize I_{BL} for analog multiply-accumulate operations, the impact of BPD is more significant than in conventional memory operation. A more-BPD-tolerant weight transfer method for accurate NAND-based SNNs is proposed and evaluated by using experimental data.

I. INTRODUCTION

Due to the immense computational costs required in deep neural networks, interest in spiking neural networks (SNNs) has notably increased. Off-chip learning, which converts well-trained weights to SNNs, demonstrates high performance and low-power operation. Charge-trap flash (CTF) memory is a powerful candidate, especially NAND flash array, due to its capability for precise weight control and high integration density. Nevertheless, the background pattern dependency (BPD), where the state of the selected cell fluctuates due to the changes in surrounding cell states during verify and read operations, is pervasive in NAND arrays [1], [2]. SNNs that utilize bit line current (I_{BL}) as synaptic weight might experience significant distortion due to the BPD, as depicted in Fig. 1 [3]. In this manuscript, we examined the impact of weight transfer errors (WTE) caused by the BPD with a fabricated NAND CTF array. Furthermore, we proposed a BPD-tolerant two-step weight transfer method for more accurate SNNs verified through an array measurement.

II. RESULTS AND DISCUSSION

A TANOS (TiN/Al$_2$O$_3$/Si$_3$N$_4$/SiO$_2$/Si) CTF NAND array was fabricated, with its process flow outlined in Fig. 2(a). The top view of the NAND array with ten word lines (WLs) and three strings is illustrated in Fig. 2(b), and the TEM image of the gate stack is shown in Fig. 2(c). To determine the optimal read and pass biases (V_{read}, V_{pass}), we applied various program pulses to the cell with its V_{th} adjusted to 2 V. As shown in Fig. 2(d), while a V_{pass} of 7 V was obtained based on the V_{th} sensing, a lower V_{pass} of 5.5 V was determined when considering I_{BL} sensing as depicted in Fig. 2(e)

(V_{read}=3.3 V). As even slight V_{th} variations induce substantial changes in I_{BL}, the V_{pass} of the NAND array for SNN synapse is limited, exacerbating the BPD issue. A Two-phase program method to address BPD in neuromorphic applications has been previously proposed, but it tends to result in significant fluctuations for the lowest weight [4]. To mitigate this, we proposed the distinct two-step transfer method, which begins by clustering all cells near the largest weight, followed by sequentially transferring according to the ascending order of weight values as shown in Fig. 3. To evaluate our strategy, we compared the weight transfer results of WL-position based (A), the method in [4] (B), and our method (C) as demonstrated in Fig. 4. Assuming weight quantization to 8 levels from 300 nA (W_0) to 1 μA (W_7), we conducted the weight transfer of the fabricated NAND array for 160 cells. The WTE values for each weight can be observed in Fig. 5(a), and average WTE values are illustrated in Fig. 5(b). While method B can reduce the WTE for larger weights (W_6, W_7) compared to method A, WTE for the smallest weight (W_0) dramatically increases due to its transfer sequence. However, method C was able to reduce the BPD across all weights. The average WTEs of total cells for the three methods are 5.44, 4.50, and 2.01 %, respectively, demonstrating that the proposed method achieves the more accurate weight transfer results.

III. CONCLUSION

This paper examines the background pattern dependency (BPD) problem in NAND flash array-based hardware spiking neural networks (SNNs). As SNNs employ multiply-accumulate operation with bit line current (I_{BL}) sensing, the NAND arrays are more vulnerable to BPD than regular memory operation, which could affect the performance of off-chip SNNs. Before proceeding with the weight-value dependent sequential transfer, a simple approach that congregates all cells near the largest weight value demonstrated more BPD-tolerant results in a fabricated NAND array. Through this method, we achieve more accurate weight transfer results in NAND flash-based SNNs.

ACKNOWLEDGMENTS

This work was supported in part by the IITP Grant funded by MSIT (2020-0-01294), in part by the NRF funded by MSIT (NRF-2022M3I7A1078544: PIM Semiconductor Technology Development Program) in part by the Brain Korea (BK) 21 Four Program of the Education and Research for Future ICT Pioneers at Seoul National University in 2022.

REFERENCES

[1] S. Lee, J.-y. Lee, I.-h. Park, J. Park, S.-w. Yun, M.-s. Kim, J.-h. Lee, M. Kim, K. Lee, and T. Kim, "7.5 A 128Gb 2b/cell NAND flash memory in 14nm technology with tPROG= 640μs and 800MB/s I/O rate," in *IEEE International Solid-State Circuits Conference (ISSCC)*, 2016, pp. 138-139.

[2] W.-C. Chen, H.-T. Lue, K.-P. Chang, Y.-H. Hsiao, C.-C. Hsieh, Y.-H. Shih, and C.-Y. Lu, "Study of the programming sequence induced back-pattern effect in split-page 3D vertical-gate (VG) NAND flash," in *Proceedings of International Symposium on VLSI Technology, Systems and Application (VLSI-TSA)*, 2014, pp. 1-2.

[3] J.-W. Back, H.-N. Yoo, J. Kim, M.-K. Park, W. Y. Choi, and J.-H. Lee, "String Current Compensation Method in VNAND Flash for Hardware-Based BNNs," *IEEE Transactions on Electron Devices*, vol. 69, no. 12, pp. 6717-6721, 2022.

[4] M. Kim, M. Liu, L. R. Everson, and C. H. Kim, "An embedded NAND flash-based compute-in-memory array demonstrated in a standard logic process," *IEEE Journal of Solid-State Circuits*, vol. 57, no. 2, pp. 625-638, 2021.

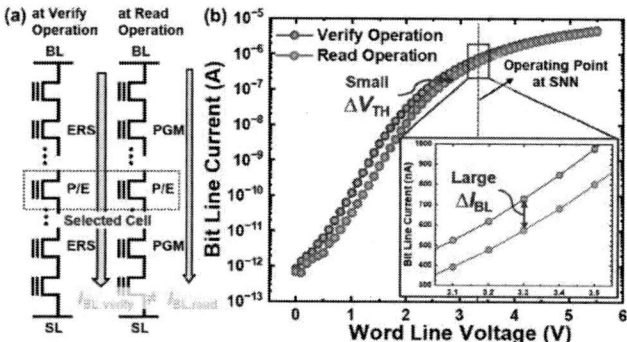

Fig. 1. (a) Background pattern dependency problem in NAND array. (b) The change in I_{BL}-V_{WL} curves is caused by BPD, which highlights the differences in ΔV_{TH} and ΔI_{BL}.

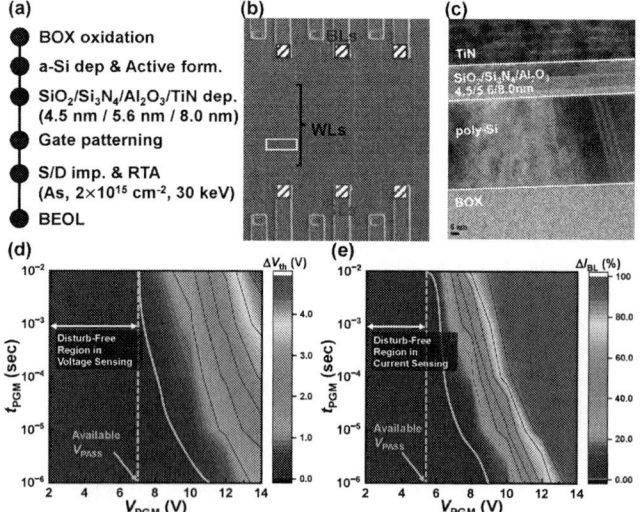

Fig. 2. (a) Process flow of charge-trap flash cell. (b) Top view of NAND array. (c) Cross-sectional TEM view of gate stack. Contour plot of V_{TH} (d) and I_{BL} (e) changes in various program pulse conditions.

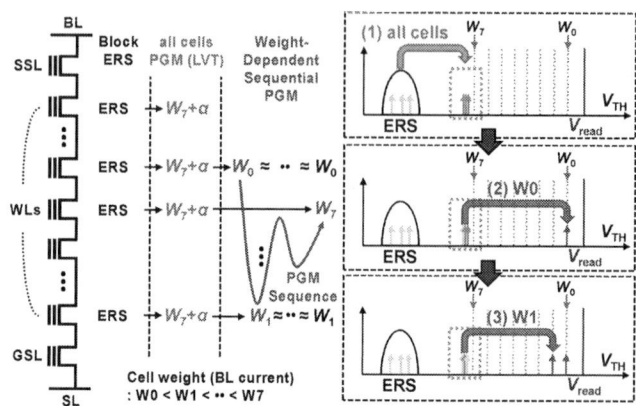

Fig. 3. Proposed BPD-tolerant two-step weight transfer method: 1) Congregate all cells to $W_7+\alpha$, 2) Weight transfer in ascending order of weight values.

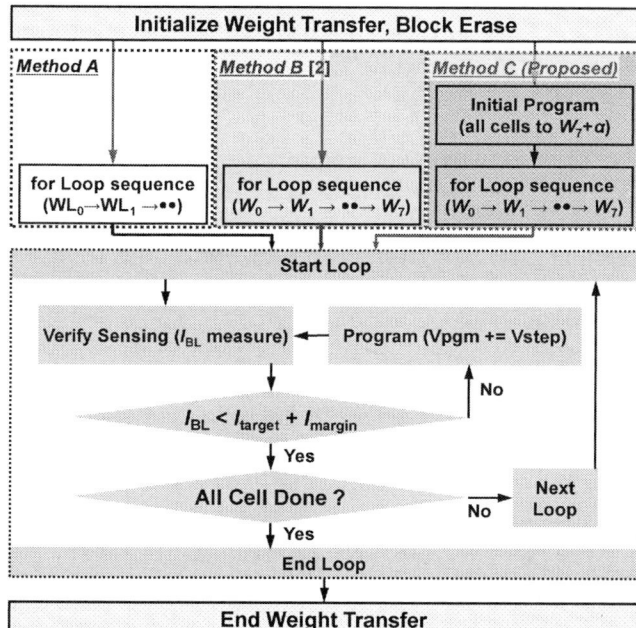

Fig. 4. Comparison of three weight transfer experiments; the sequence of weight transfer is as follows: (A) according to WL position, (B) the method from reference [4], and (C) the proposed method as in **Fig. 3**.

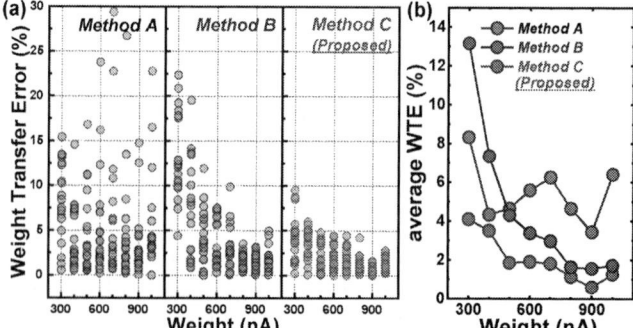

Fig. 5. For 160 cells, (a) all results of WTE measured for each weight value and (b) the average of WTE for each weight value.

979-8-3503-9164-0/24 $31.00 © 2024 IEEE

Artificial VO₂ Spiking Neurons with Protective Mechanism for Enhancing Resilience of Spiking Neural Network Against Adversarial Attacks

Chaoyi Ban[1], Linbo Shan[1], Gaoqi Yang[1], Jiajun Gao[1], Lindong Wu[1], Zongwei Wang[1,2], Yimao Cai[1,2], Ru Huang[1,2]

[1]School of Integrated Circuits, Peking University, Beijing, 100871, China.
[2]Beijing Advanced Innovation Center for Integrated Circuits, Peking University, Beijing, China
Email: {wangzongwei, caiyimao}@pku.edu.cn

Abstract — **Spiking Neural Networks (SNNs) have demonstrated significant potential in power-efficient edge computing. However, SNNs are vulnerable to adversarial attacks which seriously affects the accuracy and thus limits its application. In this study, a defense mechanism against adversarial attacks is demonstrated based on VO₂ spiking neuron. The spiking neuron circuit (SNC) can mimic the protective mechanism of neurons to effectively filter adversarial perturbations among input signals. Furthermore, we evaluate the ability of SNC-based SNNs against adversarial attacks with MNIST task. The SNC-based SNNs can effectively withstand adversarial attacks, achieving an 18.87% improvement in accuracy compared to the ReLU method under 30% disturbance. The results demonstrate that the SNC method has significant advantage in enhancing the SNN system's ability to resist adversarial attacks. This research provides an innovative solution to tackle the security issue in highly reliable neuromorphic computing.**

Keywords: SNNs, memristor, spiking neuron, adversarial attacks

I. INTRODUCTION

Spiking Neural Networks (SNNs), inspired by biological systems, have significant potential in the field of edge computing due to their low-power consumption, multitasking capabilities, and event sparsity [1-2]. However, SNNs have a weak ability to withstand adversarial attacks caused by adversarial perturbations, which can decrease the network accuracy and reliability [3-5]. Although several studies have focused on improving the resilience of SNNs against adversarial attacks, these methods still encounter significant integration challenges because of complex circuit structure and scalability issues [6]. Herein, we introduce a defense mechanism against adversarial attacks utilizing IMT (insulator-metal-transition) memristor-based spiking neuron circuit (SNC). The neuron only fires spikes within a certain range of voltage, which can be used to filter the disturbance in input signals. The SNN system, based on the SNC, can effectively confront adversarial attacks, achieving an 18.87% increase in accuracy compared to the ReLU method under 30% disturbance, enhancing the ability of SNN system to withstand adversarial attacks.

II. EXPERIMENTS

The memristor fabrication process was described as follows: a 30 nm VO₂ layer was deposited through atomic layer deposition on 300nm SiO₂ substrate. Then the planar Ti/Au electrodes were formed by Ebeam evaporator after Electron beam lithography (EBL) and patterned through a lift-off process. **Fig.1** shows the scanning electron microscope (SEM) image and schematic illustration of the planar Au/Ti/VO₂/Ti/Au device. The Raman spectrum shows that the as-fabricated VO₂ has peaks at 190.0, 224.8, 308.5, 394.3, and 611.5 cm-1, indicating that the VO₂ is in the M1 phase (**Fig.2**). The resistance-temperature curve of the VO₂ is depicted in **Fig.3**, which indicates more than three orders of magnitude

III. RESULTS AND DISCUSSION

Fig.4 shows the typical switching curves and cycle endurance test of the VO₂ memristor for 300 switching cycles. The device demonstrates IMT characteristics, which can be utilized to achieve neuronal oscillations. The distribution of positive and negative threshold and holding voltages, including V_{th}, V_{hold}, $-V_{th}$, $-V_{hold}$, in 300 repeated cycles is shown in **Fig.5**. By applying AC pulses, the measured delay time of the device is ~30ns (**Fig.6**). **Fig.7** displays the circuit diagram of spiking neuron, consisting of a load resistance (R_L) and parasitic capacitance. **Fig.8** shows the typical spiking waveforms of the artificial spiking neuron. **Fig.9** shows the spiking frequency of the artificial spiking neuron when adopting varied R_L under a constant input voltage of 5 V, demonstrating that the frequency gradually decreases as R_L increases. **Fig.10** demonstrates the relationship between the spiking waveforms and input voltage when R_L is fixed at 80 kΩ. And the dependence of spiking frequency on pulse amplitude, is demonstrated in **Fig.11**. The spiking frequency initially increases and then decreases with the rise in pulse amplitude due to the prolonged discharge time at excessively high voltage. Such characteristics can be utilized to mimic the protective mechanism of neurons, preventing neurons from aggressive stimulation.

Based on the protective mechanism of neurons, we design a fully connected neural network with spiking SNC to enhance the ability of SNN system against adversarial attacks (**Fig.12**). **Fig.13** demonstrates the SNN accuracies based on ReLU and SNC under disturbance. Under 30% disturbance, SNN achieves an 18.87% increase in accuracy compared to the ReLU. The results indicate that the SNC method demonstrates ability to enhance the performance of the SNN system in defending against adversarial attacks, owing to its filtering property.

IV. CONCLUSION

In summary, we introduce a spiking neuron circuit utilizing the IMT memristor, which enhances the resilience of the SNN system against adversarial attacks. The spiking neuron circuit mimics the protective mechanism of neurons, which can effectively filter the noise in input signals. The SNN based on SNC achieves an accuracy increase of 18.87% under 30% disturbance, demonstrating obviously advantage in confronting adversarial attacks.

Acknowledgment: C. Ban and L. Shan contributed equally. This work was supported by the NSFC (62025401, 62322401, 62304006, 92164205, 61927901), the 111 project (B18001), and the China Postdoctoral Science Foundation under Grant (2023M740053).

REFERENCES: [1] Wang Z., et.al. *ISSCC*, 2021, pp.436. [2] Zhang W., et.al. *Nat. Electron.*, 2020, vol.3, no.7, pp.371. [3] Shan L., et.al. *Adv. Intell. Syst.*, 2022, pp.2100264. [4] Zhang H., et.al. *Neural Netw.*, 2023, vol.165, pp.164. [5] Zheng R., et.al. *ECCV*, 2022, pp.175. [6] Wu Z., et.al., 2024, vol.71, no.3, pp.1446.

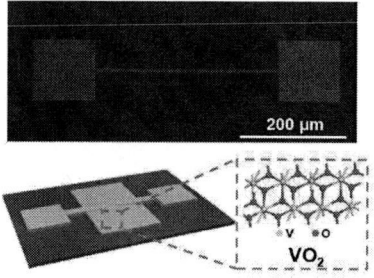

Fig.1. SEM image and schematic illustration of the in-plane Au/Ti/VO$_2$/Ti/Au device.

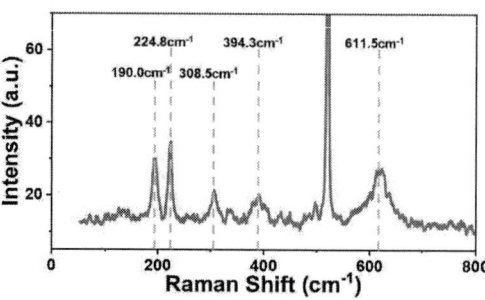

Fig.2. Raman spectroscopy of VO$_2$. The peaks at 190.0, 224.8, 308.5, 394.3, and 611.5 cm^{-1} indicate the M1 phase.

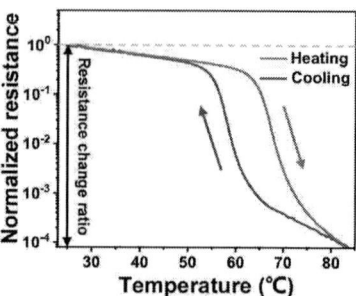

Fig.3. The resistance-temperature curve of the VO$_2$ film.

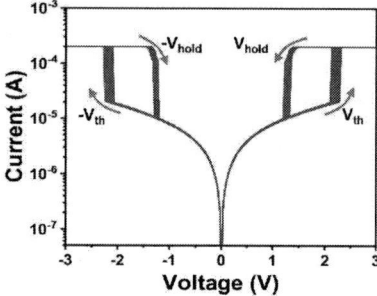

Fig.4. The typical switching curves and cycle endurance test of the device for 300 switching cycles.

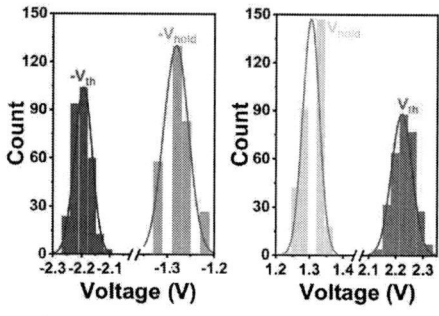

Fig.5. Statistical histogram of threshold and hold voltages under negative and positive sweeping.

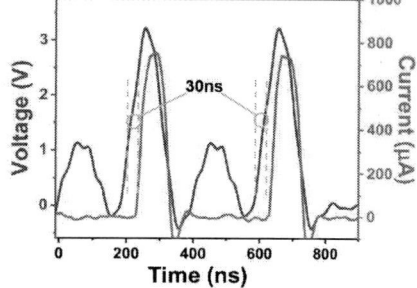

Fig.6. The AC response of the in-plane VO$_2$ device with a delay time of ~30 ns.

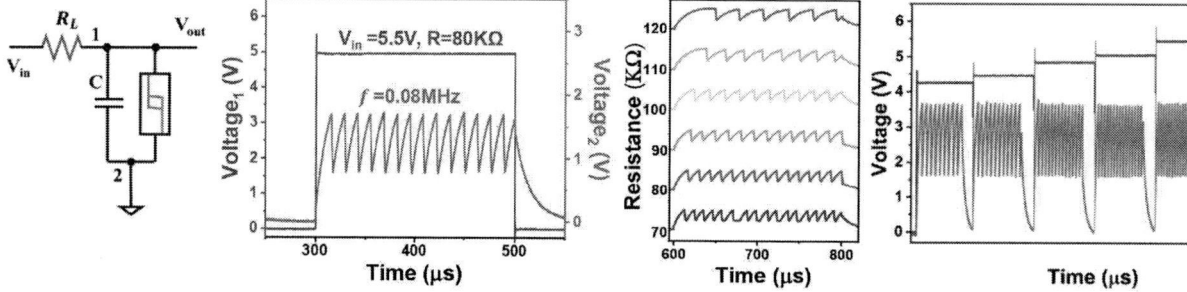

Fig.7. The circuit diagram of spiking neuron, including load resistor and parasitic capacitance.

Fig.8. The input stimulus and measured output waveforms of neuron device by applying 5.5 V pulse with 200 μs width. The loading resistance value is 80 KΩ.

Fig.9. The neuron response with varied load resistance of 70, 80, 90, 100, 110 and 120 KΩ.

Fig.10. The neuron response with different input. The firing frequency first increases and then decreases with the increase in pulse amplitude.

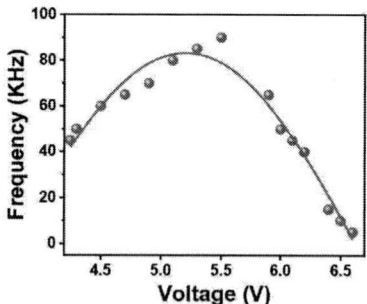

Fig.11. The correlation between input voltage and the firing frequency. The red is fitting curve by a Gaussian Function.

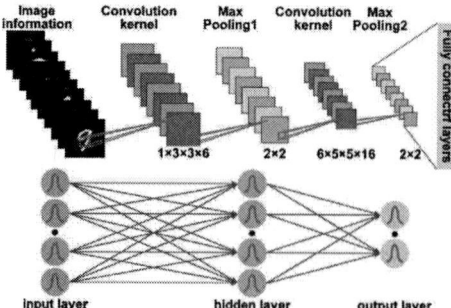

Fig.12. The schematic of the SNN system structure. and activation function in three-layer fully connected layers replaced by spiking neuron circuit.

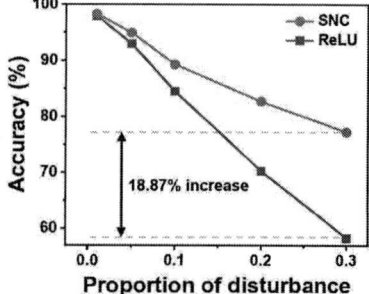

Fig.13. The comparison of accuracy between ReLU and RBSN based SNN under varying levels of adversarial attacks.

979-8-3503-9164-0/24 $31.00 © 2024 IEEE

Flash-based Computing-in-Memory (CiM) Towards Stochastic Computing with Low Power-consumption and High Noise-immunity

Hai Wang, Yang Feng, Xuepeng Zhan*, Maoying Bai, Pengpeng Sang, Jixuan Wu, Qianwen Wang, and Jiezhi Chen

School of Information Science and Engineering, Shandong University, P. R. China,

*Email: zhanxuepeng@sdu.edu.cn

Abstract- Flash-based In-Memory-Computing (IMC) scheme with low power-consumption and high noise-immunity is proposed to proceed with stochastic computing (SC). The logical AND function and number counter operation in the SC system are hardware implemented in individual flash memory cells and the current accumulation is processed in the array, respectively. By adopting the SC algorithm and matrix multiplexing scheme (MMS), the convolution operation is achieved with advantages in array scaling and high robustness. This work could be of great importance for developing low-power IMC systems.

Keywords- Stochastic Computing, In-Memory-Computing, NOR Flash, Robust, Convolution.

I. INTRODUCTION

As a vital part of convolutional neural networks, an efficient hardware implementation of convolution operation is of great importance [1]. This operation involves both addition and multiplication functions, often necessitating extensive numerical processing for intricate networks and applications. In the conventional Von-Neumann architecture, challenges such as memory wall bottlenecks result in substantial power dissipation and latency issues. To tackle these challenges, a data-centric In-Memory-Computing (IMC) architecture has been proposed as a compelling solution, leveraging emerging nonvolatile memories [2]. Among these memory technologies, flash memory stands out as a promising candidate due to its well-established processing technology and cost-effectiveness [3]. Nevertheless, the recognition accuracy is susceptible to significant degradation in the presence of noise during large-scale image pixel processing. Additionally, using large array sizes inevitably leads to pronounced power consumption and reliability concerns [4].

In this work, we design an individual flash memory cell to implement stochastic computing (SC), improving the noise immunity of flash arrays. According to the topology of the convolution kernel, a matrix multiplexing scheme (MMS) is proposed. This method can significantly reduce the redundant hardware cost and network power consumption.

II. METHOD

SC is a kind of probabilistic computing that converts binary or other base numbers to random number bit-stream (RNBS) of '0' and '1'. The RNBS performs logic operation bitwise leading to another RNBS as a result. The '1's of the resulting RNBS are counted by a counter and converted to a binary or other base number [5][6]. The calculation diagram is shown in Fig. 1. The logic used in convolution are AND function for multiplication and multiplexers for addition. An individual NOR flash memory cell is adopted to achieve AND function as shown in Fig.2(a). Fig.2(b) shows the corresponding truth table. When V_{Gate} is 1.2V, the '1' cell with a low threshold voltage (V_{TH}) is on, and the '0' cell with a high V_{TH} is off. When V_{Gate} is 0V, both cells are off. Fig.2(c) shows the NOR flash array, in which the source lines of the flash array act as counters to sum up the logic results of transistors connected in parallel. The diagram of convolution is shown in Fig. 3. The input image is stretched into a column vector and converted to RNBS as the gate voltage (V_{Gate}), and the kernel is repeatedly stored in the NOR flash memory as the V_{TH}.

III. SIMULATION AND RESULTS

The impact of the length of RNBS and V_{Gate} noise on accuracy is evaluated in Fig.4, which shows the numerical and SC convolution results. Fig.5(a) shows the accuracy loss of SC when V_{Gate} fluctuates from 0 to 10%. The comparison of SC and HEX solutions is shown in Fig.5(b). When the V_{Gate} fluctuates by 10%, the calculation accuracy drops by 64.58% in the conventional convolution method. As a comparison, the accuracy of the proposed structure only drops by 17.33% under the same V_{Gate} fluctuations. By alleviating the effect of V_{Gate} noise among multiple flash memory cells, stable results are acquired. MMS is proposed to reduce the growing array size and energy consumption caused by an increase in the input image pixels. Fig.6(a) shows that the convolutional kernel matrix is repeatedly stored in the unnecessary flash array. In this work, only the necessary part is stored to save the area. The complete result is obtained through multiple operations illustrated in Fig.6(b). In this way, the array size reduces significantly ($\sim 1/n^2$) and the number of operations only becomes n times without any accuracy loss, in which 'n' refers to the number of pixels in one dimension of the image. For one convolution operation, this method can reduce the energy consumption from 17.7nJ to 1.66nJ. The average energy consumption of a flash memory cell can be as low as $\sim 6.47fJ$. The benchmark is listed in Fig.7. This work exhibits low power consumption with a small array size.

IV. CONCLUSION

A new convolution scheme based on NOR flash is proposed. Benefitting from the IMC-based SC, the proposed convolution scheme shows lower power consumption and higher noise tolerance compared with the traditional method. Moreover, MMS is adopted to reduce the array size and power consumption significantly. This work may be essential for neural network and edge computing which requires low power and high reliability.

Acknowledgment

This work was supported by China Key Research and Development Program under Grant (No. 2023YFB4402500), National Natural Science Foundation of China (Nos. 62034006, U23B2040, 92264201), National Natural Science Foundation of Shandong Province (No. ZR2023LZH007, tsqn202306059) and Program of Qilu Young Scholars of Shandong University.

979-8-3503-9164-0/24 $31.00 © 2024 IEEE

6.5

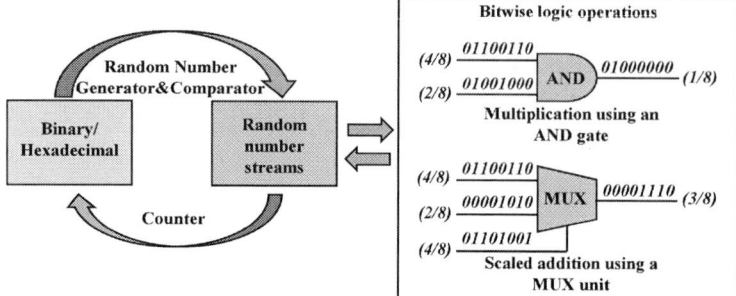

Fig.1 The schematic diagram of stochastic computing.

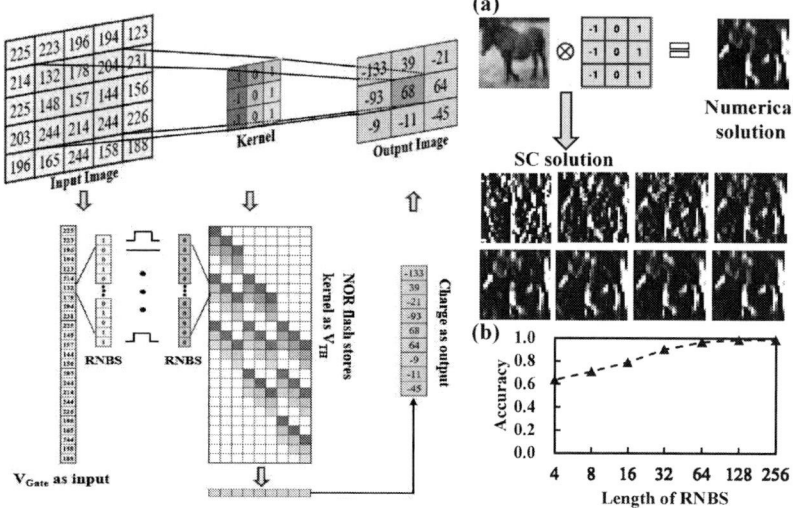

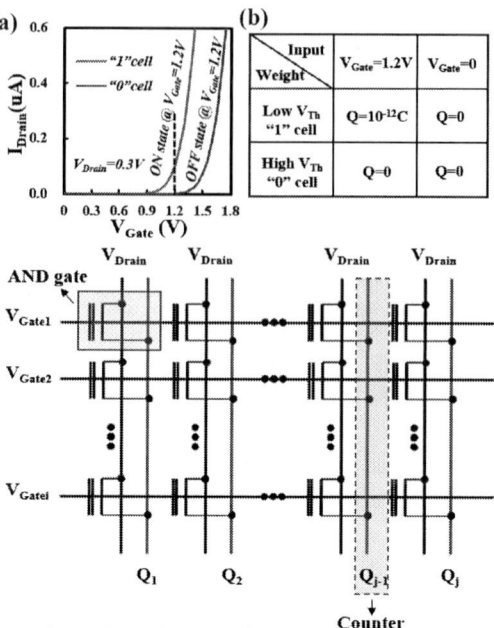

Fig.2. (a) The transfer characteristic curves of the flash memory cells with different V_{TH}. (b) The truth table of flash-based AND function. (c) Flash memory cells and array to perform multiplication and accumulation.

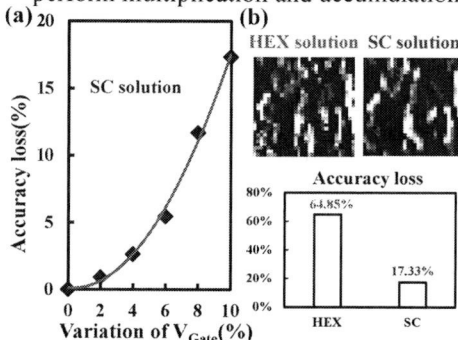

Fig.5. (a) The accuracy loss with different V_{Gate} fluctuates from 0 to 10%. (b) The results of HEX and SC solutions and their accuracy loss when V_{Gate} fluctuates by 10%.

Fig.3 The schematic of convolution by NOR flash. The input image is converted to a column vector and mapped to V_{Gate}; Kernal is converted to a matrix and stored as a V_{TH} in the flash array. The charge accumulated on the source line is the result of matrix multiplication.

Fig.4 (a) The numerical SC solution and (b) accuracy with various RNBS lengths from 4 to 256. The results are derived by using the Prewitt edge detection operator. The accuracy with the length of RNBS from 4 to 256, indicating that BSL=64 gets satisfactory accuracy and considerable array reduction.

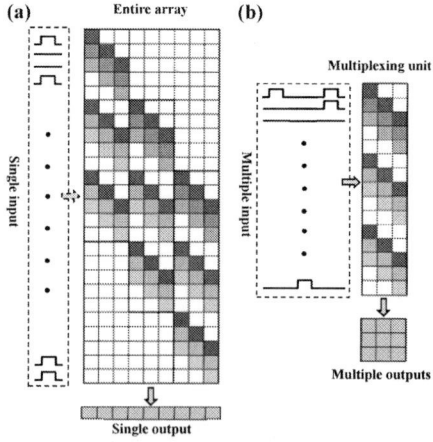

Fig.6. (a) The topology of the convolutional kernel storing in the flash. (b) The convolutional kernel is stored one time and computed multiple times.

	This work (w/matrix multiplexing)	This work (w/o matrix multiplexing)	TED 2020[7]	TCAS-I 2019[1]
Technology node	65nm	65nm	65nm	130nm
Image size	32*32	32*32	28*28	256*256
Energy Consumption/image	1.66nJ(convolution only)	17.7nJ(convolution only)	30.85uJ(inference)	5.44uJ(convolution only)
Array size	0.37M	118M	9.92M	8.46G

Fig.7. The benchmark with reported SC works. Our scheme shows low power consumption and high area efficiency.

References [1] R. Han, *et al.*, IEEE TCAS-I, 66(5):1692-1703, 2019. [2] Q. Xia, *et al.*, Nature Elec., 18(4): 309-323, 2019. [3] Y. Feng *et al.*, IEDM 2021,12.1.1-12.1.4. [4] D. Zhang *et al.*, IEEE EDL, 42(11): 1603-1606, 2021. [5] Y. Zhang, *et al.*, IEDM 2017, 6.6.1-6.6.4. [6] A. Alaghi, *et al.*, DAC 2013, 1-6. [7] Y. Xiang, *et al.* IEEE TED, 67(6): 2329-2335, 2020.

979-8-3503-9164-0/24 $31.00 © 2024 IEEE

Multi-Vt Gate Stack Technologies for Nanosheet and CFET Devices

H. Arimura, J. Franco, L.-Å. Ragnarsson, A. Vandooren, S. Brus, W. Maqsood, T. Conard,
G. Alessio Verni[1], J. W. Maes[2], B. Kannan[3], M. Givens[2], and N. Horiguchi

imec, Leuven, Belgium, [1]ASM, Helsinki, Finland, [2]ASM, Leuven, Belgium, [3]ASM, Phoenix, USA

Email: Hiroaki.Arimura@imec.be

Abstract — In this paper, we present our recent studies on multi-Vt gate stack technology elements required for Nanosheet and CFET devices, which are WFM scaling and interface dipole engineering. MoN has been proposed as a high-WF p-metal as well as a scalable barrier under nWFM. As for the interface dipole engineering, dipole-first scheme has been proposed as the alternative of dipole-last, and a Vt fine-tunable novel n-type shifter has been studied for dipole-first scheme. These material innovations are the driving force for the NS/CFET multi-Vt technology.

I. INTRODUCTION

Multi-Vt is the key to optimize the power and performance of System-on-Chip technology. On the conventional FinFETs, WFM stacks with nWFM and barrier layers with varied thickness has been used to provide multiple Vt's [1]. Entering to the nanosheet era, however, the tight space between nanosheets has become the strictest space constraint for the gate stack design (**Fig. 1**). Therefore, while the continuous scaling of WFM thickness is desired, interface dipole engineering needs to be greatly exploited as it can shift Vt without consuming much space [2]. As for the dipole integration, a typical scheme uses a high temperature drive-in anneal to cause the diffusion of shifter element into high-k film to form dipoles at high-k/SiO$_2$ interface. However, when considering a Complementary FET (CFET) [3], a low-temperature RMG solution is required for monolithic CFET when RMG module is processed after a contact module or for the top tier of sequential CFET. Thus, in our recent studies, we have evaluated MoN as the alternative scalable pWFM possibly replacing TiN [4]. Then, we have proposed "dipole-first" integration scheme as the scalable and thermal-budget-flexible dipole integration option compatible with CFET. A novel n-type shifter is shown as a suitable material for the new dipole scheme [5].

II. WFM SCALING USING MoN

ALD MoN layer shows a higher WF value than ALD TiN according to Ultraviolet Photoelectron Spectroscopy (UPS) (**Fig. 2**), thus having a potential to provide a lower pMOS |Vt| than TiN when it is used as pWFM. In addition, the film closure of ALD MoN was found to occur ~5 Å earlier than that of ALD TiN (**Fig. 3**). This is an indication of a superior scalability as a barrier under nWFM compared to TiN. Using a short-loop MOS capacitor test vehicle, MoN barrier under TiC nWFM is confirmed to maintain a high EWF down to a thinner layer thickness than TiN barrier (**Fig. 4**). Also, for a low EWF suitable for low-Vt nMOS, although a 5-Å-thick MoN barrier shows EWF as low as that with TiN, some

benefit in gate leakage reduction was observed, which could be attributed to the earlier film closure of MoN [4]. On long-channel planar RMG pMOSFETs, 80 mV |Vt| reduction was demonstrated with MoN as compared to TiN reference. Moreover, when nWFM was deposited on top, although pMOS |Vt| was increased by 170 mV with TiN barrier, the impact was kept negligible by using MoN, resulting in ~200 mV lower |Vt| (**Fig. 5**). Maintained hole mobility and device performance have confirmed the absence of major penalty [4].

III. INTERFACE DIPOLE ENGINEERING

In the proposed "dipole-first" scheme, shifter is deposited and patterned on SiO$_2$ IL whereas it is done on high-k in dipole-last. Since a more space is remaining at the multi-Vt patterning step, dipole-first has a better scalability as compared to dipole-last. Moreover, it does not require a high-temperature drive-in anneal (**Table 1**) [5]. Dipole-last is suitable for Vt fine-tuning as it can be tuned by shifter thickness, drive-in anneal condition and insertion of buffer layer. Therefore, it can be used together with band edge WFM for nMOS. In contrast, dipole-first is advantageous in largely shifting Vt as it can maximize the areal density of dipole (**Fig. 6**), thus can be even combined with mid gap WFM for nMOS. However, it has a difficulty in obtaining a small Vt shift, thus, not applicable to pMOS w/o a very high-WF p-metal. Here, a novel shifter forming a weak individual dipole is evaluated on MOS capacitors and compared with LaO, which is the typical n-type shifter (**Fig. 7**). Whereas the dipole-first LaO has no process window for a small EWF shift w/o EOT increase, EOT-penalty-free 100 meV EWF reduction and subsequent gradual EWF shift have been achieved using 1 and 2-3 cycle novel ALD shifters, respectively. This will be a promising low thermal budget-compatible fine-tunable multi-Vt option applicable to future CFET devices.

IV. CONCLUSION

Evaluation of novel WFM as well as dipole forming materials are conducted to enable multi-Vt of NS / CFET devices. MoN is a promising option as a scalable barrier under nWFM and high-WF p-metal, while a novel shifter provides a fine-tunable EWF shift at a maintained EOT even in a low thermal budget-compatible dipole-first shifter-kept integration scheme. Combination of these options can provide multi-Vt solutions for advanced device architecture within a given space and thermal budget constraints.

REFERENCES

[1] S. Hung, IEDM 2017, short course. [2] R. Bao *et al.*, IEDM 2018, p. 648. [3] J. Ryckaert *et al.*, VLSI 2018, p. 141. [4] H. Arimura *et al.*, VLSI 2023, T6-3. [5] H. Arimura *et al.*, IEDM 2021, p. 290. [6] T. Kamioka *et al.*, JJAP **57**, 076501 (2018).

979-8-3503-9164-0/24 $31.00 © 2024 IEEE

7.1

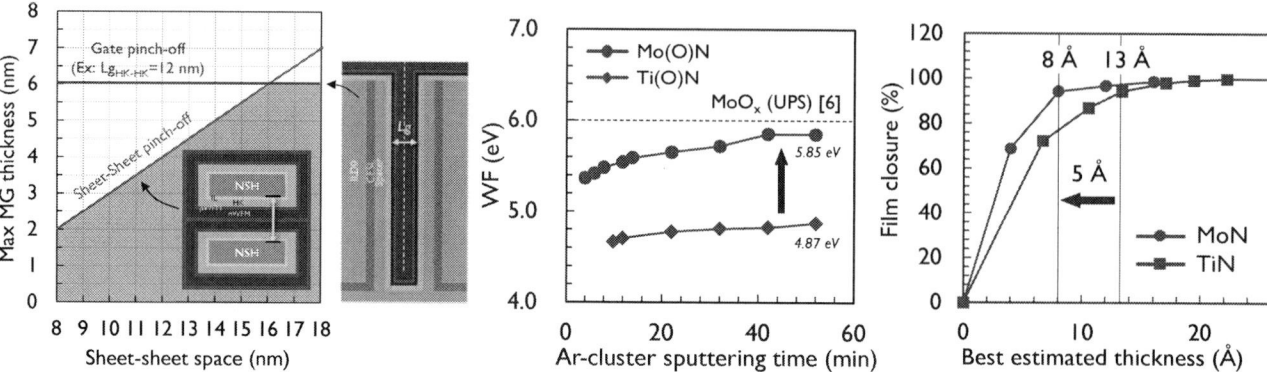

Fig. 1 Maximum metal gate layer thickness applicable to the nanosheet FET. The sheet-to-sheet space rather than L_g is expected to limit the available WFM thickness in the realistic scaled sheet-pitch devices.

Fig. 2 WF measured by UPS on 10 nm pWFM with repeated Ar-cluster sputtering, showing WF of 5.85 eV for Mo(O)N and 4.87 eV for Ti(O)N after a long surface cleaning.

Fig. 3 Film closure of MoN and TiN deposited on a thick HfO_2 using Low Energy Ion Scattering (LEIS). More than ~5 Å scaling benefit can be seen with MoN as compared to TiN.

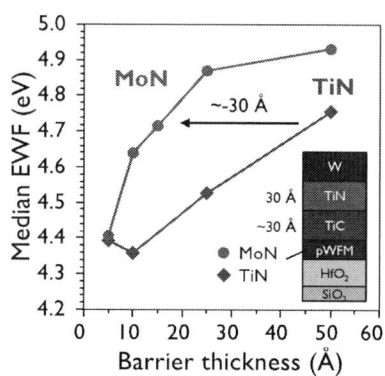

Fig. 4 EWF of gate stack using MoN or TiN barrier under ~3 nm TiC nWFM. MoN barrier maintains a high EWF with a thinner barrier layer than TiN, resulting in a superior scalability over TiN. The steeper EWF vs barrier thickness trend for a given nWFM also shows EWF tunability within the limited space.

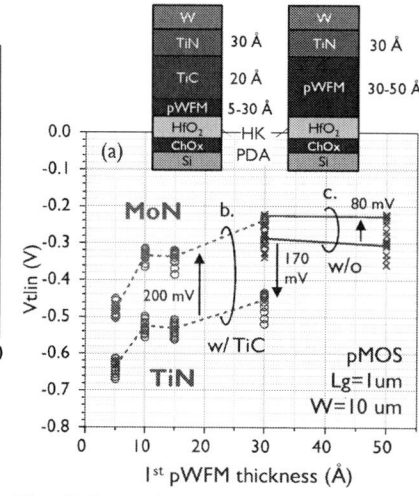

Fig. 5 Long-channel Vt of pFETs using MoN or TiN as the single pWFM or barrier under ~2 nm TiC nWFM. Reduced pFET |Vt| with single MoN pWFM as well as suppressed impact of TiC nWFM on top using MoN barrier were demonstrated.

Table. 1 Comparison of dipole-last and dipole-first shifter-kept multi-Vt options. Dipole-first-kept has advantage in low thermal budget-compatibility and scalability. With the novel shifter material, it can be used for pFET as the EWF of WFM w/o dipole can be around Ev.

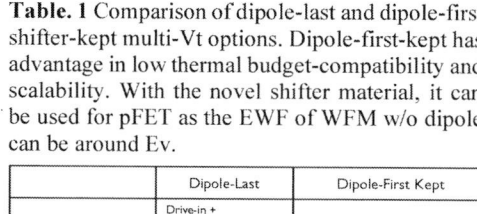

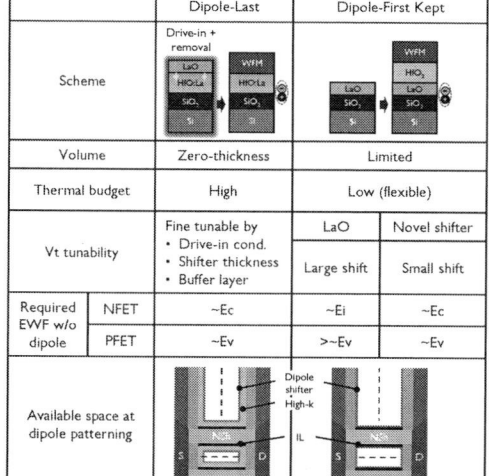

		Dipole-Last	Dipole-First Kept	
Scheme				
Volume		Zero-thickness	Limited	
Thermal budget		High	Low (flexible)	
Vt tunability		Fine tunable by • Drive-in cond. • Shifter thickness • Buffer layer	LaO	Novel shifter
			Large shift	Small shift
Required EWF w/o dipole	NFET	~Ec	~Ei	~Ec
	PFET	~Ev	>~Ev	~Ev
Available space at dipole patterning				

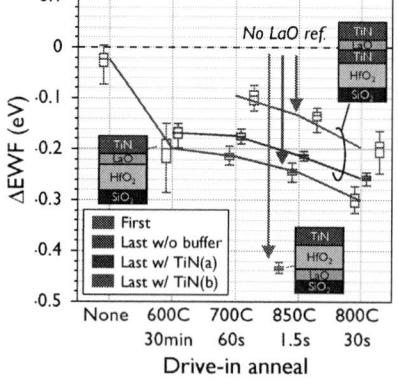

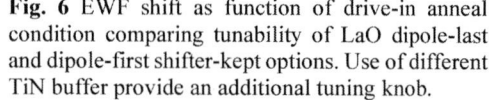

Fig. 6 EWF shift as function of drive-in anneal condition comparing tunability of LaO dipole-last and dipole-first shifter-kept options. Use of different TiN buffer provide an additional tuning knob.

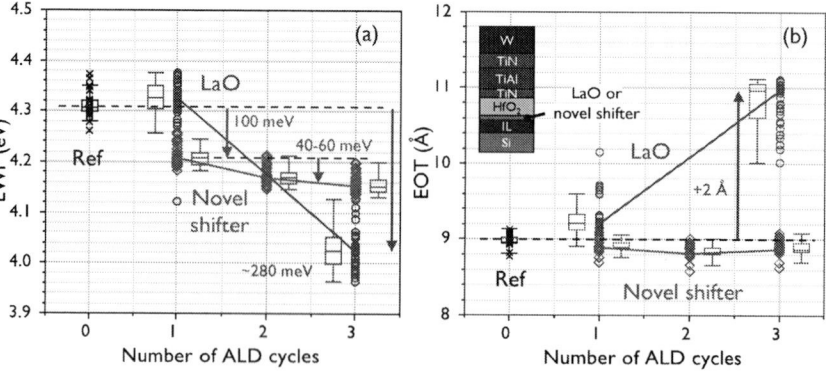

Fig. 7 (a) EWF and (b) EOT of dipole-first shifter-kept low-T gate stacks comparing ALD LaO and a novel ALD n-type shifter. The novel shifter provides the more modest and gradual EWF shift than LaO, with no clear EOT penalty. The target ~50 meV EWF shift is available between 1 and 2-cycle of novel ALD shifter.

979-8-3503-9164-0/24 $31.00 © 2024 IEEE

48

Investigation of Sheet Width Dependence on Hot Carrier Degradation in GAA Nanosheet Transistors

Zixuan Sun[1#], Zirui Wang[1#], Runsheng Wang[1,2*] and Ru Huang[1,2]

[1]School of Integrated Circuits, Peking University, Beijing 100871, China, [2]Beijing Advanced Innovation Center for Integrated Circuits, Beijing 100871, China ([*]Email: r.wang@pku.edu.cn) [#]These authors contributed equally.

Abstract — In this paper, we investigate the sheet width (SW) dependence of hot carrier degradation (HCD) in gate-all-around (GAA) nanosheet. We observe that HCD intensifies with an increase in SW, displaying a more pronounced width dependency in nGAA compared to pGAA. The analysis indicates that the worsening of HCD can be attributed to an increase in carrier energy due to more severe self-heating effect (SHE), which are caused by the increased current driven by wider sheets. It is elucidated that the width dependency of HCD is less pronounced in pGAA devices because they have lower mobility and higher threshold voltages, leading to a weaker dependence of their thermal power on SW. Consequently, under the same HCD stress, the increase in SW has a lesser impact on self-heating in pGAA than in nGAA, resulting in a weaker SW dependence of HCD in pGAA devices.

Keywords: hot carrier degradation, gate-all-around, sheet width dependence, self-heating effect.

I. INTRODUCTION

With the CMOS technology scaling down, stacked nanosheets in a gate-all-around (GAA) structure have become the predominant choice for the 3nm nodes and beyond, due to their superior electrostatic control [1,2]. Hot carrier degradation (HCD), a critical reliability issue in transistors, has previously been extensively studied in both planar transistors and FinFETs [3-7]. Recently, HCD in GAA has begun to receive increasing attention in order to provide a comprehensive assessment of the reliability of GAA devices [8,9]. Among them, the correlation between HCD and sheet width (SW) in GAA devices has been reported, but there is a lack of analysis of the physical reasons behind this phenomenon [10,11].

In this paper, the SW dependence of HCD in GAA nanosheet is studied. The deterioration of HCD can be attributed to the more severe self-heating effect (SHE) caused by the increase in SW. It is clarified that the width dependency of HCD in pGAA devices is less marked, attributable to their lower mobility and higher threshold voltages, which result in a diminished thermal power response to changes in SW. Therefore, at the same HCD stress, the increase in SW impacts self-heating less in pGAA than in nGAA, leading to a less pronounced width dependency of HCD in pGAA devices. The results are beneficial for enhancing the understanding and assessment of HCD in GAA.

II. RESULTS AND DISCUSSION

The GAA nanosheet geometry is shown in Fig.1. In this work, we studied the HCD in nGAA devices with a channel length of 60nm and widths of 20nm, 35nm, and 40nm, as well as in pGAA devices with widths of 20nm, 45nm, and 60nm. As shown in Fig. 2, the HCD in nGAA worsens with increasing SW, under V_g=1.7V, V_{ds} =1.5V stress conditions. A similar phenomenon was also observed in pGAA devices. Fig. 4 compares the SW dependency of HCD under the same stress voltage between nGAA and pGAA devices. The results indicate that HCD in nGAA devices shows a more pronounced SW dependency compared to pGAA.

Through simulations of self-heating effect and HCD, the underlying physical mechanisms behind this phenomenon were revealed. To accurately simulate the self-heating effect in GAA devices, the model parameters in the "Sentaurus" TCAD were calibrated using experimental I_{ds}-V_{gs} data. Using the power density input obtained from TCAD, the self-heating behaviour of nanosheet GAA devices was evaluated by the COMSOL tool. A thermal simulation structure, incorporating the back-end interconnect layers and substrate, was constructed based on actual GAA dimensional parameters. This simulation considers the temperature and size dependence of the thermal conductivity of material. As shown in Fig. 6, as the SW increases, the thermal resistance of the devices gradually decreases, and due to the use of SiGe material, the thermal resistance of pGAA is significantly higher than nGAA. However, pGAA devices have a lower mobility and a higher threshold voltage (V_{th}) compared to nGAA, resulting in significantly lower thermal power in pGAA, and the rate of change in thermal power with SW is also significantly lower than in nGAA, as shown in Fig. 7. Therefore, under the same stress voltage, nGAA exhibits more severe self-heating effect, and the SW dependency of self-heating effect is more pronounced, as shown in Fig.8. Fig. 9 illustrates the temperature distribution caused by self-heating effect at different SW.

Based on the differences in self-heating effect, we simulated the distribution of carrier energy at different SW, as shown in Figs. 10 and 11. Drawing from the theory of resonant scattering, the HCD is determined by the integral of EDF and DOS of Si-H state. In nGAA devices, as the SW increases, the energy of high-energy carriers also increases, exacerbating HCD. In contrast, in pGAA devices, the energy of high-energy carriers remains relatively constant with increasing SW, with mainly a rise in the energy of some low-energy carriers. However, due to the lower density of Si-H states for PMOS, carriers whose energy lower than 1eV are less likely to interact with Si-H bonds and form interface defects. Therefore, HCD in nGAA exhibits a more pronounced SW dependency.

III. CONCLUSION

This study reveals that the worsening of HCD is driven by an increase in carrier energy, which results from enhanced self-heating effect associated with larger SW. It further clarifies that the difference in SW dependency of HCD between nGAA and pGAA is due to the differences in SW dependency of thermal power.

ACKNOWLEDGMENTS

This work was partly supported by NSFC (62125401, 61927901) and the 111 Project (B18001).

REFERENCES

[1] J. Jeong et al., VLSI, 2023, pp. 1-2. [2] N. Loubet et al., VLSI, 2017, pp. T230-T231 [3] Z. Wang et al., IEDM, 2023, pp. 1-4. [4] R. Wang et al., IEDM, 2021, pp. 31.2.1-31.2.4. [5] Z. Yu et al., IEDM, 2017, pp. 7.2.1-7.2.4. [6] H. J. Huang et al., VLSI, 2011, pp. 154-155. [7] A. Bravaix et al., IEDM, 2011, pp. 27.5.1-27.5.4. [8] C. Gupta et al., TED, 2020, vol. 67, no. 1, pp. 4-10. [9] M. Vandemaele et al., IRPS, 2023, pp. 1-10. [10] Choudhury et al., TED, 2022, vol.69, no.12, pp. 6576-6581. [11] Choudhury et al., TED, 2022, vol.69, no.7, pp. 3535-3541.

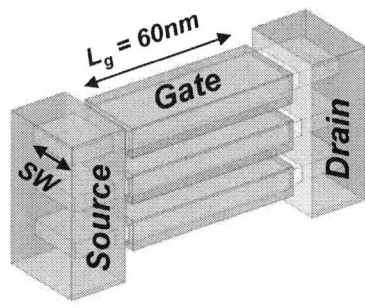

Fig. 1. The schematic of the GAA geometry. The gate length is 60nm with different sheet width.

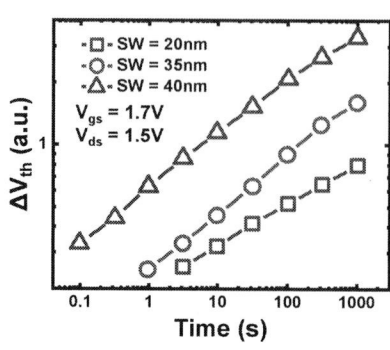

Fig. 2. Experimental results on V_{th} degradation under HCD stress across devices with different sheet widths in GAA device. Data from Ref. [10].

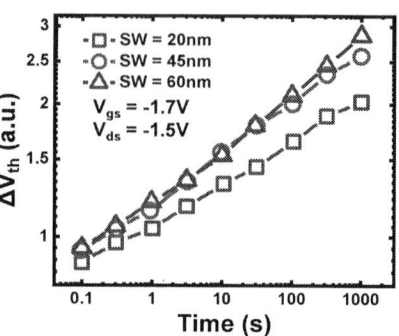

Fig. 3. In pGAA, experimental results on V_{th} degradation under HCD stress across devices with different sheet widths. Data from Ref. [11].

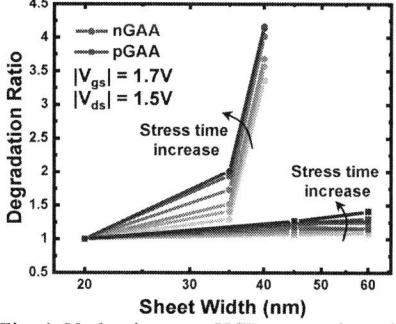

Fig. 4. Under the same HCD stress, the ratio of V_{th} degradation ($\Delta V_{th\_SW} / \Delta V_{th\_SW=20nm}$) varies with sheet width in n and pGAA. The HCD in nGAA exhibits a more pronounced sheet width dependency. Data from Ref. [10,11].

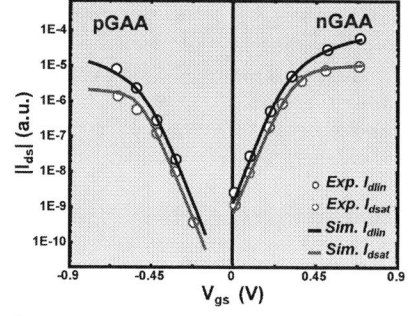

Fig. 5. TCAD calibration of I_{ds}–V_{gs} characteristics with experimental results. The TCAD results closely align with the experimental results. Experimental data from Ref. [2].

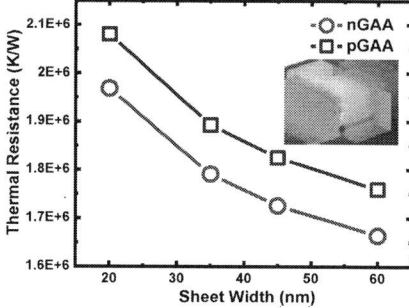

Fig. 6. Thermal resistance of GAA devices with different sheet widths. pGAA exhibit higher thermal resistance compared to nGAA.

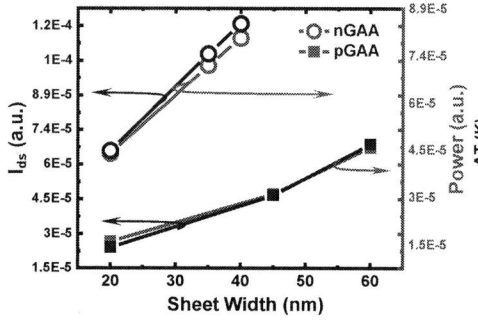

Fig. 7. I_{ds} varies with sheet width at the same voltage. The lower mobility of pGAA leads to a slower rate of change in I_{ds} with sheet width.

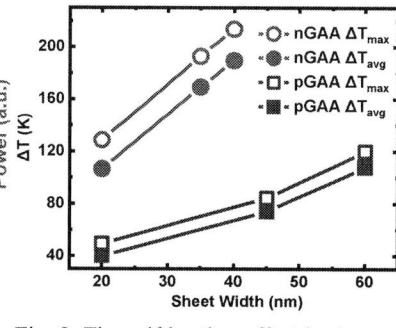

Fig. 8. The self-heating effect leads to an increase in maximum temperature rise and average channel temperature under different sheet widths.

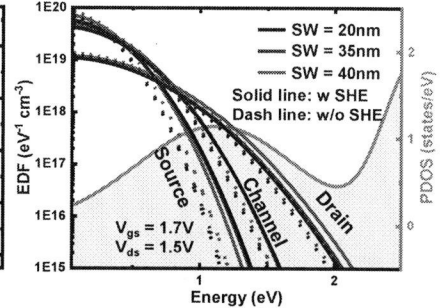

Fig. 10. Energy distribution difference of different locations in nGAA with different sheet width under HCD stress condition with and without SHE.

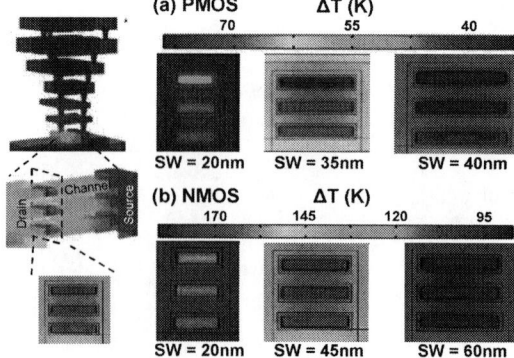

Fig. 9. Temperature distribution maps of the self-heating effect under different sheet widths. (a) NMOS, (b) PMOS.

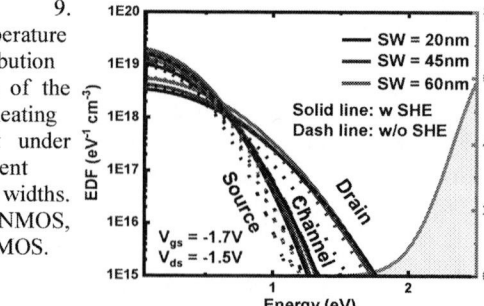

Fig. 11. Energy distribution difference of different locations in pGAA with different sheet width under HCD stress condition. with and without SHE.

979-8-3503-9164-0/24 $31.00 © 2024 IEEE

Gap in pagination due to unavailable paper.

Pages 51-52

First demonstration of SRAM transistor based on 3-dimensional stacked FET with back side interconnection structure beyond 1nm node

Mingyu Kim[1,2], Jaehyun Park[2], Sungil Park[2], Jejune Park[2], Juhyun Kim[1], Daewon Ha[2] and Hyungcheol Shin[1,3]

[1]Inter-University Semiconductor Research Center, Development of Electrical and Computer Engineering, Seoul National University, Seoul 08826 South Korea, [2]Semiconductor R&D Center, Samsung Electronics, Co. Ltd., Hwaseong, South Korea, [3]Integra Semiconductor, Ltd. E-mail: mg4721@snu.ac.kr

Abstract

For the first time, we report SRAM transistor demonstration based on 3-Dimensional Stacked FET (3DSFET) with Back Side Interconnection (BSI) structure. The use of NMOS as top and PMOS bottom device structure provides advantage of preserving the bottom PMOS channel stress engineering. However, from the SRAM perspective, PG transistor changes from NMOS to PMOS because bottom device can't be removed. By employing the BSI technology to reduce the cell height and Back-End-of-Line metal resistance, we propose using NMOS device for the PG as usual, achieving a 29% higher scalability for the SRAM bitcell, along with the most competitive SRAM performance with 28% higher IREAD and 3% WRM improvement.

Introduction

Recently, 3DSFET has emerged as one of the most promising candidates post MBCFET to drive the CMOS technology scaling beyond 1nm node thanks to its outstanding scalability [1]. To further enhance scalability, leveraging the wafer backside for interconnects, including powers and/or signals, enables effective scaling of both logic standard and SRAM cell height without compromising parasitic resistance [2]. As shown in Fig. 1 and 2, in SRAM area scaling point of view, 3DSFET presents an innovative approach to counter the traditional scaling trend due to its vertically stacked structure. So several researchers have reported and presented 3DSFET SRAM bitcell process flow and layout design [3-5]. In only 3DSFET, PG transistor should be changed from NMOS to PMOS. This results in significant change in SRAM read (from PG-PD to PU-PG) and write (from PG/PU to PG/PD) operation as shown in Fig. 3 and 4, expecting worse IREAD and WRM. To address the challenges posed by the PG PMOS SRAM bitcell, we propose an innovative structure of SRAM layout design and device fabrication.

SRAM Layout Design

Fig. 5 shows the traditional and 3DSFET on BSI SRAM layout. The notable feature of the SRAM layout is that it has both bottom and top SRAM design layout adopting the 3DSFET and it utilizes only two nanosheets by folding the NMOS on PMOS in six transistor configuration. Morevoer, to achieve the BSI structure, top bottom connection contact (VV) plays a significant role in connecting the NMOS device to VIA in CT layer. From the PD and PG transistor perspectives, both VSS and BL contacts are connected by CA-VV-VIA-M1 layer. On the other hand, the PU transistor is supported by BSI structure such as bCA-VIA-M1 layer for VDD. Consequently, BSI structure is expected to play a key role not only overcome the Back-End-of-Line resistance but also to providing a solution for PG transistors

to function as a NMOS device. Subsequently, a key SRAM layout design parameter to achieve the smallest high-density bitcell is to reduce the CT width including VV layer.

Device Fabrication

Fig. 6 and 7 illustrate the process flow of 3DSFET with BSI structure for SRAM bitcell. 3DSFET process flow, from multi-channel growth to replacement metal gate, was initially presented in [6] and we adhere to the same process fabrication. Following the replacement metal gate process, top bottom connection VV contacts are formed in advance to connect between top CA and VIA and then processing the top CA. Eventually, to achieve the BSI structure, a wafer flip process is required. After removing the wafer body region, bottom CA,VIA and M1 process are sequentially carried out to form BSI structure. This proposed process flow still remains to address the process challenges, such as the removal of the PU device when the wafer is turned over.

Result and Discussion

Fig. 8 displays the TEM image referring to fabrication process based on the SRAM layout design. In order to characterize the electrical properties of the SRAM transistor 3DSFET with BSI structure, we demonstrated, for the first time, the I_d-V_g characteristics of the bottom device, the SRAM PU transistor, exhibiting a saturation current of 245uA/um and a good subthreshold swing of 68mV/dec as shown in Fig. 9. Furthermore, to evaluate the PG device impact on SRAM bitcell performance, the high density SRAM bitcell simulation comparison of the IREAD and WRM between PG PMOS and PG NMOS is presented in Fig. 10. The results clearly indicate that PG NMOS SRAM shows both 28% higher IREAD and 3% WRM improvement. As a result, this innovative structure provides us not only the most competitive SRAM in terms of bitcell performance but also outstanding high density SRAM bitcell area scalability.

Conclusion

This work proposed the SRAM bitcell process flow and redesigned layout based on the 3DSFET along with BSI structure. Owing to the BSI technology in 3DSFET, we further report the compatible architecture between logic transistor performance and SRAM bitcell design. Accordingly, this structure give us the SRAM cell performance improvement, opening up new possibilities for monolithic 3DSFET combined with BSI integration.

Reference [1] D. Ha et al., IEDM, 2022 [2] R. Chen, IEDM, 2021 [3] Ryckaert J et al., VLSI, 2018 [4] M. Gupta et al., TED, 2021 [5] H. Liuet et al., TED, 2023 [6] J. Park et al., IEDM, 2023

979-8-3503-9164-0/24 $31.00 © 2024 IEEE

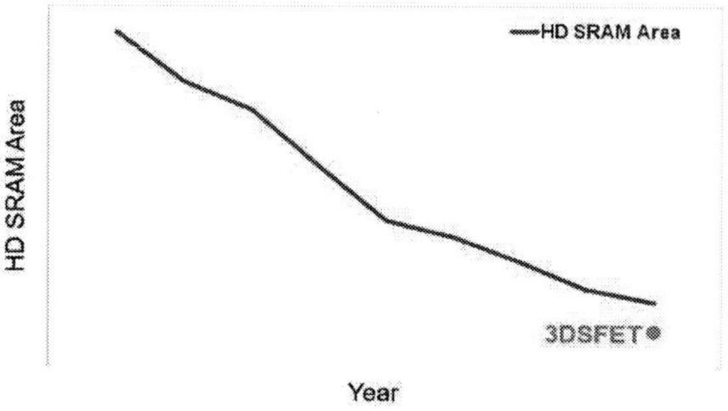

Fig.1. HD SRAM area trend

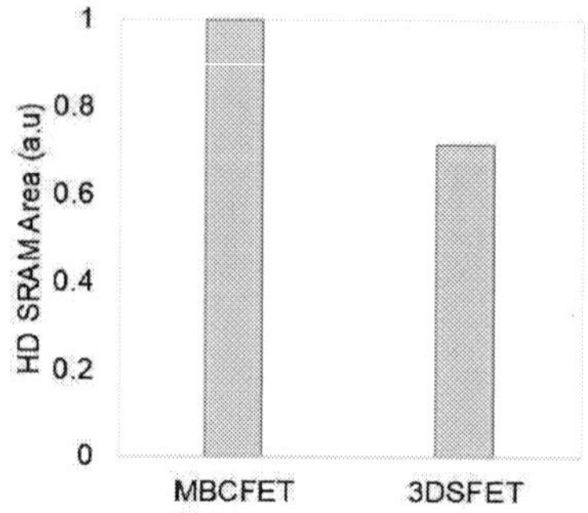

Fig.2. HD SRAM area comparison

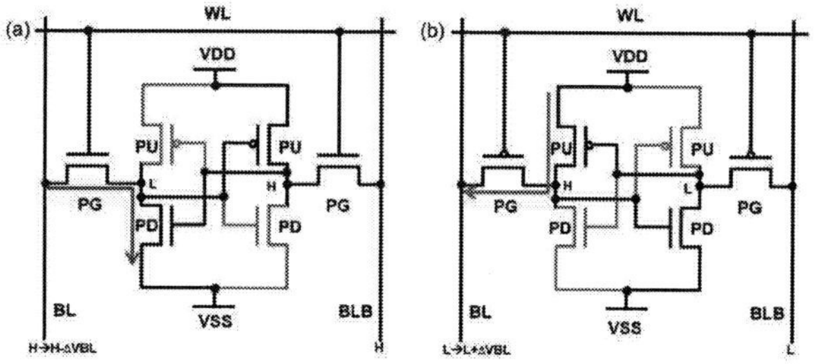

Fig.3. SRAM read operation comparison (a) Traditional (PG NMOS) read operation (b) PG PMOS read operation

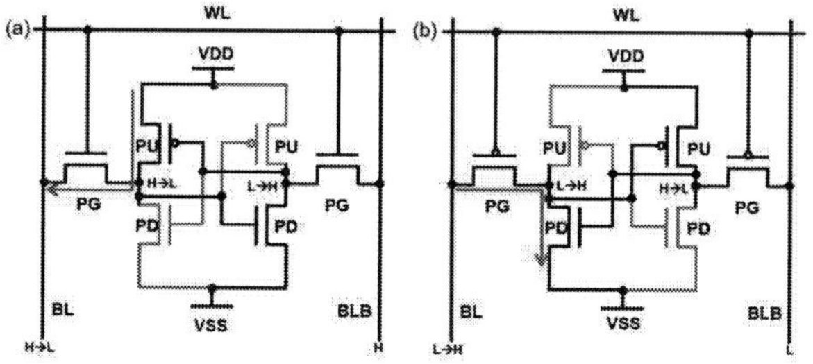

Fig.4. SRAM Write operation comparison (a) Traditional (PG NMOS) write operation (b) PG PMOS write operation

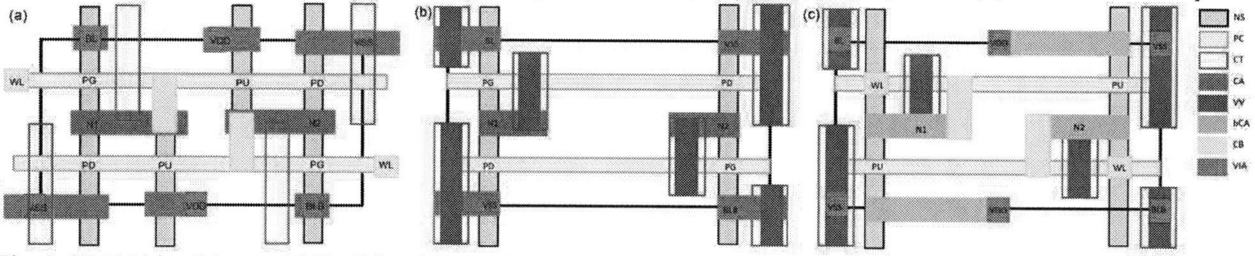

Fig. 5. SRAM bitcell layout (a) Traditional SRAM layout (b) 3DSFET Bottom layer (Front-side) SRAM layout (c) 3DSFET Top layer (Back-side) SRAM layout

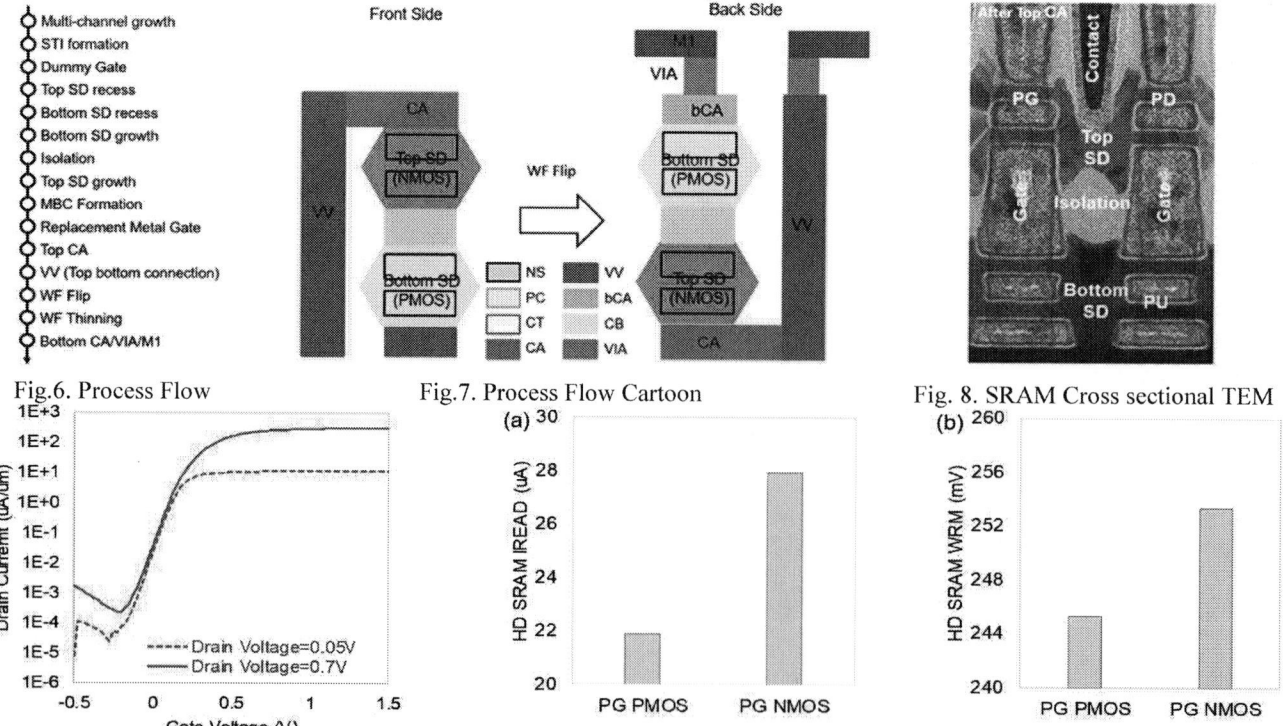

Fig.6. Process Flow

Fig.7. Process Flow Cartoon

Fig. 8. SRAM Cross sectional TEM

Fig.9. SRAM PU transistor I_d-V_g characteristics

Fig.10. Comparison between PG PMOS and PG NMOS (a) IREAD (b) WRM

979-8-3503-9164-0/24 $31.00 © 2024 IEEE

54

Gap in pagination due to unavailable paper.

Pages 55-56

Steep Slope Device "N-type Gate-Controlled Carrier-Injection SOI-Transistor": Suppression of Hysteresis by Ar-ion Implantation and Possibility of CMOS

Haruki Yonezaki, Takayuki Mori, Jiro Ida

Kanazawa Institute of Technology, 7-1 Ohgigaoka, Nonoichi, Ishikawa, 921-8501 Japan
E-mail: ida@neptune.kanazawa-it.ac.jp

Abstract

In this study, we report the detailed device characteristics of the "n-type gate-controlled carrier-injection silicon-on-insulator transistor (GCCI SOI-Tr)". It demonstrated no hysteresis steep subthreshold slope (< 60 mV/dec) by Ar Implantation and indicated that an ultralow power complementary metal-oxide-semiconductor (CMOS) can be realized by the GCCI SOI-Tr.

Keywords: CMOS, floating body effect, SOI, and steep subthreshold slope.

Introduction

Ultra-low power devices are needed to achieve the Internet of Things. The steep slope devices have been researched as ultralow power devices [1-2]. We already have proposed the steep slope device "gate-controlled carrier-injection silicon-on-insulator transistor (GCCI SOI-Tr)", which has a steep subthreshold slope (SS = 14 µV/dec) characteristics with a low drain voltage of 0.1 V and also a leaky-integrate-and-fire operation and amplification ability for the neuromorphic device [3-4]. However, we showed only a p-type device and it has the problem of hysteresis. We also reported Ar-ion implantation improving the hysteresis window on our previous device [5]. In this study, we show the measured n-type GCCI SOI-Tr basics characteristics, for the first time. We also confirmed that Ar-ion implantation again improved the hysteresis window on GCCI SOI-Tr. We show a proposal of CMOS configuration because both n-type and n-type were confirmed.

Device Structure

Fig. 1 shows the structure of the n-type GCCI SOI-Tr. The source of the NMOS and the base of the PNP bipolar transistor are connected, forming a single component. The dummy gate was used for making of PNP BJT. It achieves steep SS due to the floating body effect (FBE) by the carrier injection from the base region to the body region. The Ar-ion was implanted into the device, as shown in Figs. 1 (b) and (c). The Ar-ion implantation was conducted with an Ar dose of 2×10^{14} cm^{-2} at 7° tilted angles after making all gates. Ar-ion implantation reduces the FBE due to reduces the carrier lifetime [6]. Therefore, we consider that Ar-ion implantation suppresses the hysteresis. This study shows the measured results of 2 types of devices without Ar and with Ar implantation. The devices were fabricated using a LAPIS semiconductor 0.2 µm SOI CMOS process.

Results and Discussions

Fig. 2 shows the comparison of the Ar-ion implantation on the I_d–V_g characteristics of GCCI SOI-Tr. When V_b = -0.8 V, the hysteresis was not found in both I_d–V_g characteristics with Ar and without Ar. However, the hysteresis with Ar was smaller than that without Ar at V_b = -1 V. Specifically, the SS

is very little different between both devices, as shown in Fig. 3. The subthreshold slope was below 60 mV/dec between 1 pA to 100 nA at V_b = -0.8 V. Similarly, it was below 60 mV/dec between 1 pA to 1 µA at V_b = -1 V. It noted that the no was realized the no hysteresis steep SS characteristics with Ar. Fig. 4 shows the Ar implantation effect of transconductance. Both transconductances are like a spike, coming from steep turn-on. The peak transconductance without Ar was larger than with Ar.

Fig. 5 shows the I_d–V_g characteristics dependent on the V_d with the constant V_b = -0.8 V with Ar. The SS of GCCI SOI-Tr with every V_d is still below 60 mV/dec However, an increase of the leakage current was observed at high V_{dd} (= 0.8 V).

Fig. 7 shows I_d–V_d characteristics. It indicates saturation characteristics like MOSFET, as same as previously reported p-type [4]. However, it needs attention that the bias condition is different from MOSFET.

Fig. 8 shows the I_d / I_b – V_g characteristics dependent on the V_b with the constant V_d = -0.1 V with Ar. The β is defined as I_d divided by I_b. β is almost constant with V_b. However, the amplification ability of the n-type is lower than that of the p-type GCCI SOI-Tr [3-4] due to using PNP BJT.

Fig. 9 shows the p-type, reported on EDTM2024 [4] and n-type double sweep I_d–V_g characteristics. It shows that the ultra-low voltage swing (< 0.2 V) will realized with I_{off} (< 1 pA) and I_{on} (> 1 µA). We confirmed p-type and also n-type GCCI SOI-Tr, therefore, we think it is possible to realize CMOS configuration with GCCI SOI-Tr, as a proposal is shown in Fig. 10.

Conclusion

This paper demonstrated the n-type GCCI SOI-Tr, for the first time. We show it has the I_d–V_d characteristics of SS below 60 mV/dec between 1 pA to 1 µA (p-type at V_b = 1 V and n-type at V_b = -1 V) and the Ar implantation is possible to suppress the hysteresis. As a result, the n-type and p-type [5] GCCI SOI-Tr have achieved the steep SS (< 60 mV/dec). Therefore, we believe that GCCI SOI-Tr can realize steep SS CMOS with the ultralow V_{DD} (e.g. 0.2 V).

Acknowledgements

This work is the result of collaboration with the High Energy Accelerator Research Organization (KEK) and LAPIS Semiconductor Co., Ltd. This work was supported in part by JST-CREST Grant Numbers JPMJCR16Q1 and JPMJCR20Q1. This work was also supported through the activities of VDEC, The University of Tokyo, in collaboration with Cadence Design Systems and Mentor Graphics.

References

[1] A. M. Ionescu, and H. Riel, Nature, 479, pp. 329, 2011. [2] K. -S. Li et. al., IEDM, 2015, pp. 22.6.1. [3] H. Yonezaki, T. Mori, and J. Ida, JJAP, 63, 02SP83, 2024. [4] H. Yonezaki, T. Mori, and J. Ida, EDTM, 2024, pp. 1. [5] H. Matsushita, T. Mori, J. Ida, VLSI-TSA, 2024, pp. 1. [6] T. Ohno, et. al, IEEE TED, 45, 1071, 1998.

8.4

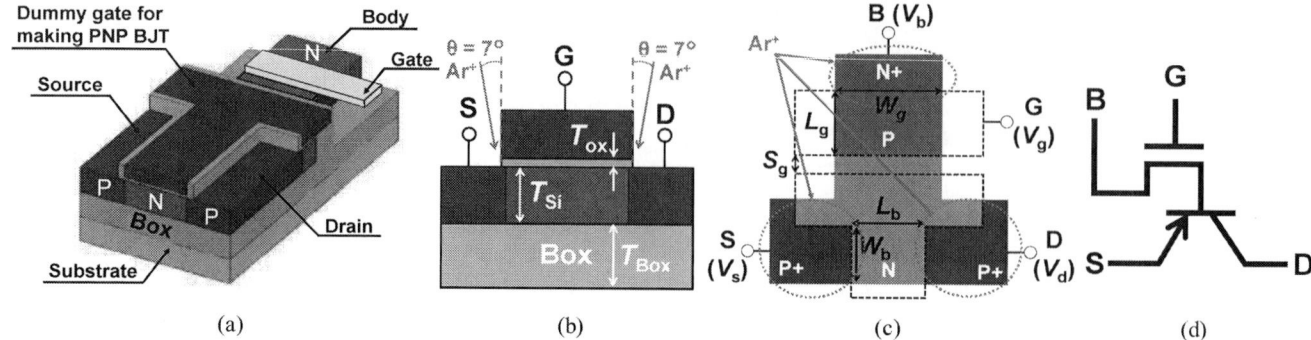

Fig. 1 GCCI SOI-Tr device structure. (a) Bird eye's (b) Front view (c) Plane view (d) circuit schematic.

TABLE I
Device Parameter

Gate Oxide	T_{OX}	4.4 nm
SOI Thickness	T_{Si}	40 nm
Buried Oxide Thickness	T_{BOX}	145 nm
Gate Length	L_g	1 µm
Gate Width	W_g	0.5 µm
Gate Space	S_g	0.32 µm
Base Length	L_b	0.2 µm
Base Width	W_b	1 µm

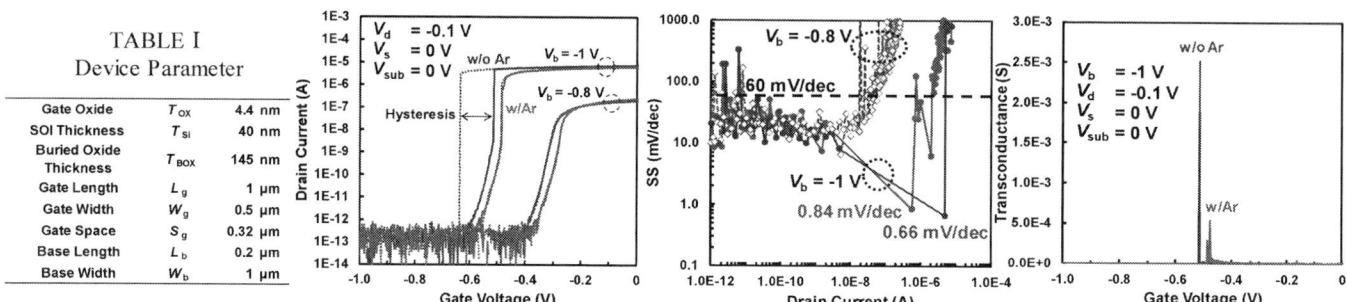

Fig. 2 Measured Ar implantation effect of I_d–V_g characteristics. Solid line: forward scan, Dotted line: backward scan.

Fig. 3 SS as a function of I_d from Fig. 2.

Fig. 4 Transconductance as a function of Fig. 2 at $V_b = -1$ V.

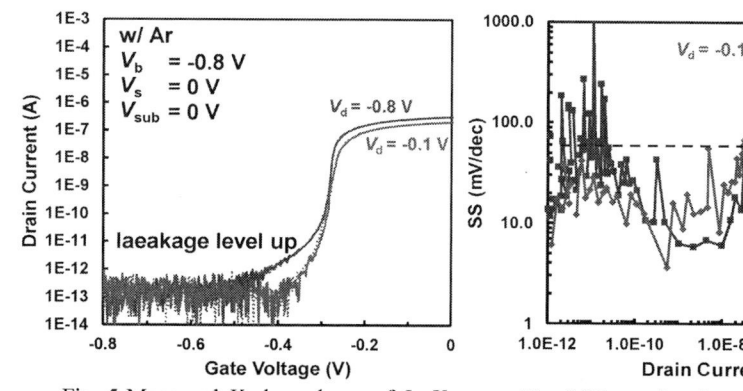

Fig. 5 Measured V_d dependence of I_d–V_g characteristics. Solid line: forward scan, Dotted line: backward scan.

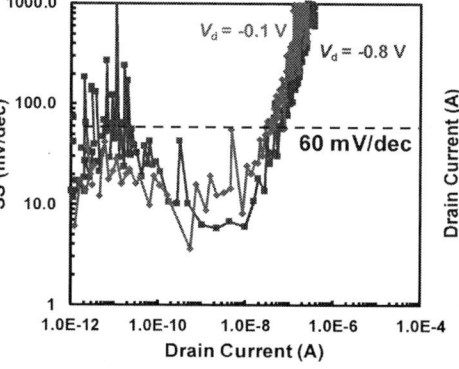

Fig. 6 SS as a function of I_d from Fig. 4.

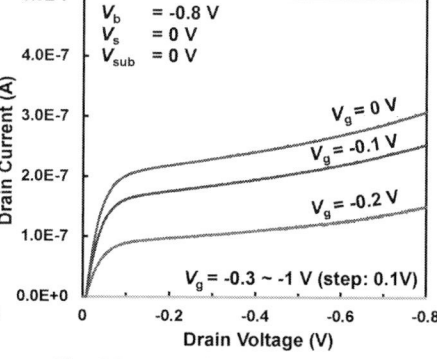

Fig. 7 Measured V_g dependence of I_d–V_d characteristics.

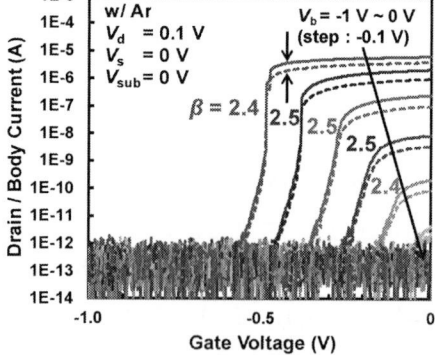

Fig. 8 Measured V_b dependence of I_d–V_g characteristics. Solid line: drain current, Dotted line: body current.

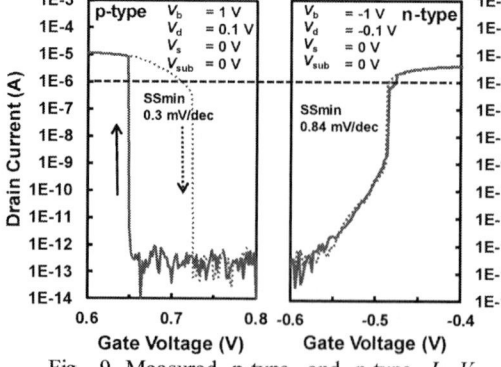

Fig. 9 Measured p-type and n-type I_d–V_g characteristics in same device parameter. Solid line: forward scan, Dotted line: backward scan.

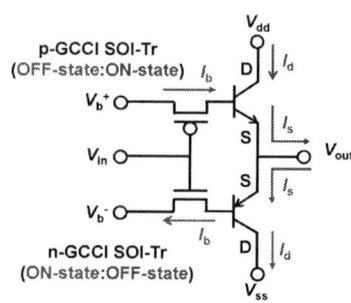

Fig. 10 GCCI SOI-Tr CMOS inverter circuit schematic.

979-8-3503-9164-0/24 $31.00 © 2024 IEEE   58

Gap in pagination due to unavailable paper.

Pages 59-60

A Nanosheet Oxide Semiconductor FET Using ALD InZnOx Channel

Sung-hun Kim[1], Kaito Hikake[1], Zhuo Li[1], Takuya Saraya[1], Toshiro Hiramoto[1], and Masaharu Kobayashi[1,2]

[1]Institute of Industrial Science, The University of Tokyo, [2]d.lab, The University of Tokyo. Email: sh-kim@nano.iis.u-tokyo.ac.jp

We have fabricated and characterized nanosheet InZnOx (IZO) FETs by ALD for monolithic 3D integration. Thermal stability and composition/thickness dependence on device characteristics are systematically investigated. IZO FETs show high thermal stability at 400 °C and trade-off among mobility, V_{th}, and V_{th} shift under bias stress. IZO FETs show higher mobility, lower V_{th}, and smaller V_{th} shift than previously reported ALD InGaOx (IGO) FETs. This can be due to the difference in oxygen dissociation energy between IZO and IGO. These findings provide insights for process optimization and device design to realize high-performance and high-reliability monolithic 3D integration.

I. INTRODUCTION

As the demand for AI technology continues to grow, the need for high-performance computing accelerates. However, 2D scaling is near the physical limit and monolithic 3D integration is a promising alternative solution for high density and energy-efficiency [1, 2]. In the monolithic 3D integration, upper-layer FETs must be fabricated at low temperature (<400 °C) to ensure the integrity of BEOL interconnects and the property of lower-layer devices. Oxide semiconductor (OS) FETs have excellent properties such as high mobility, low leakage, and high thermal stability with fabrication process less than 400 °C [3]. ALD enables precise control of OS composition with high uniformity and high conformality for 3D structures such as sidewalls and deep trenches for M3D BEOL and 3D OS FETs [4]. Although studies on individual OS compounds such as InO, IZO, IGO, and IGZO were reported, systematic study on nanosheet OS FETs with comparison among various OS compounds is needed.

In this paper, we fabricate and characterize nanosheet OS FETs with ALD IZO studying characteristics trade-off, and compare IZO FETs to IGO FETs to understand the effect of atomic species in InO-based OS material.

II. DEVICE FABRICATION

We developed IZO deposition process by ALD with alkyl, diethyl-based precursors and O_3 at 250 °C for channel material. Zn provides structure stability in amorphous phase to InOx. ALD IZO layer remains amorphous up to 600 °C anneal as shown in GI-XRD spectra in **Fig. 1**. We fabricated IZO FETs by the process flow in **Fig. 2**. Channel width and length of the FETs are 50 μm and 50 μm, respectively. **Fig. 3** shows the typical I_d-V_g curves of the fabricated IZO FET.

III. RESULTS AND DISCUSSIONS

A. Post-deposition anneal (PDA)

First, the thermal stability of FET characteristics after PDA was investigated. IZO FET characteristics were maintained well up to 400 °C as shown in **Fig. 4 and 5**, which is compatible to BEOL process for LSI application. Note that characteristics were hindered by high gate leakage above 500 °C due to HfO$_2$ gate oxide crystallization [5].

B. Zn-concentration dependence

Second, we studied Zn atomic concentration (Zn%) dependence to find optimum composition. As Zn% increases, threshold voltage (V_{th}) increases but effective mobility (μ_{eff}) decreases while subthreshold swing (SS) is almost maintained as shown in **Fig. 6 and 7**. Oxygen vacancy (V_o) is reduced, donor concentration decreases and thus V_{th} increases by adding Zn to InOx. ZnOx interrupts InOx conduction path and the mobility decreases [6,7]. Post-metallization anneal (PMA) greatly helps to annihilate V_o, raises V_{th}, and lowers SS. In:Zn=2:1 has relatively high V_{th} and high μ_{eff} among others.

C. Thickness dependence

Third, we studied IZO thickness dependence for tuning V_{th} tuning and studying thickness scalability. As thickness decreases, V_{th} increases due to the TFT operation principle, while SS and mobility are maintained down to 4-5 nm as shown in **Fig. 8 and 9**. 3 nm-thick IZO shows too large deviation in characteristics. 4-5 nm thickness range is a suitable choice for controlling short channel effect in scaled devices, while maintaining IZO channel property.

D. Bias-stress V_{th} instability

Fourth, we evaluated V_{th} shift (ΔV_{th}) by multiple DC V_g sweeps as shown in **Fig. 10**. ΔV_{th} is stable in the wide range of Zn% in this study at 10 nm thickness. This indicates that IZO does not suffer much from excess oxygen compared to IGO. However, when the Zn% is fixed at In:Ga=2:1 and the thickness is varied, ΔV_{th} gradually becomes large as the thickness decreases as shown in **Fig. 11 and 12**. Ultrathin IZO is prone to the exchange of oxygen with ambient and trap states in ultrathin OS film.

E. Characteristics Trade-off

Fig. 13 summarizes μ_{eff} versus V_{th} with ΔV_{th} of IZO FETs, which shows characteristics trade-off. We made comparison between IZO FETs in this work and IGO FETs in previous work [8]. **Fig. 14** compares μ_{eff} and V_{th} for the same thickness of IZO and IGO FETs. IZO FETs show higher μ_{eff} and lower V_{th} at the same thickness and same composition ratio, which is related to the difference in oxygen dissociation energy. **Fig. 15** compares the thickness dependence of ΔV_{th} at the same composition ratio. ΔV_{th} is less sensitive to thickness scaling in IZO FETs than in IGO FETs. This can be because IZO takes less excess oxygen which causes electron traps and ΔV_{th} than IGO. As we faced in IGO FETs, however, it is still a challenge to achieve high mobility, high V_{th} for normally-off operation, and high bias-stress reliability, simultaneously [9].

IV. CONCLUSION

We fabricated and characterized nanosheet IZO FETs by ALD. IZO FETs show high thermal stability at 400 °C and trade-off trends among mobility, V_{th}, and V_{th} shift under bias stress. IZO FETs show higher mobility, lower V_{th}, and higher reliability than ALD InGaOx (IGO) FETs. This can be due to the difference in oxygen dissociation energy between IZO and IGO. These findings provide insights for process optimization and device design for monolithic 3D integration.

979-8-3503-9164-0/24 $31.00 © 2024 IEEE

9.2

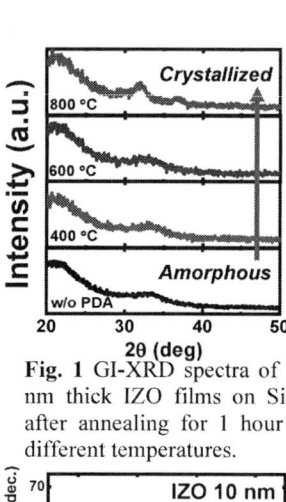

Fig. 1 GI-XRD spectra of 10 nm thick IZO films on SiO$_2$ after annealing for 1 hour at different temperatures.

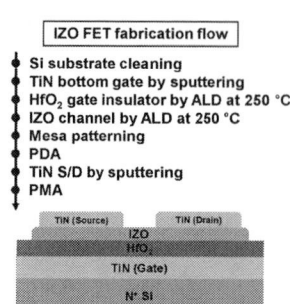

Fig. 2 Fabrication process flow and schematics of fabricated ALD IZO FETs. Bottom gate structure is used.

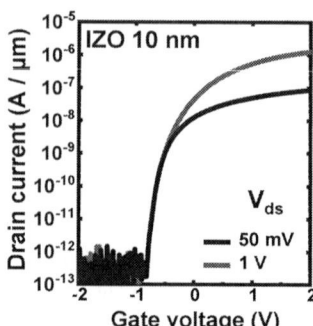

Fig. 3 Measured I$_d$-V$_g$ curves of the fabricated IZO FET with In:Zn=2:1 and 10 nm-thick IZO at V$_{ds}$=50 mV and 1 V.

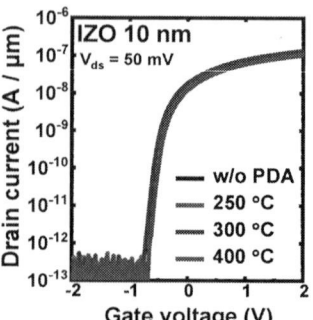

Fig. 4 Measured I$_d$-V$_g$ curves of IZO FETs with HfO$_2$ gate oxide. IZO was annealed at different PDA temperature conditions.

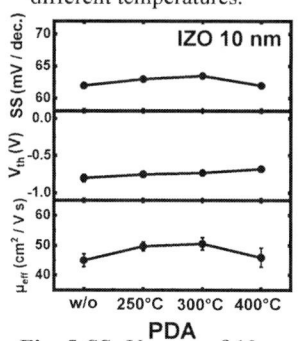

Fig. 5 SS, V$_{th}$, μ_{eff} of 10 nm-thick IZO FETs with HfO$_2$ gate oxide from Fig. 4 as a function of PDA temperature.

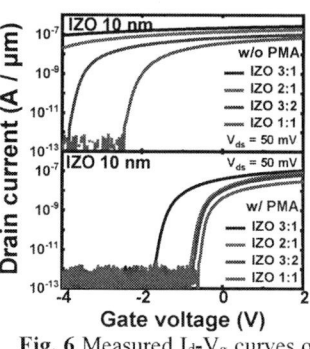

Fig. 6 Measured I$_d$-V$_g$ curves of the fabricated IZO FETs w/ and w/o PMA. Zn% for 10 nm-thick IZO channel is varied.

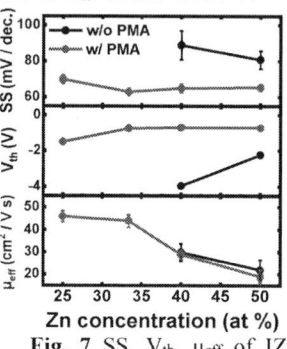

Fig. 7 SS, V$_{th}$, μ_{eff} of IZO FETs as a function of different Zn% from Fig. 6 with and without PMA.

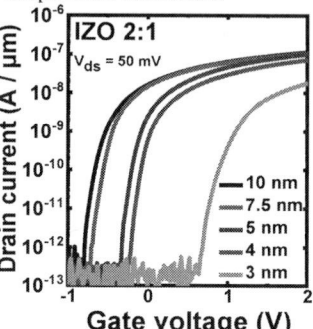

Fig. 8 Measured I$_d$-V$_g$ curves of the fabricated IZO FETs with different IZO thicknesses for In:Za=2:1.

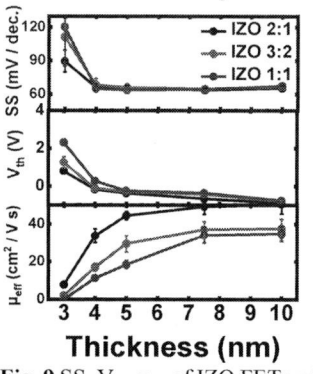

Fig. 9 SS, V$_{th}$, μ_{eff} of IZO FETs with different IZO thickness (3-10 nm) for In:Zn = 2:1, 3:2, 1:1.

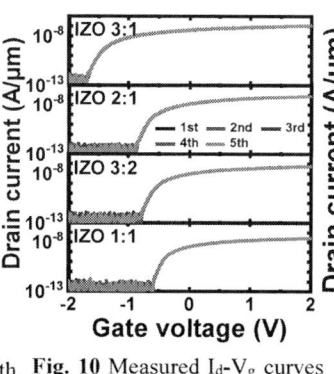

Fig. 10 Measured I$_d$-V$_g$ curves of the fabricated IZO FETs by multiple V$_g$ sweeps with different Zn%.

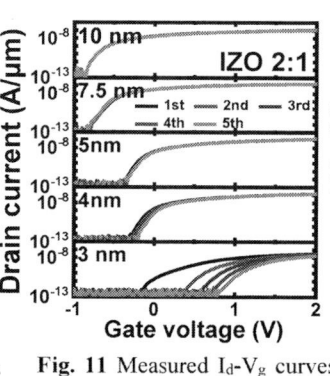

Fig. 11 Measured I$_d$-V$_g$ curves of the IZO FETs by multiple V$_g$ sweeps with different IZO thickness for In:Za=2:1.

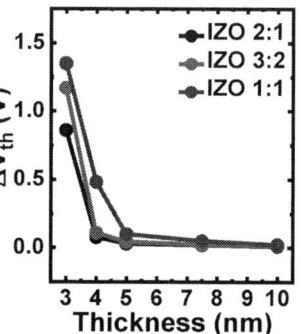

Fig. 12 V$_{th}$ shift of the fabricated IZO FETs with different IZO thickness extracted from Fig. 10 and 11.

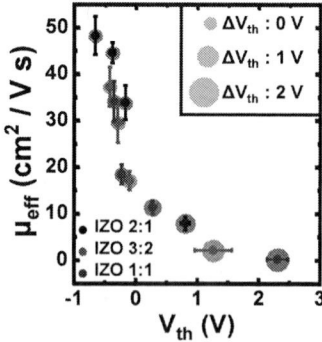

Fig. 13 Trade-off among μ_{eff}, V$_{th}$, and V$_{th}$ shift in IZO FETs with different Zn% and IZO thickness.

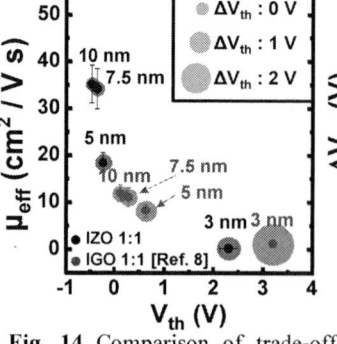

Fig. 14 Comparison of trade-off between IZO and IGO FETs [8] at 1:1 ratio with different thickness.

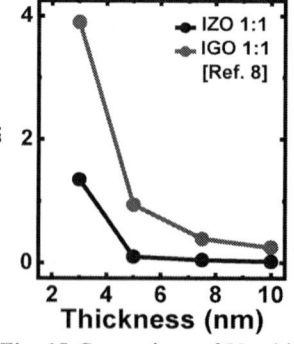

Fig. 15 Comparison of V$_{th}$ shift between IZO vs IGO FETs [8] at 1:1 ratio with different thickness.

Acknowledgements: This work was supported by JST CREST (23830112), JST ASPIRE (23836464), JSPS KAKENHI (21H04549), TSMC Advanced Semiconductor Research Project.

References: [1] Bishop, M.D. et al, IEEE Micro, 39(6), pp.16-27, (2019), [2] W. Gomes, IEDM, 15-5, (2023), [3] K. Nomura et al., Nature,432, 25, 488 (2004), [4] Kim, H.M. et al, IJEM. (2023), [5] Murdzek, J.A. et al, Journal of Vacuum Science & Technology A, 38(2). (2020), [6] Jeon, H. et al, Journal of The Electrochemical Society, 158(10), p.H949. (2011), [7] Liang, Y.K. et al, IEEE TED, 70(3), pp.1067-1072. (2023), [8] K. Hikake et al., VLSI Symp. T14-1, (2023), [9] Kim, H.M. et al, ACS Applied Materials & Interfaces. (2024)

979-8-3503-9164-0/24 $31.00 © 2024 IEEE

Investigation of In-Sn-Zn Composition on the Characterization of Submicron Channel Length Ultra-Thin Atomic Layer Deposited InSnZnO Channel Transistors

Yan-Kui Liang [1,2], Li-Chi Peng [1], Yu-Lon Lin [1], June-Yang Zheng [1], Dong-Ru Hsieh [1], Tsung-Te Chou [3]
Huai-Ying Huang [4], Yu-Ming Lin [4], Yuan-Chieh Tseng [1], Tien-Sheng Chao [1], Edward-Yi Chang [1]
Kasidit Toprasertpong [2], Shinichi Takagi [2], Chun-Hsiung Lin [1, #]

[1] National Yang Ming Chiao Tung University, Taiwan; [2] The University of Tokyo, Japan; [3] Taiwan Instrument Research Institute;

[4] Taiwan Semiconductor Manufacturing Company (TSMC) #Email: chun_lin@nycu.edu.tw

Abstract — The study focuses on developing ultrathin 3 nm-thick indium-tin-zinc-oxide (ITZO) films for submicron channel length thin-film transistors (TFT) using atomic layer deposition (ALD). We aimed to enhance the performance and thermal stability of TFTs by optimizing the ITZO channel composition with a low thermal budget (250 °C). The optimized TFTs with the $In_{0.6}Sn_{0.2}Zn_{0.2}O$ channel demonstrated high performance with a threshold voltage (V_{th}) of 0.56 V, field-effect mobility (μ_{FE}) of 36 cm²/V·s, a subthreshold swing (SS) of 78 mV/dec, a low drain-induced barrier lowering (DIBL) of 88 mV/V, and an I_{ON}/I_{OFF} ratio of over 10^9. Notably, the devices with the $In_{0.6}Sn_{0.2}Zn_{0.2}O$ channel exhibited exceptional thermal stability under positive bias temperature stress (PBTS) at 85 °C, resulting in a minimal V_{th} shift of ~0.18 V at 85 °C. This study underscores the potential of ultrathin ALD ITZO films as viable candidates for BEOL-compatible transistors in monolithic 3-D (M3D) integration, offering significant advancements in the performance and durability of TFTs.

Keywords: InSnZnO TFT, ALD, and BEOL transistor.

I. INTRODUCTION

Significant research has been directed towards using amorphous oxide semiconductors (AOS) in thin-film transistors (TFTs) for monolithic 3-D integration (M3D) with CMOS technology at the BEOL [1]. Notably, indium-containing AOS such as InZnO, InWO, InSnO, InGaO, and InGaZnO are favored for their high electron mobility and material stability, making them prime for AOS-TFT applications [2]-[7]. ALD, preferred over PVD methods for AOS deposition, offers precise control over film thickness and composition, essential for 3D integration at the BEOL. InSnZnO (ITZO) is a promising quaternary compound for the ALD process due to its positive threshold voltage, high ON-current (I_{ON}), thermal stability, and high mobility, as shown in **Fig. 1(a)**. Although fundamental material studies on ITZO TFTs have been conducted [8-9], detailed reports on the practical device designs and integration technology of ITZO TFTs for BEOL applications are still lacking. Additionally, the PBTS test for submicron L_{ch} ITZO TFTs has not been investigated sufficiently.

In our study, we present 3 nm ultrathin ITZO TFTs with an L_{ch} of 500 nm, demonstrating high performance and thermal stability, with notable short channel effects, including an SS of 78 mV/dec and DIBL of 88 mV/V. We also investigate the high-temperature PBTS for devices. The quaternary ITZO devices show greater stability than ternary ITO and IZO devices under PBTS at 85 °C.

II. EXPERIMENTS

Fig. 1(b) illustrates the schematic of an ultra-thin channel TFT. **Fig. 1(c)** shows the fabrication for TFTs. P-type Si wafers with a thermally deposited SiO_2 layer served as the substrates. The TiN as the bottom gate structure was patterned and etched, followed by an 8 nm-thick ALD-HfO_2 gate dielectric. This was followed by a 3 nm-thick ALD AOS channel deposition at 250 °C. Pt source/drain

(S/D) contacts were then established using a lift-off process involving e-beam lithography for patterning and PVD for S/D deposition. **Fig. 1(d)** presents a top-view image of the channel in the TFT, confirming the L_{ch} of approximately 500 nm in these submicron devices.

III. RESULTS AND DISCUSSION

Fig. 2(a)-(d) displays the I_D-V_G characteristics of amorphous ITZO TFTs with an L_{ch} of 500 nm, measured at V_D of 0.1V and 1V. The threshold voltage (V_{th}) is determined as the gate voltage (V_G) at a constant drain current (I_D) of 10^{-2} μA/μm is maintained. The field-effect mobility (μ_{FE}) was calculated from the linear region of the transfer characteristics using the formula: $\mu_{FE} = (L \times gm) / (W \times C_{ox} \times V_D)$, which g_m represents the transconductance, W/L the channel width to channel length ratio, C_{ox} the gate dielectric oxide capacitance, and V_D the applied drain voltage set at 0.1 V. The V_{th} were recorded at -0.22 V, 0.56 V, 0.76 V, and 0.95 V for $In_{0.6}Sn_{0.4}O$, $In_{0.6}Sn_{0.2}Zn_{0.2}O$, $In_{0.6}Sn_{0.1}Zn_{0.3}O$, and $In_{0.6}Zn_{0.4}O$ TFT, respectively. The devices exhibited peak μ_{FE} of 47 cm²/V·s, 36 cm²/V·s, 20 cm²/V·s, and 13 cm²/V·s, respectively. These TFTs also demonstrated outstanding short-channel performance, with Drain-Induced Barrier Lowering (DIBL) values of 178 mV/V, 88 mV/V, 111 mV/V, and 188 mV/V, and subthreshold swing (SS) values of 95 mV/dec, 78 mV/dec, 86 mV/dec, and 89 mV/dec for compositions of $In_{0.6}Sn_{0.4}O$, $In_{0.6}Sn_{0.2}Zn_{0.2}O$, $In_{0.6}Sn_{0.1}Zn_{0.3}O$, and $In_{0.6}Zn_{0.4}O$, respectively. It was noted that the μ_{FE} of TFTs diminishes with an increase in the zinc content in ITZO. Concurrently, positive V_{th} shifts and improvements in SS were observed in ITZO TFTs.

Furthermore, **Fig. 3(a)-(d)** illustrates the output characteristics of these ITZO TFTs, exhibiting V_G from -2 V to 4 V and V_D values up to 1V, demonstrating excellent saturation and pinched-off traits. The transistors achieved I_{ON} of 205 μA/μm, 154 μA/μm, 75 μA/μm, and 37 μA/μm at a V_D of 1V and a V_G of 4V, respectively.

The positive-bias temperature stress (PBTS) test, a standard evaluation for the reliability of n-type TFTs, was demonstrated in **Fig. 4(a)-(d)**. The test was assessed at 25 °C, 55 °C, and 85°C on a TFT, stressed under a gate voltage of $V_G = V_{th} + 3V$ (equivalent to 3.75 MV/cm). The results revealed high PBTS stability for $In_{0.6}Sn_{0.2}Zn_{0.2}O$ TFT and $In_{0.6}Sn_{0.1}Zn_{0.3}O$ TFT, indicated by minimal shifts in V_{th}: an increase of only 177 mV and 203 mV from the temperature increased from 25 °C to 85 °C, after a stress period of 1000 seconds.

IV. CONCLUSION

The thin-film transistors employing an $In_{0.6}Sn_{0.2}Zn_{0.2}O$ channel exhibited superior performance characteristics, including a μ_{FE} of 36 cm²/V·s, a V_{th} of 0.56 V, a SS of 78 mV/dec, and a low DIBL of 88 mV/V. Moreover, the device demonstrated remarkable thermal stability under PBTS at 85 °C. This research highlights the efficacy of ALD-deposited ITZO films, positioning ITZO as a promising candidate for BEOL-compatible transistors in M3D integration, significantly enhancing the performance and durability of thin-film transistors.

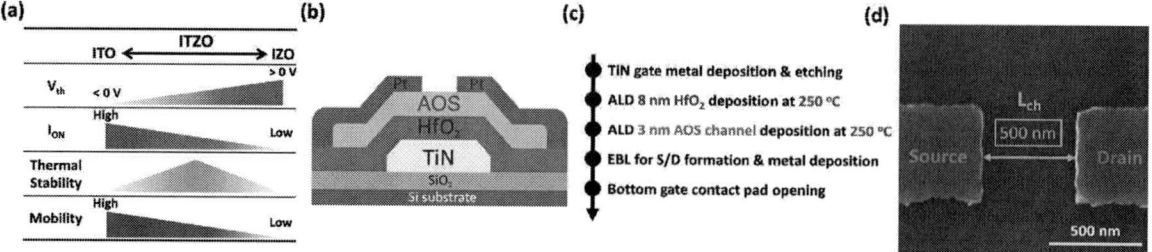

Fig. 1. (a) A characteristic trend of ITZO with different concentrations: V_{th}, I_{ON}, thermal stability, and mobility. (b) Schematic of AOS TFTs. (c) Process flow of the ITZO TFTs, and (d) Top-down SEM image of the submicron channel length (L_{ch}= 500 nm) of the ITZO TFTs.

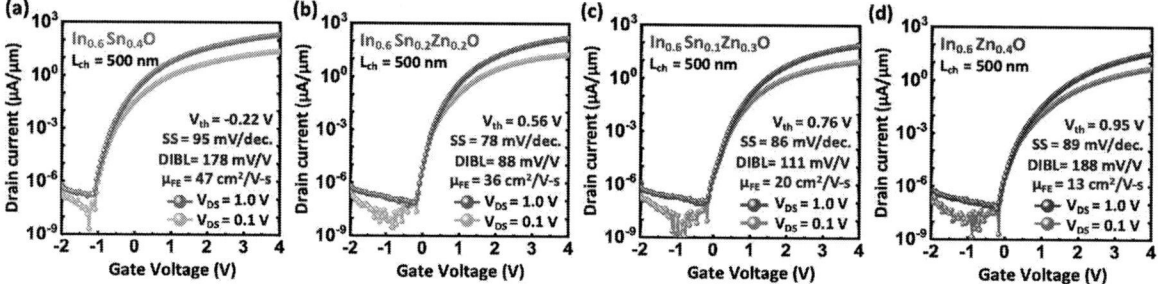

Fig. 2. I_D-V_G of the ALD ITZO TFTs. (a) $In_{0.6}Sn_{0.4}O$ (b) $In_{0.6}Sn_{0.2}Zn_{0.2}O$ (c) $In_{0.6}Sn_{0.1}Zn_{0.3}O$ (d) $In_{0.6}Zn_{0.4}O$.

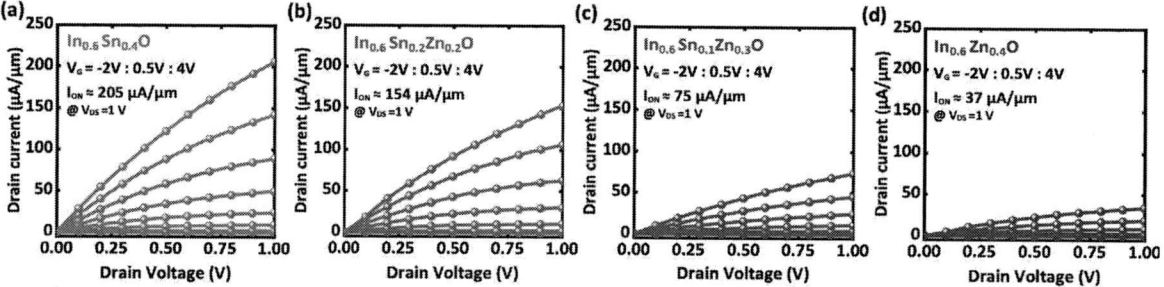

Fig. 3. I_D-V_D of the ALD ITZO TFTs. (a) $In_{0.6}Sn_{0.4}O$ (b) $In_{0.6}Sn_{0.2}Zn_{0.2}O$ (c) $In_{0.6}Sn_{0.1}Zn_{0.3}O$ (d) $In_{0.6}Zn_{0.4}O$.

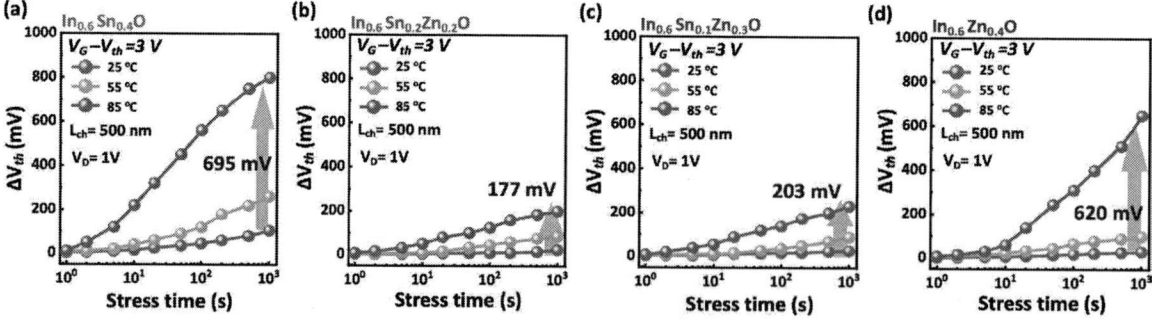

Fig. 4. ΔV_{th} for ALD ultra-thin ITZO TFTs with L_{ch} of 500 nm at 25 °C, 55 °C, and 85 °C with V_{th} +3V for 1000s. (a) $In_{0.6}Sn_{0.4}O$ (b) $In_{0.6}Sn_{0.2}Zn_{0.2}O$ (c) $In_{0.6}Sn_{0.1}Zn_{0.3}O$ (d) $In_{0.6}Zn_{0.4}O$.

ACKNOWLEDGMENTS

This work was supported in part by Taiwan Semiconductor Manufacturing Company (TSMC), JST CREST Grant JPMJCR20C3, JSPS KAKENHI Grant 21H01359, Japan and in part by the National Science and Technology Council (NSTC), Taiwan, under Grant NSTC-112-2221-E-A49 -172.

REFERENCES

[1] S. Datta, et al., *IEEE Micro*, 2019. [2] Y.-K. Liang, et al., *2023 IEEE Symposium on VLSI*, 2023. [3] W. Chakraborty, et al., *IEEE Trans. on Electr. Dev.* 2020. [4] Y.-K. Liang, et al., *IEEE Trans. on Electr. Dev.* 2024. [5] K. Toprasertpong et al., *2023 IEEE Symposium on VLSI*, 2023 [6] K. Hikake et al., *IEEE Trans. on Electr. Dev.*, 2024. [7] J. Zhang et al., *2023 IEEE Symposium on VLSI*, 2023. [8] S. Tomai, et al., *Jpn. J. Appl. Phys.* 2012. [9] Y.-S. Shiah, et al., *Nat Electron.*, 2021.

979-8-3503-9164-0/24 $31.00 © 2024 IEEE

BEOL-Compatible In₂O₃ Thin-Film Transistor with Linear Dielectric ZrO₂ Achieving Dielectric Constant over 27 and Enhanced Field-Effect Mobility Up To 89.3 cm²·V⁻¹·s⁻¹

Zehao Lin and Peide D. Ye.

Elmore Family School of Electrical and Computer Engineering, Purdue University, West Lafayette, IN 47907, USA. *Email: yep@purdue.edu

Abstract

In this work, we report development of atomic-layer-deposited (ALD) In_2O_3 thin-film transistor (TFT) using ZrO_2 as a linear higher-k dielectric. Employing ALD In_2O_3 as the capping layer for ZrO_2 and the channel semiconductor, $W/ZrO_2/In_2O_3$ stack with post-deposition annealing (PDA) results in a higher dielectric constant in ZrO_2 and mobility boost in In_2O_3 simultaneously. The capacitive equivalent oxide thickness (EOT) of 12.5-nm ZrO_2 reaches 1.77 and 1.72 nm under O_2 and N_2 PDA environments, corresponding to k value 27.5 and 28.2, respectively. PDA in O_2 enables not only a higher field-effect mobility (μ_{FE}) of In_2O_3 of 89.3 cm²·V⁻¹·s⁻¹ at channel length (L_{ch}) of 2 μm, but also negligible hysteresis, high on/off ratio over 10^{10}, low DIBL of 50 mV/V and low sub-threshold swing (SS) of 80 mV/dec at L_{ch} of 60 nm. Such advancements contribute to the TFTs with high saturation current (I_{sat}) exceeding 1.3 mA/μm at a L_{ch} of 1 μm, and maximum current (I_{max}) reaches 4.7 mA/μm at L_{ch} of 40 nm. This approach highlights an easy-to-scale methodology to boost oxide-semiconductor (OS) TFTs from higher-k dielectric engineering.

Introduction

Since the revisit of nanometer-thin oxide semiconductor unveiled the transport superiority of In_2O_3-based material over other amorphous semiconductors [1-3], extensive research has been conducted on various aspect of TFTs based on In_2O_3, ITO and IGZO due to their high mobility, high scalability, and back-end-of-line (BEOL) compatibility. While recent advancements in OS-TFTs are primarily focused on the channel material, contact and device structure [4-10], the benefits of these approaches shrink when L_{ch} scaled down to sub-50 nm, where the electrostatic gate control dominates. This demands a new, effective, and straightforward way to form higher-k dielectric with high-quality interface and lower EOT.

The tremendous effort on high-k dielectrics gives some insights. Although HfO_2 is adopted in high-k-metal-gate (HKMG) technology, the most stabilized monoclinic (m) phase of HfO_2 has a lower k value of around 15 than that over 45 in tetragonal (t) phase. The problem of t-phase is the high formation energy, unavoidability of interlayer formation in high-temperature process, and mobility degradation due to remote phonon scattering (RPS) [11]. While compared with HfO_2, t-phase in ZrO_2 is more preferable, although with a reduction in k value [12, 13]. This enables a new strategy to form higher-k metal gate with OS TFTs using ZrO_2 by proper strain engineering.

In this work, we report a $W/ZrO_2/In_2O_3$ stack with asymmetric capping profile under PDA treatment to achieve linear higher-k value of near 30 in ZrO_2, and boosted μ_{FE} in In_2O_3 to 89.3 cm²·V⁻¹·s⁻¹, simultaneously. Such enhancements boost OS-TFTs with negligible hysteresis, on/off ratio over 10^{10}, small SS of 80 mV/dec and DIBL below 50 mV/V. And the I_{sat} and I_{max} exceed 1.3 and 4.7 mA/μm at L_{ch} of 1 μm and 40 nm, respectively.

Experiments

Figs. 1 and 2 show the device schematics and fabrication process flow of higher-k ZrO_2-In_2O_3 TFTs based on ALD. 12.5 nm ALD ZrO_2 is sandwiched by sputtering W as the back gate metal and ALD In_2O_3 as the channel material. The same batch of $W/ZrO_2/In_2O_3$ stack underwent a 350°C PDA under O_2 and N_2 environment as comparison. The asymmetric capping builds up a strain distribution inside ZrO_2, which is critical to form polycrystalline t-phase in ZrO_2 under PDA with higher dielectric constant. PDA also passivates the oxygen vacancies in In_2O_3, especially in O_2 environment. The whole process is BEOL-compatible.

Results and Discussion

Figs. 3 (a) and (b) demonstrate the C_g-V response of the $W/ZrO_2/In_2O_3$ metal-insulator-semiconductor capacitor (MOS-CAP) at frequency of 10 kHz with PDA in O_2 and N_2, respectively. With the same batch of materials, MOSCAP in O_2 PDA achieves a capacitive EOT of 1.77 nm with k value of 27.5, slightly worse than that of EOT=1.72 nm and k=28.2 in N_2 PDA. However, the C-V response in O_2 PDA shows a minor hysteresis than in N_2, indicating a much reduced antiferroelectric phase co-exhibiting in a large area.

Figs. 4 and 5 show the bi-directional transfer characteristics of ALD ZrO_2-In_2O_3 TFTs with L_{ch} of 2 μm under PDA in O_2 and N_2, respectively. Like the hysteresis difference in C-V response, the TFT with O_2 PDA has a negligible hysteresis compared with a clockwise one in N_2. Besides, TFTs with O_2 PDA have a greater threshold voltage (V_{TH}), smaller subthreshold swing (SS) and higher transconductance (g_m). Since the capacitance is slightly smaller in O_2, the performance enhancement is mainly due to the mobility enhancement of In_2O_3 channel, likely due to the extra passivation from O_2 annealing on In_2O_3 and ZrO_2-In_2O_3 interface. Fig. 6 presents the μ_{FE} of the TFTs with PDA in O_2 and N_2, extracted from the g_m of Figs. 4 and 5 and C_g of Figs. 3. PDA in O_2 enables a 27.5% increase in μ_{FE} from 70 cm²·V⁻¹·s⁻¹ in N_2-PDA to 89.5 cm²·V⁻¹·s⁻¹. These results show a trade-off between a smaller EOT and higher mobility when choosing the PDA environment in higher-k-OS structure. Even though, both μ_{FE} values are generally higher than 30-60 cm²·V⁻¹·s⁻¹, a range widely reported in OS TFTs without engineering on channel. This mobility enhancement originates from the reduction of RCS potential from the higher dielectric constant [14].

The performance boost from higher-k dielectrics doesn't diminish in short-channel devices. Figs. 7 and 8 show the bi-directional transfer characteristics of ALD ZrO_2-In_2O_3 TFTs with L_{ch} of 60 nm under PDA in O_2 and N_2, respectively. Both devices show a high on/off ratio over 10^{10}. Compared with N_2 PDA, TFT with O_2 PDA has a ΔV_{TH} of 7 mV, one order smaller than that of 70 mV under N_2 PDA. Besides, the TFT with O_2 PDA has a slightly greater V_{TH}, smaller SS, smaller drain-induced barrier lowing (DIBL) and higher g_m. The DIBL of the TFT is reduced from 250 mV/V with N_2 PDA to 50 mV/V with O_2. Figs. 9(a) and (b) show the dependence of bi-directional SS value on drain current with V_{DS} of 0.1 and 0.5 V in O_2 and N_2 PDA, respectively. SS presents a negligible dependence on direction and V_{DS}. The SS with O_2 PDA is 80 mV/dec, smaller than that of 120 mV/dec with N_2 PDA. Figs. 10 and 11 show the output characteristics of ALD ZrO_2-In_2O_3 TFT with L_{ch} of 1 μm and 40 nm under O_2 PDA, respectively, utilizing ultra-fast pulse I-V (UFPIV) scheme to alleviate self-heating effect. The transistor shows a good saturation behavior with I_{sat} exceeding 1.3 mA/μm at L_{ch} of 1 μm and drive supply of 4 V. I_{max} reaches 4.7 mA/μm at V_{DS} of 1.2 V and V_{GS} of 3 V with L_{ch} of 40 nm. Both I_{sat} and I_{max} outperform other reported OS-TFTs at the same L_{ch}.

Fig. 12 benchmarks the μ_{FE} with k value of dielectrics of currently reported high-performance OS-TFTs with nanometer-thin channel [1, 4-10]. For most approaches to enable μ_{FE} over 100 cm²·V⁻¹·s⁻¹, to integrate with a higher-k dielectric is not compatible. While this work highlights a higher k value approaching 30 with a high mobility around 90 cm²·V⁻¹·s⁻¹, which leads to a good trade-off between electron transport with good electrostatic control. Fig. 13 benchmarks the DIBL, SS and I_{max} of these high-performance OS-TFTs. This work highlights the highest I_{max} with low DIBL and SS.

Conclusion

In conclusion, high-performance OS-TFTs based on higher-k ZrO_2 are demonstrated. ALD In_2O_3 is incorporated as the capping for ZrO_2 as well as the channel semiconductor simultaneously. By performing PDA under $W/ZrO2/In_2O_3$ asymmetric capping, ZrO_2 becomes a linear dielectric with the k value approaching 30. Compared with N_2 environment, PDA in O_2 enables a higher mobility of 90 cm²·V⁻¹·s⁻¹, with incremental decrease in k value to 27. These enhancements lead to a high I_{sat} exceeding 1.3 mA/μm with L_{ch} of 1 μm and I_{max} reaches 4.7 mA/μm with L_{ch} of 40 nm. Since ZrO_2 is easy to scale, the TFT performance can be further boosted by EOT scaling to sub-nm region. The work is mainly supported by Samsung Electronics and Purdue-imec Research Center.

Reference: [1] S. Li et al., *Nat. Mater.* 18, 1091–1097, 2019. [2] M. Si et al., *Nat. Electron.* 5, 164-170, 2022. [3] Z. Lin et al., *ACS Nano*, 16, p. 21536, 2022. [4] M. Si et al., *VLSI*, T2-4, 2021. [5] K. Chen et al., *VLSI*, pp.298-299, 2022. [6] Y. Kang et al., *VLSI*, T11-2, 2023. [7] S. Hooda et al., *VLSI*, T17-1, 2023. [8] Y. -K. Liang et al., *EDL*, 44, 10, pp. 1644-1647, 2023. [9] K. Han et al., *EDL*, 44, 12, pp. 1999-2002, 2023. [10] J. Zhang et al., *TED*, 70, 12, pp. 6651-6657, 2023. [11] M. Fischetti et al., *J. Appl. Phys.* 90, p. 4587, 2001. [12] X. Zhao et al., *Phys. Rev. B*, vol. 65, 075105, 2002. [13] C-K. Lee et al., *Phys. Rev. B*, vol. 78, 012102, 2008. [14] Z. Lin et al. *VLSI*, 2024.

979-8-3503-9164-0/24 $31.00 © 2024 IEEE

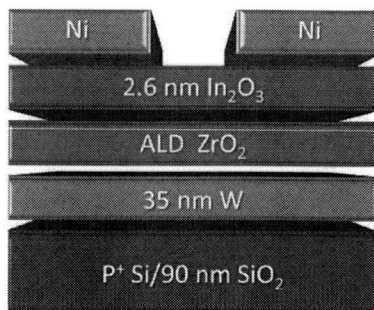

Fig. 1. Device schematics of higher-k ZrO_2-In_2O_3 TFTs.

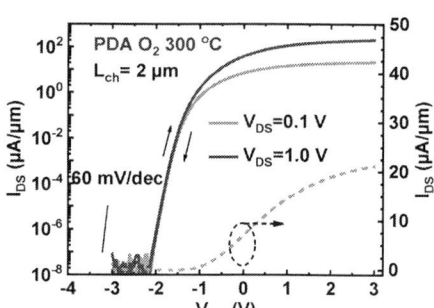

Fig. 2. Fabrication process flow of the higher-k ZrO_2-In_2O_3 TFTs.

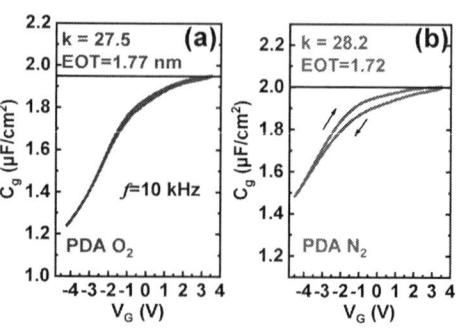

Fig. 3. C-V response of the W/ZrO_2/In_2O_3 MOS stacks frequency of 10 kHz under different PDA gas environment.

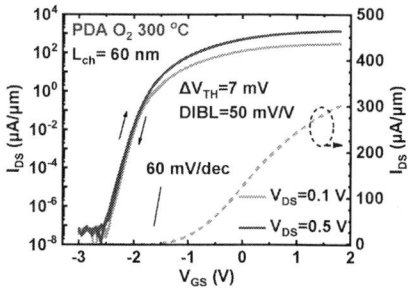

Fig. 4. Bi-directional transfer characteristics of 2-μm ZrO_2-In_2O_3 TFTs under PDA in O_2 with V_{DS} of 0.1 V and 1 V.

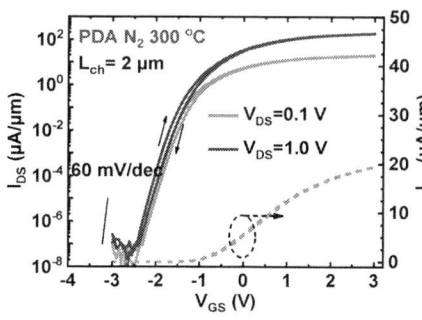

Fig. 5. Bi-directional transfer characteristics of 2-μm ZrO_2-In_2O_3 TFTs under PDA in N_2 with V_{DS} of 0.1 V and 1 V.

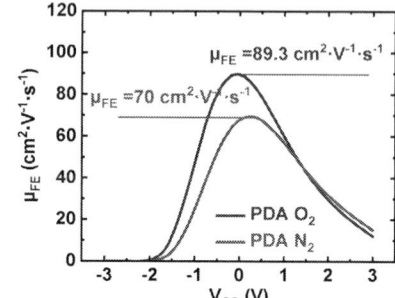

Fig. 6. Extracted μ_{FE} of ZrO_2-In_2O_3 TFTs at L_{ch} of 2 μm under PDA in O_2 and N_2, calculated from g_m of Fig. 4 and 5, respectively, and C_g of Fig. 3.

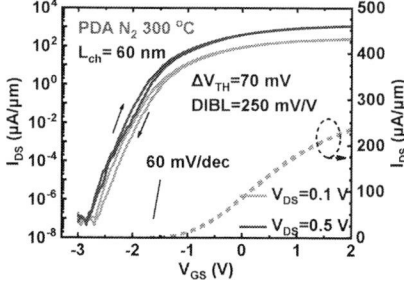

Fig. 7. Bi-directional transfer characteristics of short-channel 60-nm ZrO_2-In_2O_3 TFTs under PDA in O_2 with V_{DS} of 0.1 V and 0.5 V.

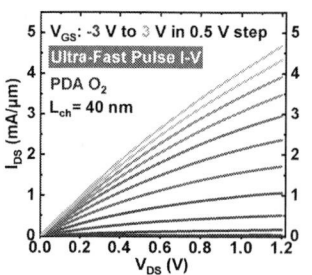

Fig. 8. Bi-directional transfer characteristics of short-channel 60-nm ZrO_2-In_2O_3 TFTs under PDA in N_2 with V_{DS} of 0.1 V and 0.5 V.

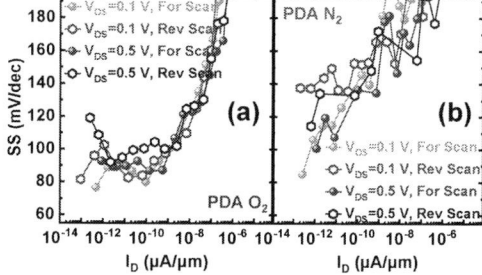

Fig. 9. Extracted bi-directional SS with V_{DS} of 0.1 V and 0.5 V of 60-nm ZrO_2-In_2O_3 transistors with PDA in (a) O_2 and (b) N_2, from Fig. 7 and Fig. 8, respectively.

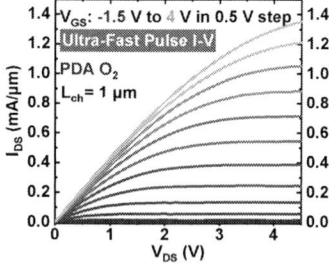

Fig. 10. UFPIV output characteristics of a ZrO_2-In_2O_3 TFTs with L_{ch} of 1 μm with PDA in O_2.

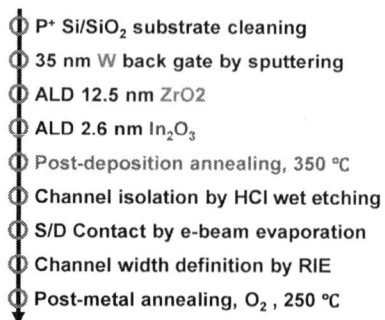

Fig. 11. UFPIV output characteristics of a ZrO_2-In_2O_3 TFTs with L_{ch} of 40 nm with PDA in O_2.

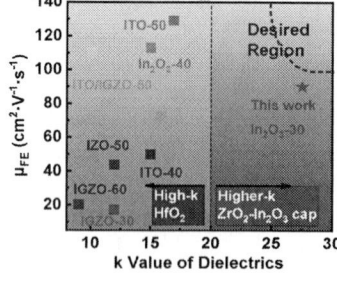

Fig. 12. Benchmarks of the μ_{FE} and k value of reported OS-TFTs with nanometer-thin channel [1, 4-10].

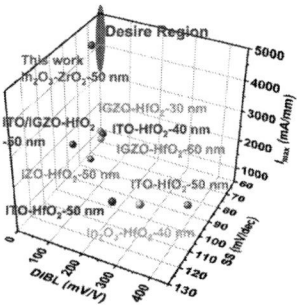

Fig. 13. Benchmarks of the DIBL, SS and I_{max} value of reported OS-TFTs with nanometer-thin channel.

979-8-3503-9164-0/24 $31.00 © 2024 IEEE

Gap in pagination due to unavailable paper.

Pages 67-68

A Novel Ternary Transistor with Nested Source Design Incorporating Hybrid Switching Mechanism for Low-Power and High-Performance Applications

Shaodi Xu[1], Tianyang Luo[1], Jin Luo[1], Ru Huang[1,2]* and Qianqian Huang[1,2]*

[1]School of Integrated Circuits, Peking University, Beijing 100871, China.
[2]Beijing Advanced Innovation Center for Integrated Circuits, Beijing 100871, China.
*Email: {hqq, ruhuang}@pku.edu.cn

Abstract — **In this work, we propose and experimentally demonstrate a novel ternary transistor with a nested source design through hybrid switching operation of band-to-band tunneling and thermionic emission mechanism. Measured ultra-low off-state leakage current (I_{OFF}) of below $10^{-7}\mu A/\mu m$ and steep subthreshold swing (SS) show its great potential for low power applications. A flat current region is also experimentally obtained for the application of ternary state. In addition, for the on state, the thermionic emission mechanism largely enlarges I_{ON}, which is promising for a high-speed ternary inverter.**

I. INTRODUCTION

Implementing ternary logic can significantly increase the information density (N) and reduce the complexity of the computational system to $\log_3 N/\log_2 N \approx 63.1\%$ [1], which can efficiently reduce the power density of the chip. Among various approaches of ternary inverter implementations, the ternary-CMOS (T-CMOS) method stands out due to the same compact configuration as binary inverters (Fig. 1a) and good wafer-level integration potential. Several Tunnel FET (TFET) -based ternary transistor designs have been proposed [2], [3] to enable steep subthreshold swing (SS) for low power, but they incur a speed penalty due to fundamental limitations in the drive current of band-to-band tunnelling (BTBT) mechanism (Fig. 1b). Therefore, advancing ternary transistors with both steep SS and high drive current is necessary to exploit the full potential of ternary logic.

In this work, we propose and experimentally demonstrate a novel ternary transistor with hybrid switching mechanism which integrates BTBT mechanism and thermionic emission mechanism together through a nested source configuration. The device features ultra-low I_{OFF} and steep SS for low power operations. The hybrid switching mechanism generates a flat current region for the third logic state and also allows high I_{ON} due to the dominant thermionic emission mechanism in on-state, showing its great potential for high-speed ternary logic.

II. DEVICE DESIGN AND WORKING MECHANISM

Fig. 2 shows the schematic illustration of the proposed p-type ternary transistor with a nested source region where an N^{++} region is nested with a P^+ source, creating fully depleted (FD) region in potion of P^+ source. It should be noted that different from the HamFET that we previously proposed in [4] for binary transistor, an additional lateral underlap between the undepleted P^+ region and channel surface is introduced in the proposed ternary transistor of this work. As shown in Fig. 2(a), the novel nested source configuration integrates BTBT current (I_{BTBT}) and thermionic emission current (I_{PN}) together, and more importantly, a larger V_G is required for I_{PN} to exceed I_{BTBT} due to the lateral underlap design. This generates a flat current region in the transfer curve (Fig. 2b), which is critical for the third logic state. Sentaurus TCAD simulation results in Fig. 3-5 further illustrate and verify the hybrid switching mechanism introduced by the nested source configuration.

III. DEVICE FABRICATION

The proposed ternary transistor is fabricated on a bulk Si substrate using CMOS-compatible process (Fig. 6). A two-step implantation is applied for fabrication of the nested source and P^+ drain, and the implantation window of N^{++} source is defined by gate and photoresist. The implantation condition is carefully designed for the vertical and lateral underlap design. The N^{++} source and P^+ source share the same source contact (Fig. 7).

IV. RESULTS AND DISCUSSION

The measured transfer characteristics of the fabricated ternary devices is shown in Fig. 8. An ultralow off-current below $10^{-7}\mu A/\mu m$ is achieved thanks to the underlap design of the P^+ source, which is about $10\times$ lower than conventional MOSFET in the same wafer [4] and desirable for ultralow power applications. Besides, the fabricated device also shows sub-thermionic minimum SS (SS_{min}) of 26mV/dec due to the dominant BTBT mechanism in the subthreshold-state. A flat current region is achieved with $|V_{GT}|$ increasing for the third logic state. Fig. 9 shows the measured output curves. At low $|V_{GT}|$, the typical super-linear onset behavior of BTBT mechanism [5] is observed, while at high $|V_{GT}|$, the drain conductance at low $|V_{DS}|$ is increased, evidencing the transition of dominant switching mechanism to thermionic emission for high drive current. Moreover, by using HSPICE, the ternary inverter characteristics is further simulated based on our developed BTBT current and thermionic emission current model [6]. As shown in Fig. 10, the ternary inverter features three stable output states. The proposed hybrid switching mechanism enables further supply voltage (V_{DD}) reduction while maintaining a high drive current at the same time, showing its great potential for low-power and high-performance applications.

From the experimental results of Fig. 11, it is worth noting that the middle-state I_{DS} increases with increasing $|V_{DS}|$ since the thermionic emission current is enhanced, and a transfer curve similar to a binary transistor is achieved at high $|V_{DS}|$. This unique characteristic can further contribute to even faster switching of the output state from high/low state to middle state due to the enlarged current difference between n/p ternary transistors, as indicated in Fig. 12.

V. CONCLUSION

In summary, a novel low-power and high-performance ternary transistor with tunneling/thermionic-hybrid switching mechanism is proposed and experimentally demonstrated. Measured ultra-low I_{OFF} and steep SS are promising for low power applications. The flat current region and the dominant thermionic emission current in the on-state facilitate a high-speed ternary inverter simultaneously.

Acknowledgments: This work was supported by National Key R&D Program of China (2018YFB2202801), NSFC (61927901, 62374009), and 111 Project (B18001).

979-8-3503-9164-0/24 $31.00 © 2024 IEEE

References: [1] S. L. Hurst, *IEEE Trans. Comput.*, pp. 1160–1179, 1984. [2] H. W. Kim et al., *IEEE TED*, pp. 4541-4544, 2020. [3] A. Gupta et al., *IEEE TED*, pp. 5305-5310, 2021. [4] Q. Huang et al., *IEEE JXCDC*, pp. 1-7, 2023. [5] L. De Michielis et al., *IEEE EDL*, pp. 1523–1525, 2012. [6] C. Wang et al., *Sci. China Inf. Sci.*, pp. 1-8, 2015.

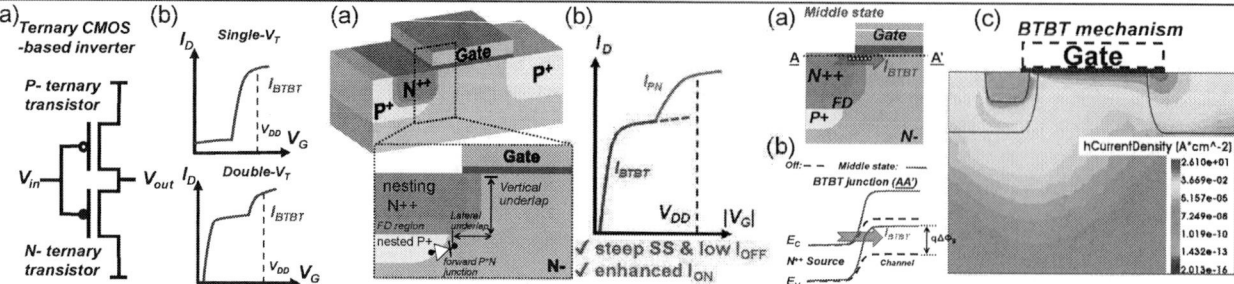

Fig.1 Illustration of (a) Ternary CMOS-based inverter and (b) Transfer curves of TFET-based ternary transistors.

Fig.2 (a) Schematic illustration of the proposed p-type ternary transistor and (b) hybrid switching mechanism realized through a nested source configuration.

Fig.3 Schematic illustration of (a) middle state current and (b) corresponding energy bandgap of the nested source; (c) TCAD simulation of hole current.

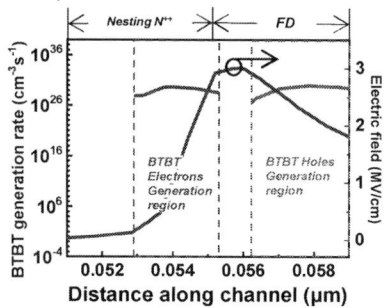

Fig.4 TCAD simulation of carriers BTBT generation rate and electric field distribution at tunnel junction.

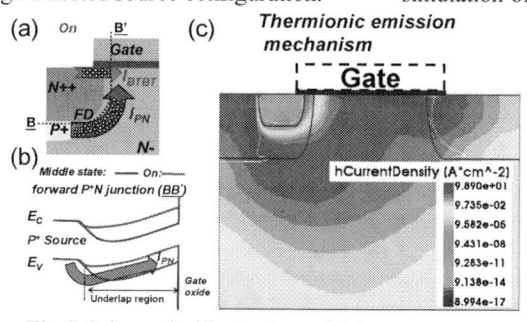

Fig.5 Schematic illustration of (a) on state current and (b) corresponding energy bandgap of the nested source; (c) TCAD simulation of hole current.

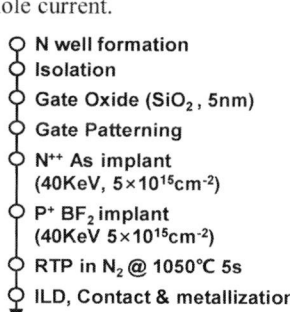

- N well formation
- Isolation
- Gate Oxide (SiO$_2$, 5nm)
- Gate Patterning
- N^{++} As implant (40KeV, 5×10^{15}cm^{-2})
- P$^+$ BF$_2$ implant (40KeV 5×10^{15}cm^{-2})
- RTP in N$_2$ @ 1050℃ 5s
- ILD, Contact & metallization

Fig.6 Process flow of the proposed ternary transistor.

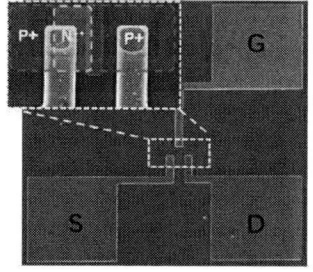

Fig.7 Top view SEM image of the fabricated ternary transistor (L$_G$=4μm, W$_G$=10μm) and illustration of implantation window.

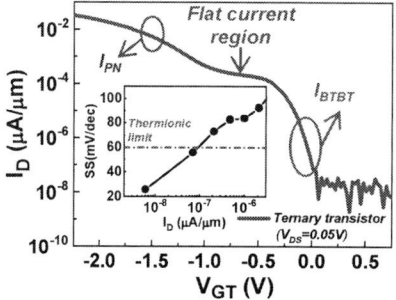

Fig.8 Measured transfer curve of the fabricated ternary transistor. Inset shows the extracted SS-I$_D$ relationship.

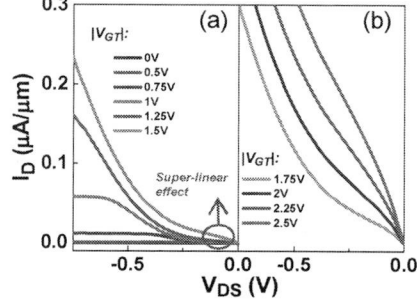

Fig.9 Measured output characteristics, evidencing (a) BTBT mechanism and (b) thermal emission mechanism.

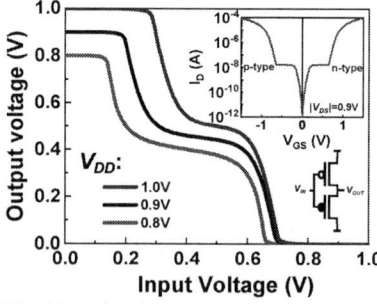

Fig.10 Simulated voltage transfer curves based on the proposed ternary transistor with hybrid switching mechanism. Inset: transfer curves of the transistors used for simulation.

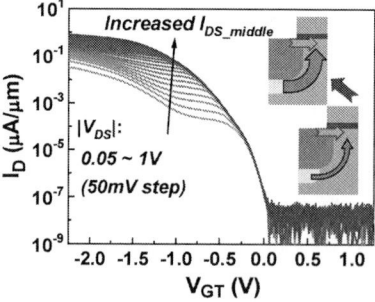

Fig.11 Measured transfer curves under different |V$_{DS}$|. I$_{DS\_middle}$ is increased with |V$_{DS}$| due to larger thermal emission current and a binary-like characteristic is realized.

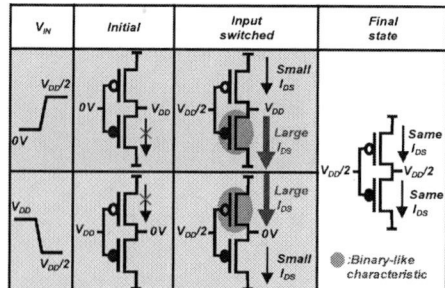

Fig.12 Schematic illustration of the output state switching from high/low state to middle state. The binary-like characteristic can enlarge the switching speed by enlarging the current difference between n/p transistors.

979-8-3503-9164-0/24 $31.00 © 2024 IEEE

Single electron charge detection in nanoscale metal double-dot:
DC Single Electron Transistor vs gate RF sensing.

Mohammad Istiaque Rahaman[1], Gergo P. Szakmany[1], Xavier Jehl[2], Alexei O. Orlov[1] and Gregory L. Snider[1]

[1]Electrical Engineering, University of Notre Dame, USA, [2]IRIG/IPHELIQS, University Grenoble Alpes, France

Email: mrahaman@nd.edu

Abstract — **We fabricated two sub-20 nm aluminum dots separated by an alumina tunnel barrier. Single electron transport from one dot to the other is controlled by applying appropriate DC potential across the dots. We utilized Single Electron Transistors (SETs) to detect the electron transport, and the radio frequency (RF) gate sensing technique was deployed to sense the state of the SET. This study leverages the amplifier action of the SET for charge sensing and alleviates the need to aligning the RF sensor probe in very close proximity to the charging object for challenging nanoscale charge sensing applications.**

Keywords: Single Electron Transistor (SET), RF sensing, metal double dots

I. INTRODUCTION

Single electron charge sensing is crucial for reading the bistable charge states in transistor-less computing platforms like Quantum-dot Cellular Automata (QCA) and for charge qubit detection in quantum computing. Alongside single charge sensing, determining the spatial localization of the charges is equally essential. We have previously reported the detection of single electron charge transfer between two sub-20 nm metal dots separated by a tunnel barrier (DD) driven by a differential voltage applied to the DD gates [1, 2]. The detection was performed using Single Electron Transistors (SETs) coupled to one of the dots [1] and radio frequency (RF) gate reflectometry at one of the gates with no SET present [2]. However, RF gate reflectometry of DD without SET [2] is only capable of detecting single charge transitions but fails to sense the spatial localization of electrons before and after transfer.

Here, we employ a structure consisting of a DD and SET similar to [1], with an added RF sensor connected to one of the DD gates. We conducted combined low-frequency ("DC") and high-frequency (RF) measurements to capture single-electron transport in the DD. The experiment reveals that RF gate transmission measurements yield a signal proportional to the SET conductance in DD detection. In contrast to the work of [2], in this study, we detect the RF response of the SET induced by the charge transfer in the DD. This approach has two main advantages: first, it increases the signal-to-noise ratio (SNR), and second, the presence of the SET enables the RF sensor to detect the spatial localization of electrons before and after transfer (proportional detection) rather than functioning solely as a transition detector. By leveraging the signal amplification attributes of the SET, we mitigate the need for very tight lithographic alignment requirements in RF-only (i.e., without a SET) measurements [3].

II. DEVICE FABRICATION AND MEASUREMENT

The colorized scanning electron micrograph illustrating two SETs, three gates, and DD is depicted in Fig. 1. The structure was fabricated by thin film deposition of Al with in situ oxidation using the Niemeyer-Dolan bridge technique [4]. After fabrication, gate-A was bonded to a printed circuit board (PCB) containing a resonant LC matching network, as shown in the inset of Fig. 1. The measurements for this study were performed at approximately 300mK inside a dilution refrigerator in a 1T magnetic field to suppress the superconductivity of Al. One channel of a lock-in amplifier (UHF LIA from Zurich Instruments) was used to measure RF response, and the SET conductance was measured with a FEMTO current amplifier connected to its second channel.

III. RESULTS AND DISCUSSION

With a positive DC potential applied to gate-A and a negative potential applied to gate-B of Fig. 1, whenever the electrons of dot #2 overcome the Coulomb repulsion of electrons of dot #1, a single electron transfer occurs from dot #2 to dot #1, and vice versa. This transfer process of an electron gives rise to dynamic capacitance, C_{dyn} [5] near transition points. A small change in C_{dyn} at points of single electron transfer slightly detunes the matching network, resulting in a measurable change in the output RF signal. In order to measure this change, the gate needs to be placed very close to the DDs and coupled asymmetrically to each dot [2]. Note that, in this structure, the SET island extension is placed closer to the DD than the gates, and hence, SETs detect the electron transfer within the dots, and the output RF signal measures the charge state of the SET.

For utilizing SET as a detector, SET-A was biased at the middle of the rising slope of the Coulomb blockade peak, where the sensitivity for detecting the electron transfer within DDs is the highest. A differential bias is applied across the DDs by ramping up the DC voltage in gate-A while ramping down the DC potential in gate-B. This application of the differential bias across the DDs causes periodic single electron transfer, yielding voltage oscillations across the DD that result in conductance oscillations of the SET in a sawtooth fashion [1]. The average differential conductance, $G = dI_{ds}/dV_{ds}$ (μS) of SET-A is plotted with respect to the differential voltage, $V_{diff} = V_{gate-A} - V_{gate-B}$ in Fig. 2. The conductance plot reveals two peaks in sawtooth, which represents a transfer of two electrons with a period of approximately 600 mV.

979-8-3503-9164-0/24 $31.00 © 2024 IEEE

For performing RF gate transmission measurements, the complex RF signal from the LC resonator was measured at its resonant frequency of f =358.975MHz. The in-phase, X_{RF}, and quadrature-phase, Y_{RF}, components of the averaged complex RF signal are plotted with respect to the V_{diff} in Fig. 3(a) and (b) respectively. As can be seen, the shape of the RF signal in Fig. 3 correlates well with the shape of conductance oscillations in Fig. 1. This correlation is expected in the case of gate-coupled SET reflectometry. Therefore, we conclude that the single electron transfer between the DD modulates the pre-set bias point of the SET, and the detected RF signal stems from SET detection rather than direct detection of the DD. One significant advantage of this technique is that it offers much wider bandwidth (about 15 MHz in this work) compared to DC sensing.

IV. CONCLUSION

We have experimentally demonstrated a new method of detecting single electron transfer in a nanoscale DD structure, where a SET detects the single electron transfer within DD and an RF sensor probe senses the SET. The capacitive coupling from DD to the gate is weaker than the coupling from DD to the SET and that's why SET is affected stronger than gate sensor alone. The detection outcome by the RF sensor also correlates with the DC detection by the SET.

ACKNOWLEDGMENTS

The US National Science Foundation Grant DMR-1904610 supported this work.

REFERENCES

[1] Rahaman, M. I., Szakmany, G. P., Orlov, A. O., & Snider, G. L., "Dipole charge detection: towards the readout of bi-stable charge states in Molecular QCA," *IEEE Sensors Letters*, 2023.

[2] Rahaman, M. I., Orlov, A. O., & Snider, G. L., "Determination of Direction of Single Electron Charging In An Isolated Metal Double-Dot System Using RF Reflectometry," *IEEE Transactions on Nanotechnology*, in press.

[3] Orlov, A. O., Fay, P., Snider, G. L., Jehl, X., Lavieville, R., Barraud, S., and Sanquer, M, "Study of charged island formation in nanoscale Si single-electron transistors using dual port reflectometric spectroscopy." *Silicon Nanoelectronics Workshop (SNW)*, pp. 1-2, IEEE, 2015.

[4] Dolan, G. J., "Offset masks for lift-off photoprocessing." *Applied Physics Letters* 31, no. 5 (1977): 337-339.

[5] Zirkle, T. A., Filmer, M. J., Chisum, J., Orlov, A. O., Dupont-Ferrier, E., Rivard, J., Huebner, M., Sanquer, M., Jehl, X., & Snider, G. L., "Radio frequency reflectometry of single-electron box arrays for nanoscale voltage sensing applications," *Applied Sciences*, 10(24), p.8797, 2020.

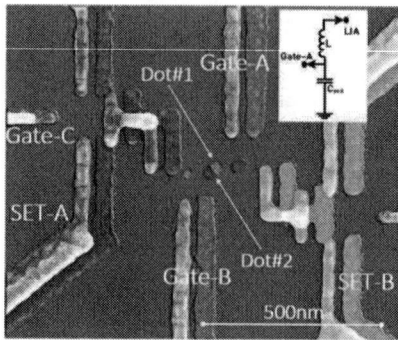

Fig. 1: Colorized SEM micrograph of the structure. SET-B was found open due to a fabrication issue. An LC resonator is connected to gate-A, as shown in the inset.

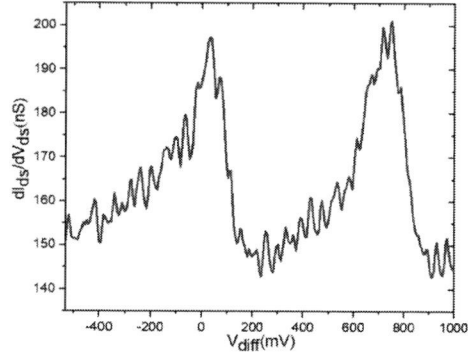

Fig. 2: Average differential conductance, dI_{ds}/dV_{ds} (μS) of SET-A of 1024 scans is plotted against differential gate voltage, V_{diff}. The peaks of the sawtooth oscillation signify the single electron transfers between the DD.

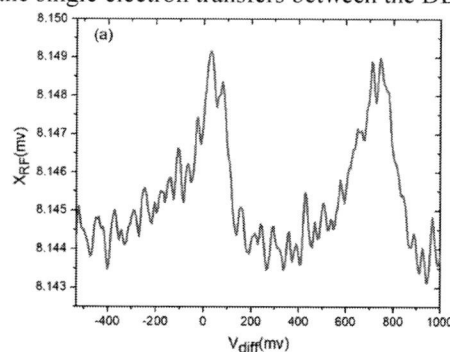

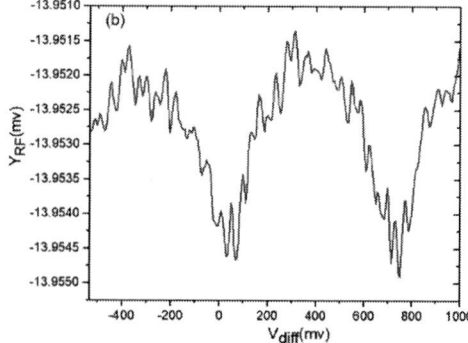

Fig. 3: Average complex RF signal of 1024 scans measured at the resonance frequency of LC resonator and plotted across differential voltage, (a) in-phase and (b) quadrature-phase component.

Ferromagnetic manganese silicide nanoparticles formed by ion implantation in silicon

R. Ohsugi[*], M. Kawano, Y. K. Wakabayashi, Y. Krockenberger, H. Sumikura,
J. Noborisaka, and K. Nishiguchi

NTT Basic Research Laboratories, NTT Corporation, Japan
Email: rento.oosugi@ntt.com

Abstract — **Manganese silicide nanoparticles (NPs) embedded in silicon were fabricated by an ion implantation method with various manganese ion acceleration energies (E_{acc}). The NPs were single crystalline with a tetragonal structure and exhibited ferromagnetic properties below the blocking temperature (T_B), at which the magnetization of NPs starts to fall into a metastable state and becomes fixed. T_B increased as E_{acc} increased. On the basis of Neel's relaxation theory, we revealed that this dependence originates from the influence of E_{acc} on the particle sizes.**

Keywords: ion-implantation, silicide, magnetism

I. INTRODUCTION

Low-dimensional ferromagnetic (FM) materials, such as FM nanoparticles (NPs) offer advantages, such as higher Curie temperature and larger magnetization than bulk specimens [1]. Additionally, they exhibit fascinating functionalities for spintronic applications [2,3], including electrical control of magnetization [1]. Among FM NPs, manganese (Mn) silicide NPs can be fabricated by conventional ion implantation and annealing processes [4] and are highly compatible with silicon (Si) technologies. To date, ion-dose [4] and annealing temperature [5] dependence on their FM properties have been investigated. However, the influence of the ion acceleration energy (E_{acc}) on the FM properties has been less discussed even though E_{acc} is also an important factor to fabricate the NPs. In this study, we fabricated Mn silicide NPs by ion implantation with various E_{acc} and investigated their magnetism.

II. EXPERIMENTS

Mn silicide NPs embedded in single crystal Czochralski silicon were fabricated by Mn-ion implantation and post-annealing under the conditions shown in Fig. 1. E_{acc} were 50, 100, 150, and 300 keV. The Mn^{1+}-ion dose was 5×10^{15} cm^{-2}. After the implantation, each sample was annealed at 700°C for 30 min in N_2 atmosphere.

III. RESULTS AND DISCUSSION

Figure 2(a) shows a cross-sectional transmission electron microscope (TEM) image and the energy dispersive X-ray spectroscopy (EDS) mapping image of Mn for a sample with E_{acc} of 50 keV. These results show that the Mn-ions precipitate in the form of NP within the Mn ion-implanted area. Figure 2(b) shows high-resolution TEM and selected-area electron diffraction (SAED) images of a representative NP. The NP was single crystalline with a tetragonal structure. The diffraction pattern and the distance of the spots between (000) and (-118) or (220) correspond to those of $MnSi_{1.7}$ reported previously [6].

Figure 3(a) shows the NP-size distribution for a sample with E_{acc} of 50 keV. From log-normal fitting of the distributions for samples with E_{acc} of 50, 150, and 300 keV,

we extracted the medial radius (R_{medi}) of each sample as shown in Fig. 3(b). R_{medi} increased as E_{acc} increased. Additionally, as shown in Fig. 3(c), the number of vacancies (N_{Vac}) increased as E_{acc} increased. In general, such defects promote the formation of large NPs via various mechanisms, such as transient enhanced diffusion of ions and Ostwald ripening [7]. Thus, the larger NPs might be formed by higher E_{acc}.

Figure 4(a) shows the magnetization as a function of a magnetic field (M-H curve) measured at 3 K for the samples with E_{acc} of 50 and 150 keV. For both samples, a clear hysteresis loop was observed, indicating that the NPs behave as ferromagnets. Figure 4(b) and (c) show the magnetization as a function of temperature (M-T curve) of the NPs under zero-field cooling (ZFC) and field cooling (FC) processes. We observed a bifurcation between ZFC and FC curves. From the peak point of ZFC curves, we extracted the blocking temperature (T_B), at which the magnetization of NPs starts to fall into a metastable state and becomes fixed. Figure 4 (d) shows T_B as a function of E_{acc}. T_B increased as E_{acc} increased. According to Neel's relaxation theory, T_B can be described as follows [8]:

$$T_B = (K/25 k_B)(4\pi R_{medi}^3/3)$$

where the K is the magnetic anisotropy constant, and k_B is the Boltzmann constant. Figure 4(e) shows T_B as a function of R_{medi}^3. T_B was proportional to R_{medi}^3. From linear fitting of the data, the magnetic anisotropy constant of our samples was estimated to be $K=6.7\times10^4$ J/m^3, which is equivalent to that of typical FM NPs [9]. These results indicate that the change in the magnetic properties originates from the change in the size of Mn silicide NPs with varying E_{acc}.

IV. CONCLUSION

In this study, we fabricated FM Mn silicide NPs embedded in Si by ion implantation. We found that the ferromagnetic properties of the NPs can be varied by controlling E_{acc}. These findings will be helpful to develop Si-based spintronic devices by ion implantation, which is highly compatible with conventional Si technologies.

ACKNOWLEDGMENTS

This work was supported by Y. Ono from Research Institute of Electronics in Shizuoka University.

REFERENCES

[1] F. Xiu, *et al.*, Nat. Mater. **6**, 440 (2007).
[2] K. Y. Camsari, *et al.*, Phys. Rev. X **7**, 031014 (2017).
[3] J. Grollier, *et al.*, Nat. Electron. **3**, 360 (2020).
[4] S. Yabuuchi, *et al.*, Jpn. J. Appl. Phys. **47**, 4487 (2008).
[5] S. Zhou, *et al.*, Phys. Rev. B **75**, 085203 (2007).
[6] Y. Xu, *et al.*, Jpn. J. Appl. Phys. **50**, 11RH02 (2011).
[7] S. Mantl, *et al.*, Mater. Sci. Rep. **8**, 56(1992).
[8] L. Néel, *et al.*, Annales de géophysique **5**, 99 (1949).
[9] S. Noh *et al.*, Nano Today **13**, 61 (2017).

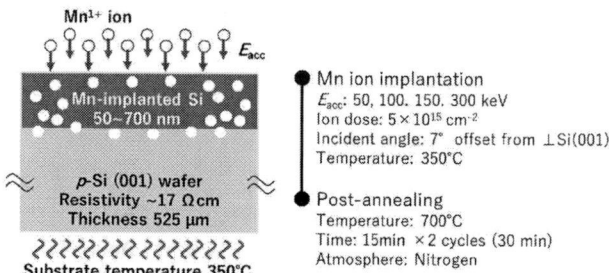

Fig. 1. The fabrication process of Mn silicide NPs in Si.

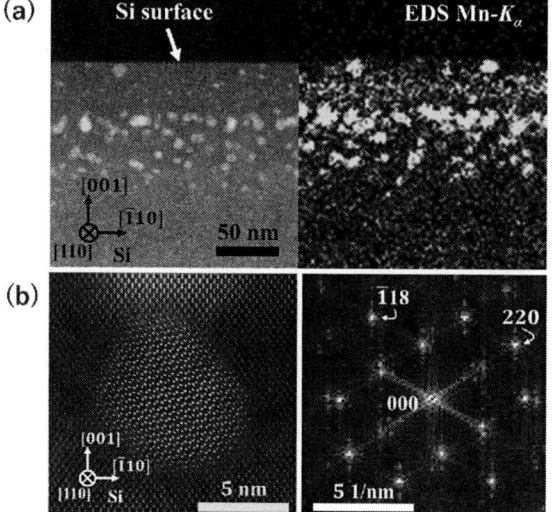

Fig. 2. (a) Cross-sectional TEM image of the Mn-implanted Si layer of the sample with E_{acc} of 50 keV (left). EDS mapping image of Mn-K_α intensity (right). (b) TEM image of Mn silicide NP (left) and its SEAD image (right).

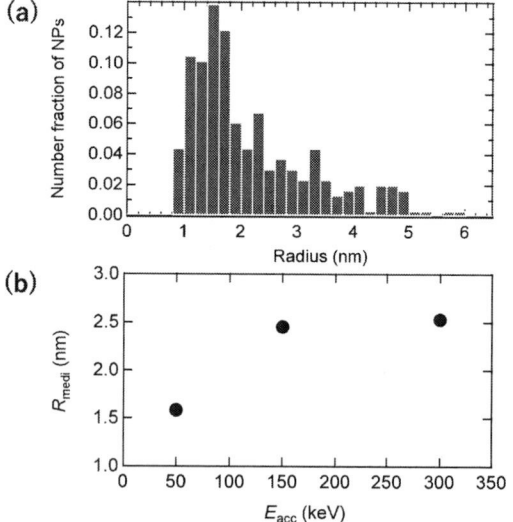

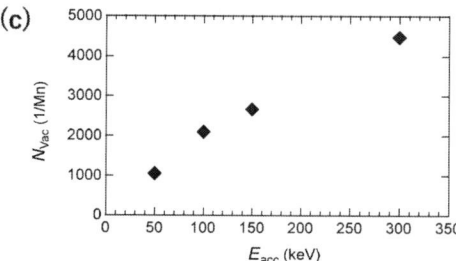

Fig. 3. (a) Particle-size distribution of the sample with E_{acc} of 50 keV. (b) E_{acc}-dependence of R_{medi}. (c) Simulated N_{Vac} assuming as-implantation state (before post-annealing) by the ion-implantation simulator: SRIM /TRIM 2008.

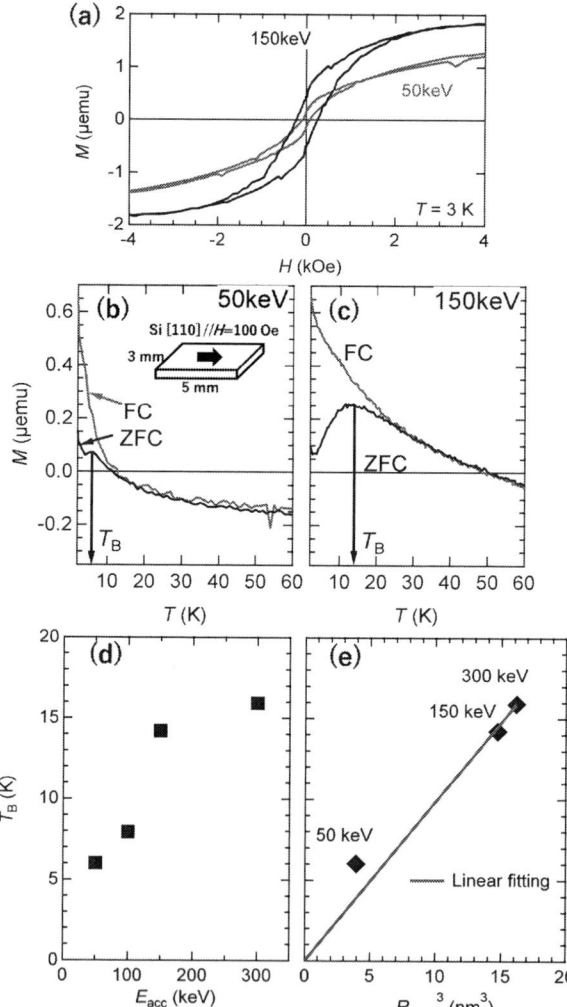

Fig.4 (a) MH curves of the sample with E_{acc} at 50 and 150 keV measured at 3 K. MT curves of the samples with E_{acc} of (b) 50 keV and (c) 150 keV. The inset in (b) shows the configuration of the applied magnetic field during the measurements. T_B is indicated by the black arrows in (c). (d) T_B versus E_{acc}, as extracted from the MT curves. (e) T_B versus R_{medi}^3.

The Simulation of Double Germanium Quantum Dots in a Ring-Shaped Quantum Structure

Cheng-En Liang and Ying-Tsan Tang

Department of Electrical Engineering, National Central University, Taoyuan, Taiwan

Email: yttang@ee.ncu.edu.tw

Abstract — **In this work, self-assembled Ge double quantum dots embedded in a ring-shaped MOS device are simulated. We utilize Poisson, Schrödinger, and Master equation models in QTCAD integrated with the open-source Gmsh tool to simulate potential changes in quantum bits within the device. By observing charge stability diagrams and variations, we discover that under the N28 technology node, SHT is more feasible than SET for realizing multi-level ring-shaped quantum dot structures.**

Keywords: quantum TCAD, quantum dot, charge stability

I. INTRODUCTION

Since the 1980s, the concept of quantum computing has spurred extensive research into photons, ion traps, superconducting circuits, and semiconductor quantum dots. Thanks to significant improvements in lithography and etching, which have facilitated VLSI technology, semiconductor quantum dots have become a focal point of recent intensive study. The introduction of long-spin-coherent atoms like Si^{28} and Ge^{74} has rapidly advanced silicon and germanium as quantum bits and related single-electron(hole) transistors (SET/SHT). Among these, a type of self-assembling quantum dot (Fig. 1(c)) attracts attention due to the harmonious interactions enabled by the incorporation of Si, Ge, and dopant. Despite the structure being implemented, gaining further physical insight into inter-bit modulation is very important, especially in understanding the relationship between quantum dots and surrounding capacitance. This study employs quantum mechanics with Schrödinger, Poisson [1, 2], and the open-source Gmsh [3] for grid optimization. Dynamic transitions are simulated using master equation in QTCAD for calculations below 100mK. Finally, charge stability diagrams and conduction/valence band edges (Ec/Ev) offer new pathways for future Si-based quantum device design.

II. DEVICE DESIGN AND MODELLING

Fig.1(a) represents a portion of the ring-shaped quantum structure with an angle of approximately 120 degrees between Source-Barrier-Drain. To facilitate the exploration of the relationships between barrier gate (BG), plunger gates (PG), and several capacitors (SiO_2, Si_3N_4, HfO_2) within the structure, we selected the simplest double-sphere unit for study inspired by Ref. [4]. After constructing the model using the open-source Gmsh, we employed the master equation suggested by Ref. [5] combined with the level-arm method [6, 7] to control the coupling capacitance for quantum TCAD simulations(Fig.2).

II. RESULTS AND DISCUSSIONS

Fig.3(a) depicts the cross-sectional profile of Ec with a double Ge spherical structure embedded in an nMOS structure with a diameter of 20nm, where the rainbow region represents the isosurface of Ec for the quantum dots. The complete spheres signify that the geometric spheres can effectively reduce leakage of quantum states outside the spherical structure. Fig.3(b)(c)(d) show the changes in the charge stability diagram as the spacing (d_{DQD}) between double spheres, is varied: 30nm, 20nm, and 10nm, respectively. The slope of the conductivity becomes steeper with decreasing d_{DQD}, indicating the transition from independent single-electron states to coupled double-electron states. This enhancement suggests that the coupling capacitance C_{SiO2} and surrounding parasitic capacitance play significant roles, emphasizing that precise control of the barrier gate is not a trivial task.

In the pMOS structure (doped with $10^{20}cm^{-3}$), the Ev profile for d_{DQD}: 20nm and 10nm is plotted in Fig.4(a). The results indicate that the 20nm hole states exhibit strong coupling and symmetric hybrid electron states compared to nMOS at the same scale. This suggests that multi-level quantum states fabricated using SHT are less susceptible to scaling pressures and process-induced defects compared to SET, reducing reliability degradation issues. It is worth mentioning that the variations in electron/hole states caused by different plugger gates can also be elucidated. This includes changes in localized wave functions, as well as transformations in the spatial distribution between the double spheres, as shown in Fig. 5. This implies the existence of spatial entanglement between hybrid quantum states, which can be manipulated by geometric structures. However, this is beyond the scope of the present study.

II. CONCLUSIONS

This work employs quantum TCAD and open-source tools to simulate the local electrical characteristics of CMOS-fabricated quantum devices. It investigates variations in Ec and Ev within real germanium ball geometries, SiO_2, HfO_2, and Si_3N_4 capacitors under multi-gate regulations. By controlling the spacing between double spheres, changes in hybrid quantum states are observed. Analysis of charge stability suggests that ring-shaped SHT fabrication of multi-level quantum dots may be more feasible than SET under N28 node technology, offering an alternative approach for future quantum device design and fabrication.

ACKNOWLEDGMENTS

We are very grateful for the valuable discussions with Dr. Félix Beaudoin and Dr. Jeremy Garaffa, as well as the valuable insights from Professor Hong Kuo at the University of Toronto and Professor P.W. Li at NYCU. This work was supported by the National Science and Technology Council (NSTC) of Taiwan (No. 112-2218-EA49-013-MBK, 112-2622-8-A49-013-SB, NSTC 113-2119-M-A49-007) and Taiwan Semiconductor Research Institute (No. JDP113-Y1-035).

REFERENCES

[1] Gao X, et al., 2013 J Appl Phys.
[2] Stopa M. 1996 Phys Rev B.
[3] Geuzaine C and Remacle JF 2009 Int J Numer Meth Eng pp.1309~31.
[4] I-Hsiang Wang et al., 2021 Nanomaterials.
[5] Félix Beaudoin et al., 2022 Physics Letters A.
[6] R. Hanson et al., 2007 Rev. Mod. Phys pp. 1217~1265
[7] Andreas Fuhrer 2003 PhD thesis ETH Zurich.

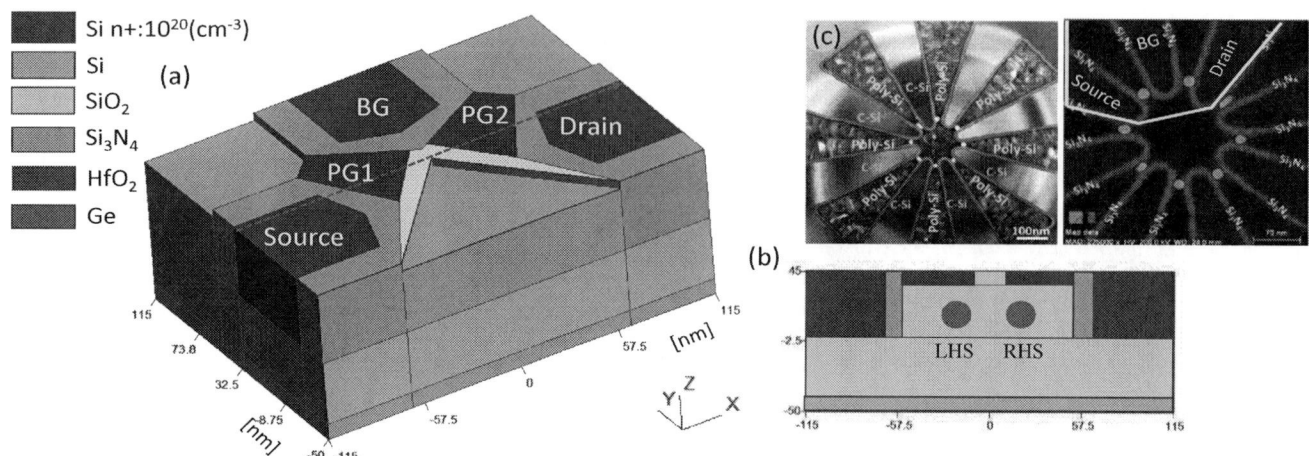

Fig. 1: Design of QDQ TCAD model (b) Slice along the red dashed line in (a); (c) TEM image of multi-QDs in Ring-Shaped Quantum Structure [4].

Non-linear Poisson $\quad -\nabla \cdot (\epsilon \nabla \varphi) = e(p - n + N_+ + N_-)$

Master equation
$$\begin{cases} \dfrac{dp_m}{dt} = -p_m \sum_{n \neq m} \gamma_{nm} + \sum_{n \neq m} p_n \gamma_{mn} = 0 \\ \gamma mn = \Gamma S(n \to m) + \Gamma D(n \to m) \end{cases}$$

Time-dependent Schrödinger's equation $\quad -\dfrac{\hbar^2}{2} \nabla \cdot [M_e^{-1} \cdot \nabla \psi(r)] + V(r)\psi(r) = E\psi(r)$

Lever arm $\quad \alpha = \dfrac{c_i}{c_j} = -\dfrac{\Delta V_i}{\Delta V_j}$

Fig. 2: Formulas utilized in QTCAD to achieve first 4 eigenstates and charge stability diagram.

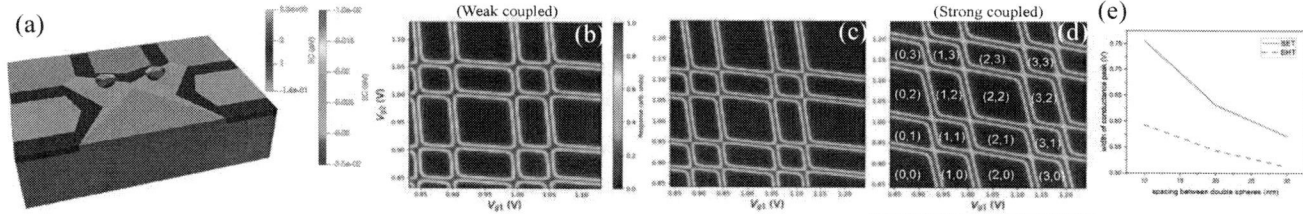

Fig. 3. (a) Ec profile of Ge-DQDs(The cross-section at z=6nm); Charge stability diagram of double quantum dots acquired by scanning plunger gates in spacing distance of (b) 30nm; (c) 20nm; (d) 10nm; (e) width of conductance peak vary with spacing (d_{DQD}) between double spheres.

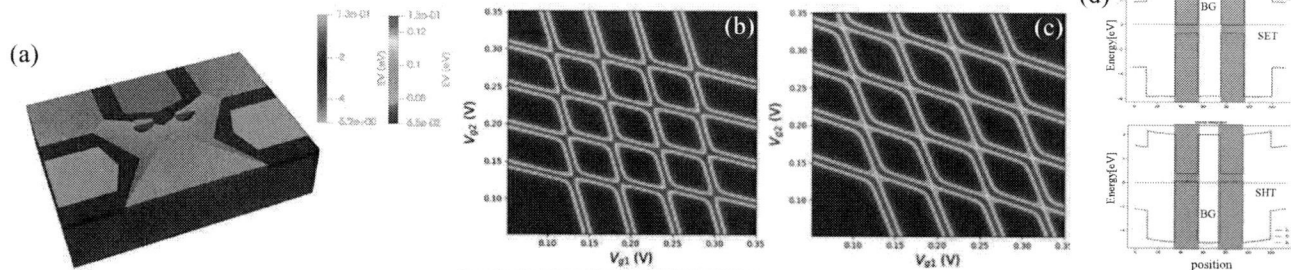

Fig. 4 (a) Valence band edge Ev profile through the system. The rainbow colour corresponds to Ev isosurfaces around Ge-DQD; Charge stability diagram of double quantum dots in spacing distance of (b) 20nm; (c) 10nm; (d) band diagrsm of SET(upper) and SHT(lower)

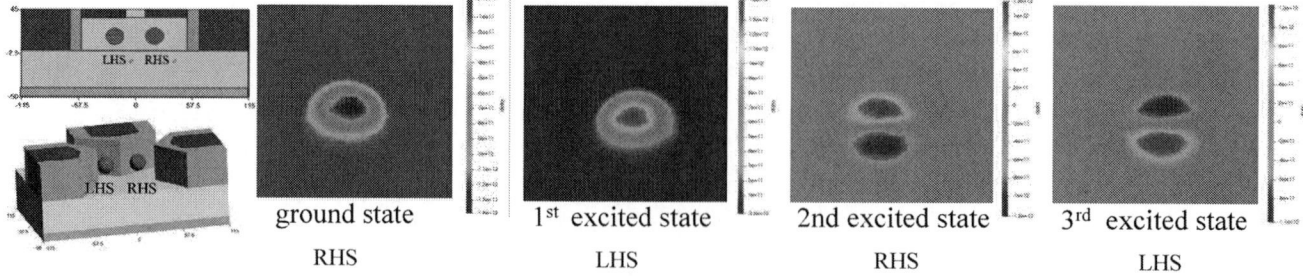

Fig. 5: The first 4 eigenstates of Ge-DQD at V_{G1}: 0.87V and V_{G2}: 0.88V in this model.

979-8-3503-9164-0/24 $31.00 © 2024 IEEE

Improved Uniformity and Excellent Endurance Characteristics of TaO_x-Based RRAM by Laser-Mediated Interface Engineering

Lindong Wu[1], Qishen Wang[1], Hongxu Liao[1], Chaoyi Ban[1], Linbo Shan[1], Zongwei Wang[1,2*], Yuan Wang[1,2], and Yimao Cai[1,2*]

[1] School of Integrated Circuits, Peking University, Beijing, China,
[2] Beijing Advanced Innovation Center for Integrated Circuits, Beijing, China
Email: {wangzongwei, caiyimao}@pku.edu.cn

Abstract — **In this work, we report a laser-mediated interface engineering method to improve RRAM performances. Compared with pristine TiN/TaO$_x$/TaN devices, a 10x uniformity improvement is achieved when a 0.4 J/cm² laser is applied to irradiate the TaO$_x$ layer. Moreover, an excellent endurance (10^6) with high on/off ratio (156) is obtained. The characterization results show that laser treatment can effectively inhibit oxygen exchange at the interface between TaO$_x$ layer and TiN electrode to improve controllability of the filament. Our study shows the potential of laser-mediated interface engineering for achieving high-performance RRAMs.**

Keywords: Laser-mediated, interface engineering, RRAM

I. INTRODUCTION

The resistive random-access memory (RRAM) has drawn great attention due to the simple structure, low power consumption and the potential of breaking Von Neumann architecture bottleneck [1-2]. However, RRAMs still suffer from poor uniformity and endurance characteristics, which poses a challenge for the applications [3-4]. Here, we propose a laser-mediated interface engineering (LMIE) method to promote the performances. The TiN/TaO$_x$/TaN devices treated by the 0.4 J/cm² laser have a great uniformity with 10x improvement compared with pristine devices and excellent endurance characteristic of 10^6 with on/off ratio of 156. The mechanism is explored by characterization methods.

II. EXPERIMENT

The fabrication process of laser-treated TiN/TaO$_x$/TaN devices is shown in **Fig. 1**. Firstly, 10 nm Ti, 60 nm TaN and 30 nm TaO$_x$ layers are deposited on the substrate. Next, a 0.4 J/cm² laser is irradiated on the TaO$_x$ surface. After etching, 20 nm TiN and 60 nm Pt layers are deposited on the TaO$_x$ layer. **Table I** shows the laser parameters adopted in the experiment. For pristine devices, the 20 nm TiN and 60 nm Pt are directly grown on the untreated TaO$_x$ layer.

III. RESULTS AND DISCUSSION

Fig. 2 shows the repeatable I-V curves of two kinds of devices for 100 cycles. During measurements, the voltage is applied on TaN electrode while keeping TiN electrode grounded. Both kinds of devices possess typical resistive switching behaviors. **Fig. 3** shows the fluctuation of high resistance state (HRS) and low resistance state (LRS). We adopt the ratio of variance (σ) and mean value (μ) of resistance as the fluctuation parameter [5]. The fluctuations of HRS (μ_h/σ_h) and LRS (μ_l/σ_l) of laser-treated devices are smaller than those of the pristine device. Especially, the μ_h/σ_h of the laser-treated device (0.016) is only 10% of that of pristine device, indicating a significant improvement in uniformity. **Fig. 4** shows the endurance characteristic of the laser-treated device under pulse mode. When the set voltage (V_{set}) is 1.7 V and reset voltage (V_{reset}) is -1.9 V, both with the width of 20 ns, an excellent endurance characteristic (10^6) with on/off ratio of 156 can be achieved.

The transmission electron microscope (TEM) and energy dispersive X-ray spectroscopy (EDS) are used to explore the mechanism. TEM results (**Fig. 5(a)** and **Fig. 5(c)**) show that there is an obvious interface between TaO$_x$ and TiN layer (red dashed box) in the laser-treated device. EDS results (**Fig. 5(b)** and **Fig. 5(d)**) show that the oxygen atomic fraction at the interface of the laser-treated device (near red dashed line) abruptly decreases compared with the pristine device. Therefore, the LMIE method can greatly improve RRAM performances by suppressing oxygen exchange at the interface, as shown in **Fig. 6**. For the pristine device, oxygen vacancies near TiN electrode randomly participate in the formation and fracture of the filament. On the contrary, the oxygen exchange at the interface in the laser-treated device is inhibited and only oxygen vacancies in TaO$_x$ layer contribute to the filament, which is more controllable.

IV. CONCLUSIONS

We propose the LMIE method to promote RRAM performances. The experiments based on TiN/TaO$_x$/TaN devices show that when a 0.4 J/cm² laser is applied to irradiate TaO$_x$ surface, the treated device has an improved uniformity (μ_h/σ_h=0.016) and an excellent endurance of 10^6 with on/off ratio of 156. These results provide a guideline for obtaining high-performance RRAMs.

ACKNOWLEDGMENTS

This work was supported by the National Natural Science Foundation of China (62025401, 62322401, 62341407, 62305002, 61927901), in part by the China Postdoctoral Science Foundation (2023M740053), in part by the 111 project (B18001), and in part by the Xiaomi Foundation.

REFERENCES

[1] H. S. P. Wong, et al., *Proc. IEEE*, 2012, 100(6): 1951-1970. [2] L. Wu, et al., *Nanoscale*, 2021, 13(6): 3483-3492. [3] Z. Yu, et al., *IEEE Electron Device Lett.*, 2020, 41(6): 940-943. [4] C. Nail, et al., *IEEE IEDM*, 2016. [5] Q. Wang, et al., *IEEE IEDM*, 2023.

979-8-3503-9164-0/24 $31.00 © 2024 IEEE

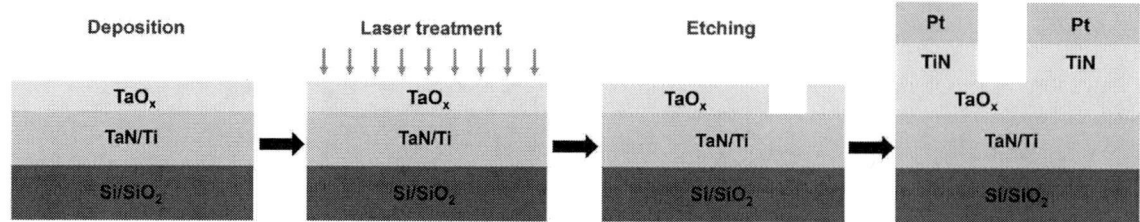

Fig. 1. The fabrication process of laser-treated RRAM devices.

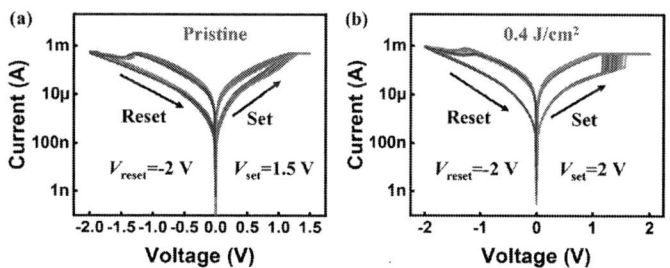

Fig. 2. The repeatable resistive switching characteristics of (a) the pristine device and (b) the laser-treated device.

Table I. The parameters of the laser adopted in the experiment.

Wavelength (nm)	527
Pulse width (ns)	150
Frequency(Hz)	500
Size (μm^2)	400×400
Energy density (J/cm^2)	0.4

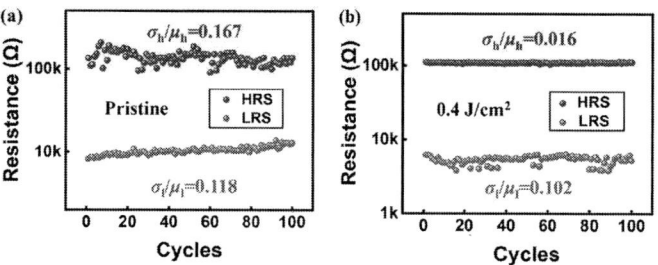

Fig. 3. The resistance fluctuations of (a) the pristine device and (b) the laser-treated device for 100 cycles. The μ_h/σ_h of the laser-treated device is only 10% of that of the pristine device.

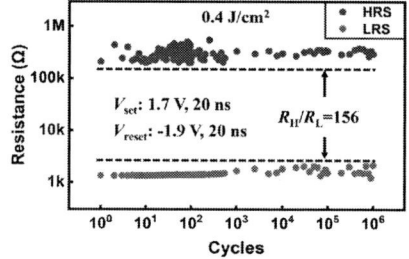

Fig. 4. The endurance characteristic (10^6) of the laser-treated device when V_{set} is 1.7 V and V_{reset} is -1.9 V under pulse mode.

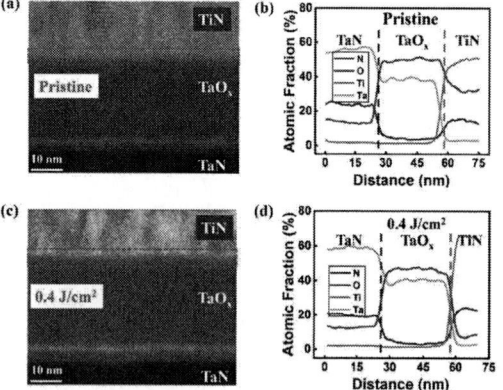

Fig. 5. The characterization results of two kinds of devices. (a-b) The TEM and EDS results of the pristine device. (c-d) The TEM and EDS results of the laser-treated device.

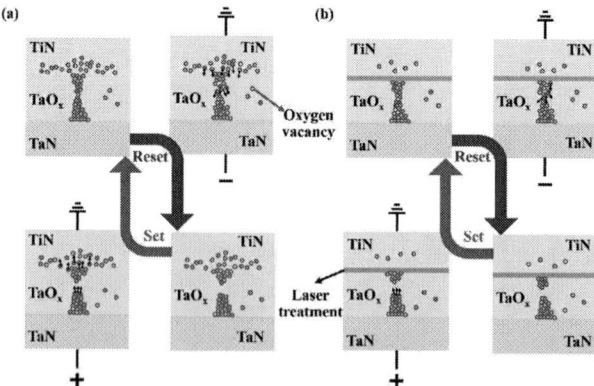

Fig. 6. The mechanism schematics of two kinds of devices. (a) There is oxygen exchange at the interface between TaO$_x$ layer and TiN electrode in the pristine device. (b) The oxygen exchange is suppressed by layer-mediated interface engineering.

979-8-3503-9164-0/24 $31.00 © 2024 IEEE

Effects of Modulation Pulse on Resistive Switching Dynamics of RRAM Devices

Cheng-Han Chien and Yeong-Her Wang*

Institute of Microelectronics, National Cheng Kung University, Tainan 701, Taiwan.

Email: q16114065@gs.ncku.edu.tw, *yhw@ee.ncku.edu.tw

Abstract —**Pulse switching is an important modulation method in addition to dc sweep on RRAM. Due to inherent randomness of conductive filament formed in RRAM, defining an optimal and reliable range for state modulation is important. OxRRAM devices operated by sub 1ms pulse-width input and possible multilevel resistance state are discussed. Compliance of forming process can induce endurance difference to a RRAM cell, with good endurance observed to be the device formed under lower compliance current.**

Keywords: RRAM, Resistive switching, endurance, pulse

I. INTRODUCTION

RRAM is seen as a good nonvolatile choice as a next-generation memory device in conventional computing or in-memory computing. But OxRRAM based on oxygen vacancy (V_O) transmitted by valence change mechanism (VCM) is unstable and inconsistent, all activities rely on the V_O-formed conductive filament (CF) in insulator. Variations can be seen between each cycle of modulation, in both DC and AC. Pulse switching is an AC method to set or reset device into low & high resistance state (LRS & HRS). By experimental results, device can behave differently depending on input waveform to the device. Forming process is generally an inevitable action needed to apply on a MIM RRAM cell for CF activation, except for the case where insulator in device is thin enough. Forming process requires an external compliance to limit the overshoot current when apply a swept positive bias. The effect of forming compliance is investigated by looking into the endurance of OxRRAM cell.

II. FAILURES BY PULSE SWITCHING

First, we demonstrate three main cases of failure observed in the Pt/HfO$_2$(5nm)/Al$_2$O$_3$(3nm)/Ti/Pt MIM stacked VCM-based OxRRAM cells, compared to the optimal case that results in best HRS/LRS ratio in **Fig. 1**, presented as transient waveform, with the responses of programming-current after applied with programming-voltage shown in **Fig. 2** showing sub-100ns current rising.

A. HRS Stuck

RRAM device can show HRS stuck if the RESET process is too dominant relative to SET programming. In the DC case of **Fig. 3**, the cell gets stuck in HRS right after first forming and reset, unable to set to LRS in initially 2.2V positive range, but it can be recovered by a larger positive range. An AC case is shown in **Fig. 4**, with a HRS stuck occurs at 1.5ms. Readjusted input SET/RESET 1.25/-1.5V can recover the resistive switching, as shown in **Fig. 5**.

B. LRS Stuck

As an opposite case to HRS stuck, a RRAM cell can show LRS stuck if SET process is too strong. This failure cannot be recovered again unlike HRS stuck. In **Fig. 6**, the cell is applied with 2/-2.6V SET/RESET pulses, showing a case that fails to reset cell back to HRS.

C. Unbalanced Write

Unbalanced write shows results of bad on/off ratio. The 0.75V SET pulse cannot induce an enough current to set device to LRS, whilst -1V RESET pulse cannot reset cell to enough low current for HRS, with eventually only an insufficient on/off ratio ($\frac{I_{LRS}}{I_{HRS}} = \frac{68.7}{45.6}$) sampled in **Fig. 7**.

III. FORMING INDUCED FAILURES

Forming process is inevitable in OxRRAM devices, and accountable for the device reliability. **Fig. 8** shows cell-wise DC forming under 1μA on crossbar array, a largest 5.76V needed to activate on of the 10 cells, and resulted in cell I-V of the array in the first cycle as shown in **Fig. 9**. From **Fig. 10**, the device formed with 100μA cannot have better endurance than the device formed under 1μA compliance, by the fact from **Fig. 9** that 1μA FORMING can function well on array with sufficient resistive switching window.

IV. RESULTS AND DISCUSSIONS

By comparing the DC result shown in **Fig. 1**, the device exhibits an average 1.1V SET transition and an smallest $|-1.5V|$ RESET transition, with an optimal 1μs pulse switching by 1.8/-1.7V SET/RESET, both programming energy of SET and RESET can certainly secure high-on/off ratio of pulse switching. Input in **Fig. 5** can recover the HRS stuck in Fig. 4 is by sufficient SET/RESET amplitude that matches the DC result, over where the SET/RESET transitions happen in I-V sweep. **Fig. 6** shows serious LRS stuck that single RESET process cannot be useful with even a high $|-2.6V|$, due to breakdown induced by strong 2V SET and overly high energy by 100μs pulse-width. Forming is a mechanism to induce soft-breakdown in the insulator of MIM OxRRAM. From **Fig. 8**, we can see that FORMING might be one of the source of RRAM reliability issue, with an extremely deviated sub-3V forming voltage from other approximately 5V formed cell. An early soft-breakdown in lower voltage might indicate that the cell is weak in its insulator, hence a lower compliance can contribute better device reliability to suppress device from breakdown.

V. CONCLUSIONS

The balance between SET/RESET programming can hugely affect the on/off ratio. Forming compliance, on the other hand, can result in different cell-wise endurance. It will be a work to do in finding the correct programming strategy regarding certain RRAM process.

ACKNOWLEDGMENTS

This work is supported by National Science and Technology Council (NSTC) of Taiwan under Contracts 112-2221-E-006-191

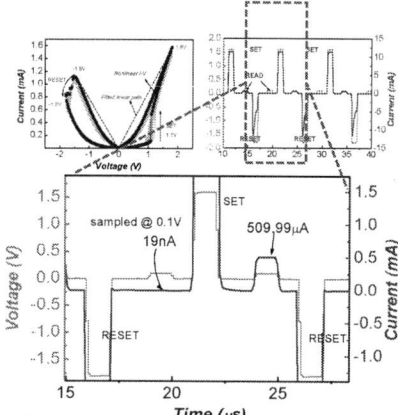

Fig. 1 Optimal case with 1.7V/-1.8V/0.1V SET/RESET/READ with $1\mu s$ pulse-width.

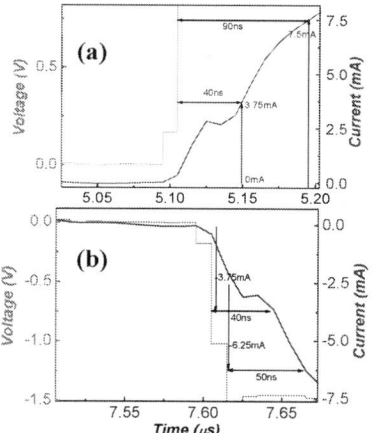

Fig. 2 Current response during (a) SET and (b) RESET process of **Fig. 1**.

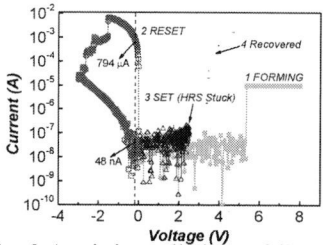

Fig. 3 A pristine cell shows failure in SET operation after first forming and RESET. (log-scale DC)

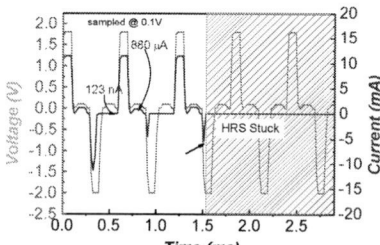

Fig. 4 HRS stuck in AC switching with 1.75/-2V SET/RESET demonstrated with $100\mu s$ pulse-width.

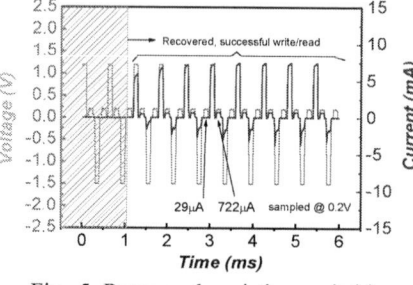

Fig. 5 Recovered resistive switching with re-adjusted SET/RESET 1.25V/-1.5V with $100\mu s$ pulse-width following the failed case in **Fig. 4**.

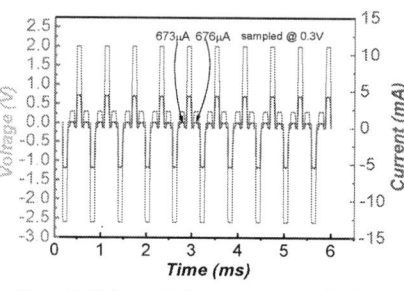

Fig. 6 LRS stuck by pulse modulation of 2/-2.6/0.3V SET/RESET/READ with $100\mu s$ pulse-width.

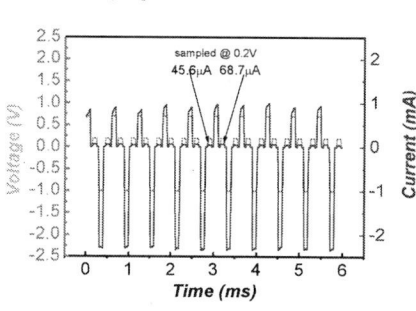

Fig. 7 Unbalanced programming with 0.75/-1V SET/RESET programming pulses with $100\mu s$ pulse-width.

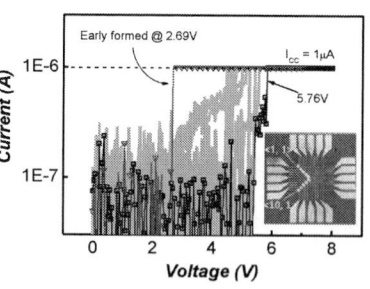

Fig. 8 Selected 10 OxRRAM-Cells formed with $1\mu A$ compliance distributes on crossbar array with cell size (crosspoint) $10\mu m \times 10\mu m$.

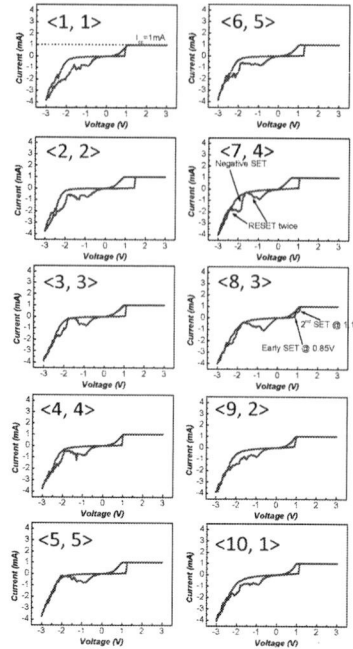

Fig. 9 First cycle of (linear-scale) I-V characteristics of OxRRAM cells on array.

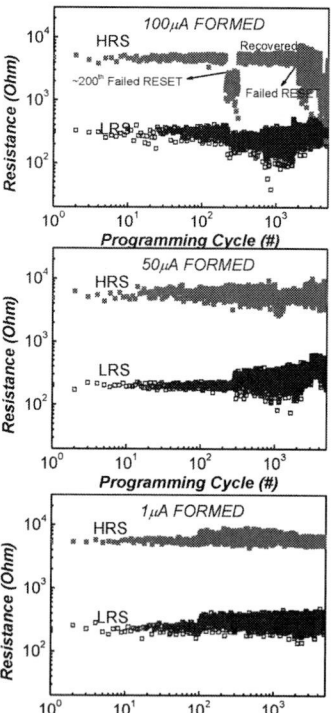

Fig. 10 Endurance of OxRRAM cell formed at $100\mu A$, $50\mu A$, and $1\mu A$ compliance respectively, all tested on fresh cell by 1.5/-1.5V SET/RESET programming pulses.

979-8-3503-9164-0/24 $31.00 © 2024 IEEE

Innovative switching mechanism, performance investigations and scaling effects in a-V_2O_3 based ReRAM devices

Killian Veyret[1,2], Leo Laborie[1], Romain Bon[1], Gabriele Navaro[1], Rachid Hida[1], Niccolo Castellani[1], Catherine Carabasse[1], Patrice Gonon[2] and Eric Jalaguier[1]

[1]CEA-Leti, Univ. Grenoble Alpes, 38000 Grenoble, France [2]Univ. Grenoble Alpes, CNRS; CEA-Leti Minatec, LTM, 38054 Grenoble, France

Email: killian.veyret@cea.fr

Abstract — **This work presents a novel mechanism for 1R a-V_2O_3 ReRAM cells fabricated with "wall" integration. Non-volatile Resistive Switching (RS) implies a Vanadium (V) depleted zone created during a unique initialisation step – highlighted by simulation, electrical and physicochemical characterisations. This new a-V_2O_3 based ReRAM features large memory window, high RS speed, excellent scalability and good endurance.**

Keywords: ReRAM, Non-volatile memory, Vanadium oxides, Depleted zone

I. INTRODUCTION

Resistive Random Access Memories (ReRAM) are considered promising candidates for future non-volatile memory, due to their scalability potential, fast read/write speed, and low write energy, which is key for emerging applications such as neuromorphic, or in-memory computing [1]. This technology usually rely on either binary metallic oxides, like HfO_2, for Oxide Random Access Memory (OxRAM), or active electrode materials, like Cu and Ag, for Conductive Bridge Random Access Memory (CBRAM) – where a conductive filament is created with respectively oxygen vacancies and metallic ions. Here, we introduce a novel ReRAM "V depleted", based on a TiN/a-V_2O_3/TiN stack with a unique initialisation mechanism, consisting in the creation and commutation of a highly resistive nanometric V depleted zone within the a-V_2O_3 thin film. Subsequent RS are surmised to follow conventional metal-oxides memory mechanisms. V depleted memory features large memory window, high RS speed, excellent scalability and good endurance.

II. DEVICE FABRICATION & METHODS

a-V_2O_3 thin films are deposited by Ion Beam Deposition (IBD) technique at room temperature using a vanadium target in an O_2 environment. Thin films of 10 nm thickness are integrated in 1R Back End Of Line test structures. TiN "wall" bottom electrode is deposited in a vertical trench and defines the cell area, varying from 10x40 nm^2 to 10x300 nm^2 (fig.1.a). Quasistatic sweeps were gathered with an Agilent 4156C parameter analyser. Endurance test were performed with Keysight B1530 and fast PIV unit. Both pristine and cycled 1R devices were investigated with Transmission Electron Microscopy (TEM) and Energy Dispersive X-ray Spectroscopy (EDX).

III. RESULTS AND DISCUSSION

Fig.2.a shows that bipolar cycling is achieved with an initialisation step, $R_{pristine}$ (~5 kΩ) to High Resistive State (HRS), at high current densities (>100 MA/cm^2). Subsequent resets occur at lower current (fig.2.a&b). Initialisation current decreases with temperature, whilst resets current remains constant with temperature, which highlight a temperature driven mechanism for the initialisation step, different from reset mechanism. Initialised device presents an altered zone (fig.3.a) within the a-V_2O_3 film. EDX mapping reveals that the altered zone is a V depleted zone, where V has migrated toward the top of the TiN "wall" (fig.4.b), creating a V rich reservoir. EDX mappings of Low Resistive State (LRS) device shows V migrating from the rich V reservoir back to the depleted zone (fig.4.c) while the depleted zone is again observed on a HRS device (fig.4.d). Electro-thermal simulations of initialisation step exhibit temperatures beyond 1600K in the altered zone, with spatial temperature profile matching the dome shaped altered zone observed in TEM (fig.3.b).

The proposed mechanisms are presented fig.5. The initialisation step consists in the creation of a highly resistive V depleted zone, concomitant to the creation of a rich V reservoir at the top of the TiN "wall. Extreme local temperatures are believed to partially dissociate the a-V_2O_3, facilitating electrically driven ionic migration. During the set operation, V migrates back towards the depleted zone, creating a conductive region composed of conductive $VO_{x\sim1}$ [2]. The highly resistive V depleted region is recreated during reset operation with electrical driven V migration, in a bulk-ReRAM manner [3].

This mechanism is highly reproducible and 1R cells where cycled up to 10k cycles on various devices sizes, ranging from 10x40 nm^2 to 10x100 nm^2 (fig.6.a). Fig.6.a&b show that device cycling is operated at low voltage (<2V) with fast pulse programming (<50ns), whilst keeping a large memory window (>100). At last, cycle-to-cycle resistive states variabilities seem to both follow lognormal distributions (fig.6.b), which hints for bulk-ReRAM-like – rather than filamentary ReRAM – operating [4].

IV. CONCLUSION

To conclude, a novel a-V_2O_3 based ReRAM has been implemented with a singular *operando* initialisation process. This initialisation process consisting in a nanometric V depleted zone creation is corroborated with both electrical and physicochemical characterisations along with electro-thermal simulation. This new ReRAM allows excellent scalability, it

979-8-3503-9164-0/24 $31.00 © 2024 IEEE

features low programming voltage (<2V), high 1R endurance with large memory window (>100), fast programming (<50ns) and good reproducibility.

ACKNOWLEDGMENTS

This work was supported by the French Public Authorities in the frame of the Important Project of Common European Interest (IPCEI) on microelectronics.

REFERENCES

[1] H. Abbas *et al.*, "Conductive Bridge Random Access Memory (CBRAM): Challenges and Opportunities for Memory and Neuromorphic Computing Applications", *Micromachines*, 13, p. 725, 2022.

[2] M. D. Banus *et al.*, "Electrical and Magnetic Properties of TiO and VO", *Phys. Rev. B*, 5, pp. 2775-2784, 1972.

[3] Y. Li *et al.*, "Filament-Free Bulk Resistive Memory Enables Deterministic Analogue Switching", *Adv. Mater.*, 32, 2020.

[4] A. Grossi *et al.*, "Fundamental variability limits of filament-based RRAM", *IEDM*, pp. 4.7.1-4.7.4, 2016

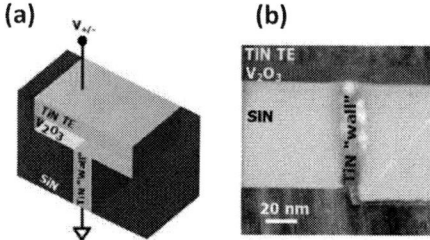

Fig 1. a. Scheme of the "wall" architecture. b. STEM picture of a pristine device (TiN "wall" damaged during sample preparation but a-V_2O_3 sheet is intact).

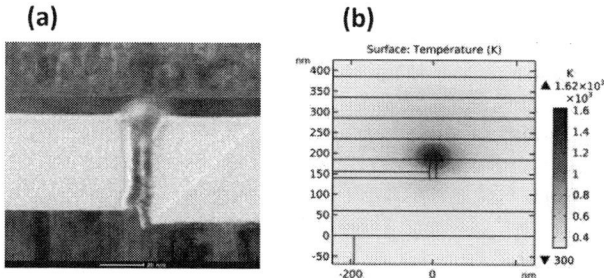

Fig.3. a. STEM picture of an initialised device. b. 2D cut plane of a 3D COMSOL simulation of device temperature during initialisation step (zoom on the "wall" vicinity).

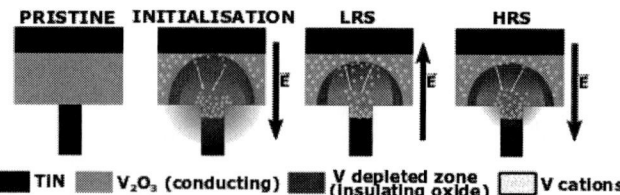

Fig 5. Proposed RS mechanisms, with pristine, initialised, LRS and HRS states.

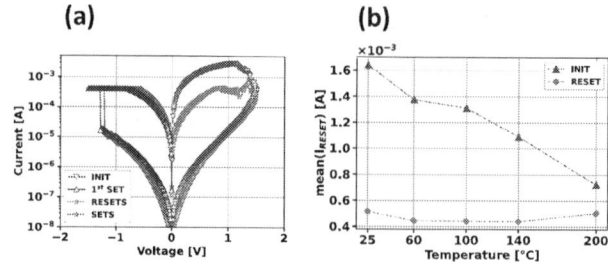

Fig 2. a. I-V characteristics of 10x80 nm² cell; initialisation (red triangles), 1st set (blue triangle), subsequent resets (light red stars) and subsequent sets (light blue stars). b. Average of reset current (5 devices) − for initialization step (red triangles) and 5th reset (light red dots) − versus temperature.

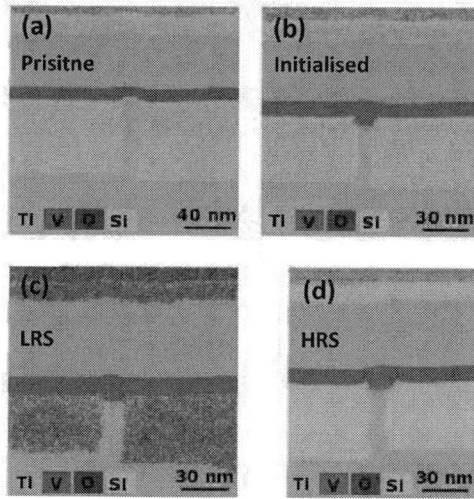

Fig 4. EDX mapping of devices a. pristine; b. initialised; c. LRS (10th set); d. HRS (5th reset) (set pulse: -1.5V/30ns; reset and initialisation: 1.7V/50ns).

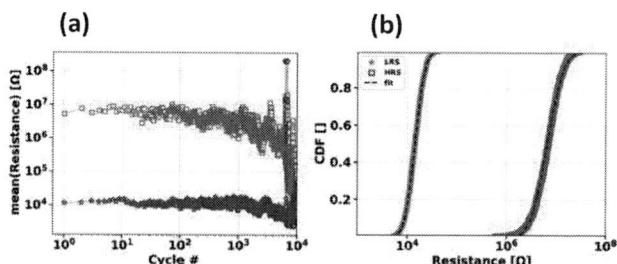

Fig 7. a. Average of 10k cycle endurance (5 devices ranging from 10x40 nm² to 10x100nm²), HRS in red, LRS in blue − reset: 1.7V/50ns pulse; set: -1.5V/30ns pulse. b. Cumulative distribution function of HRS and LRS for a 10x50nm² cell (2000 first cycles), lognormal fit for both distributions.

Gap in pagination due to unavailable paper.

Pages 83-84

High Performance HfLaO-based TiO₂-Channel FE V-NAND with High Consistency and Low Operation Voltage

Xujin Song[1], Shangze Li[2], Dijiang Sun[1], Xiaoyan Liu[1], and Jinfeng Kang[1]*

[1]School of Integrated Circuits, Peking University, China,
[2]School of Software and Microelectronics, Peking University, China
Email: kangjf@pku.edu.cn

Abstract — **A novel 3D vertical NAND (V-NAND) ferroelectric field-effect transistor (FeFET) array with TiO₂ channel and HfLaO ferroelectric layer has been demonstrated. The fabricated FeFETs exhibit excellent performances, including high device-to-device consistency of the stacked cells, low write voltage (±2V) with rapid write speed (<1μs), high on/off current ratio (>10⁶) and remarkable endurance (>10⁸). Moreover, effective string-level NAND operation of the fabricated V-NAND array has also been demonstrated.**

Keywords: FeFET, TiO₂, HfLaO, vertical channel

I. INTRODUCTION

Three-dimensional (3D) V-NAND technology has been proposed to realize continuous density and cost scaling for NAND devices [1]. For further density and energy efficiency enhancement, the HfO₂-based vertical FeFET is considered as a candidate due to its CMOS-compatibility [2-5]. However, silicon channel FeFETs encounter issue with elevated write voltage due to the interfacial layer between the FE layer and channel [5]. Although oxide semiconductor (OS) channel FeFETs eliminate this interfacial layer, they are still constrained by high erase voltage induced by insufficient hole concentration. [6]. Prior research has indicated that TiO₂ can alleviate the write voltage issue by reducing the coercive voltage and demonstrating high compatibility with HfO₂ FE layers [7]. However, integration of TiO₂ in vertical FeFETs for V-NAND memory has not been extensively explored.

In this work, we have introduced the first demonstration of a TiO₂ channel V-NAND HfLaO FeFET array. A two-cell vertical NAND array was fabricated, and the performance and consistency of stacked cells were evaluated. The low-voltage operation capability was assessed through pulse I-V testing, and string-level NAND operation was further demonstrated.

II. EXPERIMENTS

The fabrication process and schematic of the vertical channel FeFETs are depicted in Fig.1(a). Stacked TiN gate electrodes with SiO₂ spacer were deposited and etched. HfLaO and TiO₂ were then deposited using in-situ ALD. The FE-channel stack was annealed in vacuum at 600°C to enhance the ferroelectricity. Finally, channel area was etched, and Mo S/D electrodes were deposited.

Cross-sectional TEM image is provided in Fig.1(b) and (c), highlighting the uniformity of the HfLaO and TiO₂ deposition, as well as the excellent crystallinity of TiO₂ channel, which is crucial for the high performance of the vertical channel devices. The SEM image in Fig.1(d) presents an overview of the completed vertical FeFET array.

III. RESULTS AND DISCUSSION

A. Device consistency and electrical performance

Transfer characteristic curves for the two cells in Fig.2 shows high device-to-device consistency. An on-off current ratio of up to 10⁶ can also be observed. Linear plots of I_D-V_G and I_G-V_G in Fig.3 revealed the FE switching with erase peak at -1V, indicating effective reduction of erase voltage using TiO₂ channel. Consequently, drain current (I_D) exhibits a pronounced switching around -1V. Transient I-V and extracted P-V curves at 10KHz in Fig.4 confirmed these findings. The array endurance was tested with 4V, 1MHz bipolar waveforms, showing stability over 10⁸ cycles (Fig.5).

B. Low-voltage PRG/ERS operation

Pulse write operation on Gate 1 used erase (ERS) and program (PRG) pulses with varying widths (100ns-100μs) and amplitudes (0-4V), followed by a reading pulse to evaluate I_D. As depicted in Fig.6, the pulse I-V write and read curves align closely with the former reference DC I-V curve. The contour plot in Fig.7 showed the modulation of I_D by pulse width and amplitude, achieving full ERS and PRG with ±2V pulses. This low-voltage operation capability is due to the reduced coercive voltage when deposited with TiO₂ [7].

C. String-level NAND operation

String-level NAND operation configurations with varying states (all programmed, one programmed and one erased, and all erased) were examined (Fig.8). Non-destructive reading was performed at V_G=-0.5V, with the string exhibiting a distinct on-state when both cells were in the PRG state, confirming the implementation of NAND operation in the fabricated vertical FeFET array. Table 1 compares the TiO₂ FeFET with previous studies on vertical channel FeFETs [2-5]. The devices exhibit superior low operation voltage and excellent electrical performance, surpassing both widely used OS channels and traditional Si channel FeFETs.

IV. CONCLUSION

In this work, HfLaO-based V-NAND array with TiO₂ channel was demonstrated for the first time. The fabricated TiO₂ vertical FeFETs exhibit superior performances with high consistency. Low-voltage operation and string-level NAND operation were demonstrated. The presented performances of TiO₂ FeFET V-NAND array show great potential of TiO₂ as a new choice for 3D V-NAND memory applications.

ACKNOWLEDGEMENTS: This work was supported by NSFC 92064001.

REFERENCES: [1] S. Salahuddin, Nat. Electron, 2018 [2] Z. Li, EDL, 2022 [3] I. J. Kim, APL, 2022 [4] M. K. Kim, Science Advances, 2021 [5] K. Banerjee, IMW, 2021 [6] I. J. Kim, EDL, 2023 [7] X. Song, CSTIC, 2024

979-8-3503-9164-0/24 $31.00 © 2024 IEEE

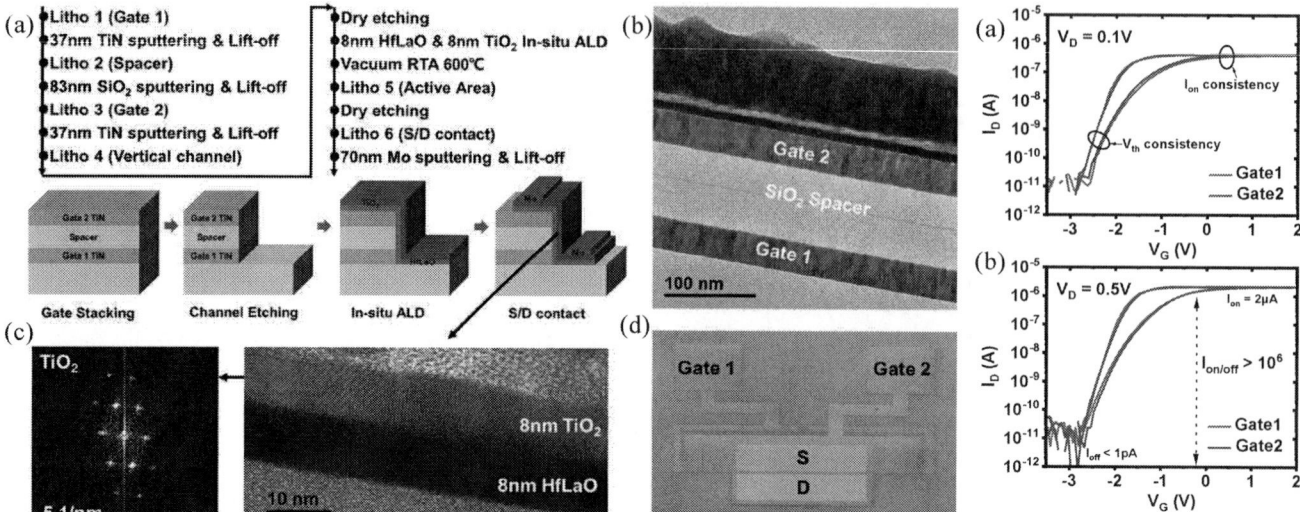

Fig. 1 (a) Schematic of the fabrication process for the V-NAND array. (b) TEM image of the stacked gate electrodes and spacer. (c) TEM image of FE-channel stack, with a FFT pattern showcasing the high-quality crystallinity of TiO_2 channel. (d) SEM overview of the completed vertical FeFET array.

Fig.2 I_D-V_G curves of Gate1/Gate2 cell with V_D= (a) 0.1V and (b) 1V. High device-to-device consistency and large $I_{on/off}$ ratio can be observed.

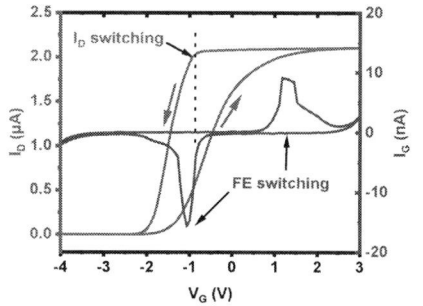

Fig.3 I_D-V_G and I_G-V_G of Gate 2 cell. Distinct FE switching peak with corresponding backward I_D switching can be observed.

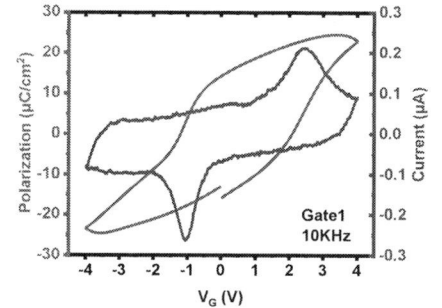

Fig.4 Transient I-V and P-V curves of Gate 1 under 10KHz triangle wave, showing similar FE behaviour with DC results of Gate 2 cell.

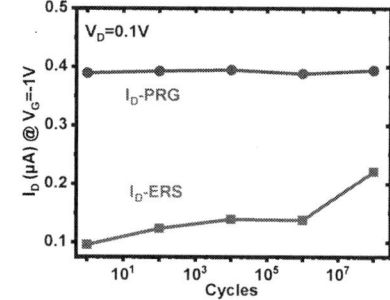

Fig.5 I_D (V_G@ -1V) during endurance test under 4V-1MHz bipolar wave cycling, presenting high endurance over 10^8.

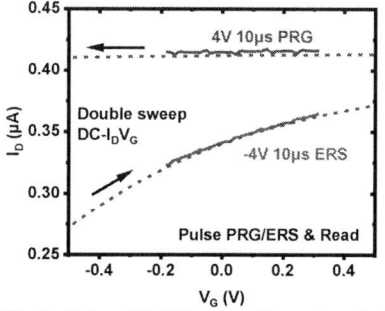

Fig.6 Pulse PRG/ERS with pulse I_D-V_G, demonstrating strong alignment with the DC I_D-V_G results (dash line).

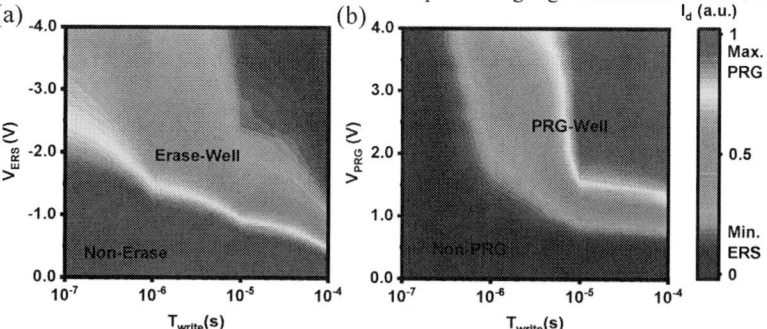

Fig.7 Contour plot of I_D (V_G@0V) under pulse width/amplitude modulation, full (a) ERS and (b) PRG state can be achieved within ±2V, demonstrating low-voltage operation capability.

Fig.8 Readout of V-NAND string, presenting on-state with all cells programmed, and off-state with any cell erased.

Table. 1 Benchmark of vertical channel FeFETs

	This work	[2]	[3]	[4]	[5]
Channel Material	TiO_2	InO_x	IZO	IZO	Si
Channel&FE thickness	16nm	17nm	44nm	44nm	60nm
PRG/ERS voltage	2V	7V	4V	6V	10V
Endurance	10^8	10^4	10^8	10^8	10^4
I_{on}/I_{off}	10^6	10^2	10^2	10	10^5

Accelerated Program Inhibition Failure by Trap-Mediated Tunneling in the Polycrystalline Floating-Channel 3-D NAND Flash Memory Array

Soomin Kim[1], Unsang Lee[2], Yeji Lee[1], Chiwook Ahn[2], Jaesung Sim[2], and Seongjae Cho[1,*]

[1]Division of Electronic and Semiconductor Engineering, Ewha Womans University, Seoul 03760, Republic of Korea
[2]SK hynix Inc., Cheongju-si, Chungcheongbuk-do 28429, Republic of Korea
*E-mail: felixcho@ewha.ac.kr

Abstract — **In this work, we have investigated the effects of trap-mediated tunneling on program inhibition in the 3-D NAND flash memory. By a series of rigorous array-level TCAD simulations, it has been found that the self-boosted channel in the unselected bitline experiences a potential drop by direct inter-band tunneling, which can be significantly accelerated by trap-mediated tunneling. It is judged both types of tunneling events are inevitable and proper methods to lessen them need to be sought since the body biasing cannot be controlled and the bitline channel is made of poly-Si having abundant trap sites in the current 3-D NAND flash memory technology.**

I. INTRODUCTION

The conventional two-dimensional (2-D) NAND flash technology encountered scaling limitations. For making a way out of it, three-dimensional (3-D) vertical NAND array architecture was introduced [1]. Unlike the 2-D NAND array where all the bitline (BL) channels are connected to the body [2], in the 3-D NAND array, they are only floating. Biasing the channel potential is also highly challenging since the channel thickness is extremely thin. For these reasons, it is important to have a closer look into the non-ideal effects occurring in performing program and erase operations in the 3-D NAND flash array and seek for the effective ways of sustaining the operation reliability. In this work, a series of technology computer-aided design (TCAD) simulations have been conducted to accurately trace the failure mechanisms in program inhibition in the unselected bitline [3], with a particular emphasis on the effects of trap-assisted tunneling.

II. EXCESSIVE CHANNEL POTENTIAL LOWERING

For the array-level simulations, a single BL with 16 cells (WL0 ~ WL15) was constructed as shown in Fig. 1. In Fig. 2, three cells numbered with WL7, WL8, and WL9 at the string center were set to erase states, whereas all the other cells are in the program states. Erase state and programmed states are defined by the distinctive threshold voltages (V_{th}'s) of the cells, which are -0.7 V (E), 0.7 V (P1), 1.4 V (P2), and 2.1 V (P3), respectively (4-level operation). Fig. 3 shows the channel potential under the WL8 cell over the verify and recovery operations depending on the states of neighboring cells while the three cells are in the erase states. With higher V_{th} of the surrounding cells (P1 < P2 < P3), the channel potential of the WL8 cell shows a faster and larger drop. Fig. 4(a) and (b) depict the WL8 channel potential for 100 ms right after the recovery is completed. It is confirmed by Fig. 4(a) and (b) that the WL8 channel potential is gradually elevated with time, and the speed of increase gets higher (P1 > P2 > P3) as less electrons are stored in the neighbor cells.

Fig. 5(a) through (c) shows the changes in electron and hole concentrations at the BL center over time with all the other cells in P1 states. After a sufficient amount of time, the difference between inter and outer carrier concentrations becomes more distinct, causing an electrical isolation of the inner island region from the outer channel. The electrically de-coupled channel potential increases as previously shown.

III. INCOMPLETE SELF-BOOSTING AND FAST LOWERING

Fig. 6(a) shows demonstrates the change in WL8 channel potential as a function of time over verify, recovery, and program operations, with neighbor cells of different V_{th}'s. The excessive channel potential lowering makes it hard for the self-boosting of the channel to take place effectively, which might lead to a program inhibition failure. Fig. 6(b) and (c) depict the channel potential as a function of time after a program voltage of 18 V is applied on the WL8 with all the neighbor cells in P1 states. In order to investigate the effects of channel crystallinity, three cases were considered: (i) Si channel with a set of drift-diffusion, recombination, and mobility models, (ii) addition of band-to-band tunneling model, and (iii) addition of trap-assisted tunneling model (to emulate the poly-Si BL channel). No significant difference is observed among the cases within 5 μs as shown in Fig. 6(b). However, as the time is extended to 200 ms, the differences become apparent. The valence electrons are tunneled into the isolated BL center and drag down the potential. If traps are introduced in the BL channel, trap-assisted tunneling further accelerates the introduction of electrons and the channel potential is drastically lowered, and an entire inhibition fail can be resulted in a short time of 1 ms as shown in Fig. 6(c).

IV. CONCLUSION

In this work, the causes for program inhibition fail in an unselected BL in the 3-D NAND have been intensively studied by a systematic series of TCAD simulations. Excessive channel potential lowering, incomplete self-boosting, and inter-band tunneling are the prominent causes and are originating from the fact that the BL channels are floating and they are made up of poly-Si in the current flash.

ACKNOWLEDGMENT

This work was supported by SK hynix through an industry-university cooperation research project.

REFERENCES

[1] R. Katsumata, *et al.*, *Proc. 2009 Symposium on VLSI Technology*, pp. 136–137, Kyoto, Japan, Jun. 2009.
[2] S. Cho, J.-D. Choi, B.-G. Park, and I. H. Cho, *Current Appl. Phys.*, vol. 10, no. 4, pp. 1096–1102, Jul. 2010.
[3] ATLAS User's Manual, Silvaco International Inc., Aug. 2017.

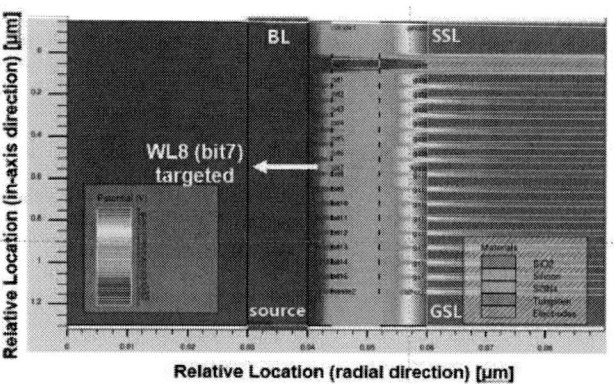

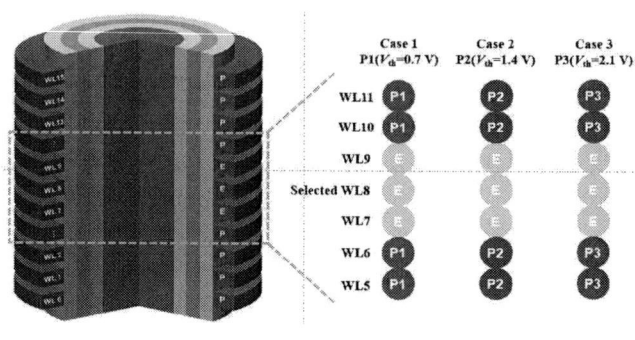

Fig. 1. Schematic of the simulated 3-D NAND unselected bitline (half cross-sectional view).

Fig. 2. Three-dimensional view of the unselected bitline. WL7 through WL9 are in the erase state and all the other cells are programmed states, P1 (Case 1), P2 (Case 2), and P3 (Case 3).

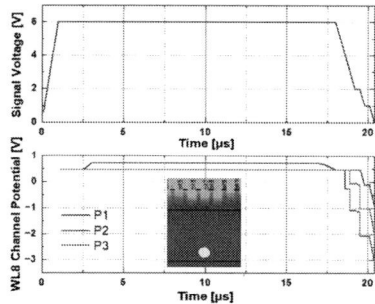

Fig. 3. Change in channel potential of the unselected cell under the WL8 as a function of time over verify and recovery operations (yellow in the inset: potential check point).

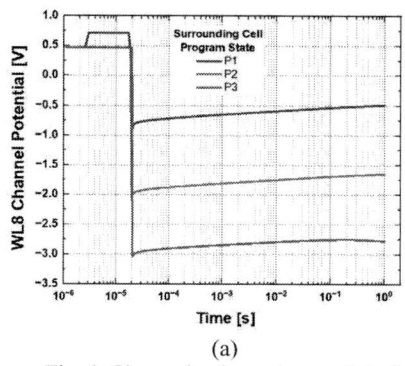

(a)

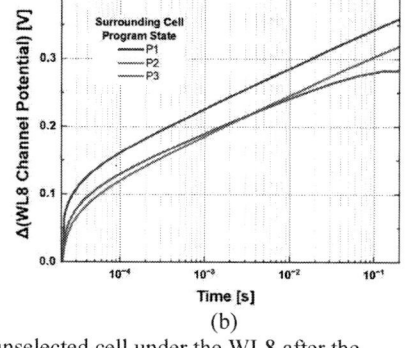

(b)

Fig. 4. Change in channel potential of the unselected cell under the WL8 after the completion of recovery operation. (a) Nominal changes. (b) Amount of change.

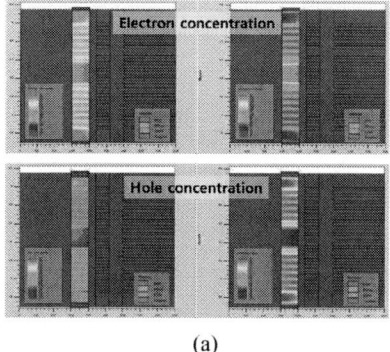

(a)

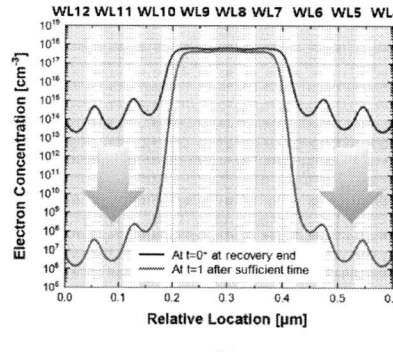

(b)

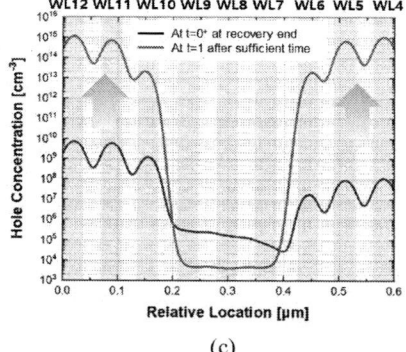

(c)

Fig. 5. Time-dependent changes in carrier concentrations. (a) Graphical analyses. Changes in (b) electron and (c) hole concentrations.

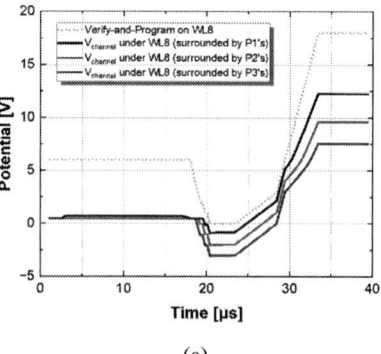

(a)

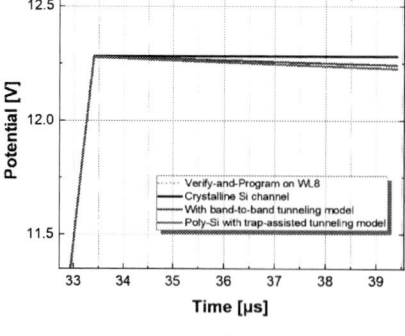

(b)

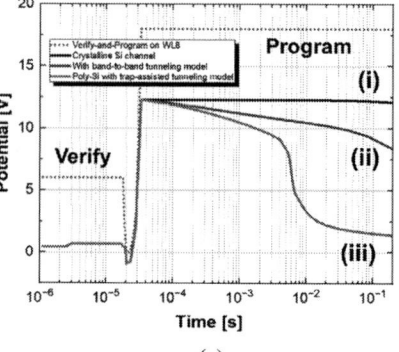

(c)

Fig. 6. Time-dependent changes in carrier concentrations. (a) Graphical analyses. Changes in (b) electron and (c) hole concentrations.

979-8-3503-9164-0/24 $31.00 © 2024 IEEE

Investigation of Row Hammer and Passing Gate Effects Based on the Work Functions of Dual Gates in DRAM Cells

Hansol Kim[1], Jisung Im[1], Jinsu Kim[1], Seungmin Woo[1], Taeseong Kwon[1],
Young Jun Yoon[2], Jong-Ho Bae[3], Sung Yun Woo[1*]

[1]School of ECE, Kyungpook National University, Korea, [2]Dept. of EE, Andong National University, Korea, [3]School of EE, Kookmin University, Korea
Email: sywoo@knu.ac.kr

Abstract — In this paper, the effect of the work function of the double metal gate inside DRAM cells on the row hammer (RHE) and passing gate effect (PGE) was analyzed. The RHE and PGE were observed to be significantly affected by the work function of the top metal, mainly due to the leakage current (GIDL). Additionally, it was found that the PGE of DRAM cells was also significantly affected by the work function of the bottom metal.

Keywords: DRAM, BCAT, TCAD simulation, row hammer effect, passing gate effect.

I. INTRODUCTION

Dynamic Random Access Memory (DRAM) is still being developed in a $6F^2$ structure with buried gates and a saddle-finFET architecture, which significantly reduces parasitic capacitance and maintains high sensing margins even under sub-20nm process conditions. However, as the size of the cell transistor decreases, the value of cell capacitance also diminishes, leading to soft errors due to gate-induced drain leakage (GIDL), junction leakage, subthreshold leakage, and gate leakage, which pose considerable threats to DRAM reliability [1]-[2]. Additionally, the phenomenon known as activation induced bitflips (AIBs), caused by the activation of adjacent word lines (WL) resulting in bitflips in victim cells, has become a major cause of retention failures. In this paper, we investigate the influence of varying the work functions (WF) of the top-metal and bottom metal on GIDL, row hammer effect (RHE), and passing gate effect (PGE) using TCAD simulation (Sentaurus, Synopsys).

II. SIMULATION RESULTS AND DISCUSSION

To investigate the RHE and PGE, the 3D-TCAD simulation based on a 2y nm node DRAM architecture was performed using the parameters summarized in Fig. 1 [1]. Fig. 2 (a) shows a diagram of the DRAM cell array where RHE and PGE occur, and the mechanisms are detailed in Fig. 2 (b). The RHE involves the movement of electrons by the neighboring gate to the storage node of the victim cell, leading to bitflips in the victim cell upon off-state. Conversely, PGE involves electrons being pulled by the passing gate from the storage node of the victim cell, causing a bitflip upon off-state in the victim cell. Fig. 3 (a) shows simulated I_{SN}-V_{WL} curves of DRAM cells as a parameter of WF_{TM}s. It shows that the GIDL current (I_{GIDL}) increases approximately fourfold as the WF_{TM} increases from 4.0 eV to 4.9 eV. Similarly, in Fig. 3.

(b), increasing the WF_{BM} results in a rise in the cell transistor's threshold voltage. In Fig. 3. (c), the leakage current of the storage node due to neighboring WLs is shown, with the higher leakage current observed with a higher WF_{TM}. Fig. 4. illustrates the BTBT generation values, with WF_{TM} of 4.0 eV resulting in a generation rate of 1×10^{13} [/cm³s], and 4.9 eV resulting in 8.5×10^{20} [/cm³s]. Simulations to investigate RHE and PGE for different WF_{TM}s and WF_{BM}s were performed respectively by applying aggressive signals (3V, 20ns) to adjacent WLs. The RHE and PGE of the victim cell were 1.2V and 0V of V_{SN}, respectively. Fig. 5 illustrates the results of RHE with different WF_{TM}s, showing a significant increase in I_{GIDL} and subsequent storage node voltage change (ΔV_{SN}) as WF_{TM} increases. Similarly, in Fig. 6(a), as WF_{TM} increases from 4.2eV, the PGE becomes more prominent due to the effect of GIDL on the storage node voltage. The RHE is strongly influenced by WF_{TM}, whereas PGE is more influenced by the carrier density in the channel region. When WF_{BM} changes while fixing WF_{TM} at 4.0 eV, ΔV_{SN} decreases as WF_{BM} increases, as shown in Fig. 6(b). Fig. 7 shows the tolerance of the RHE and PGE with the difference WF_{TM}s and WF_{BM}. Since RHE is greatly affected by the I_{GIDL}, as WF_{TM} decreases, I_{GIDL} decreases, leading to an increase in the number of aggressive signals for bit-flops. The RHE tolerance of WF_{TM} at 4.0eV is up to 3.75 times that of WF_{TM} at 4.9eV. In the case of PGE, it is up to 5.51 times higher based on the WF_{TM}.

III. CONCLUSION

The increase in WF_{TM} heightened the sensitivity of the V_{SN} to RHE, driven by additional electrons generated by GIDL between the storage node of the victim cell and the neighboring WL. For PGE, a lower WF_{BM} facilitated easier formation of the channel activated by the passing WL, leading to a more sensitive electron outflow from the storage node of the victim cell. This paper contributes to the practical design of the optimal WF of a dual metal gate structure against soft errors in DRAM arrays.

ACKNOWLEDGMENTS

This work was supported by the National Research Foundation of Korea funded by MSIT (RS-2024-00405200) The EDA tool was provided by the IC Design Education Center (IDEC), Korea.

REFERENCES

[1] H. Nam, et al., *IEEE CAL*, vol. 22, no. 2, 2023.
[2] M. Suh, et al., *IEEE TED*, vol.71, no. 4, 2024

P1-11

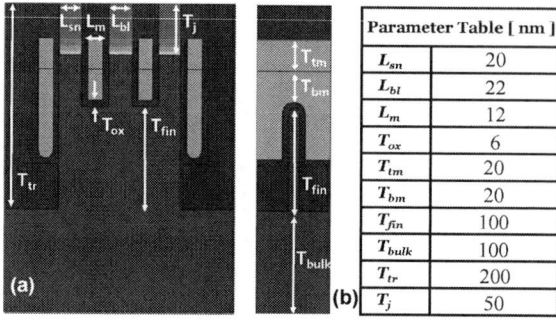

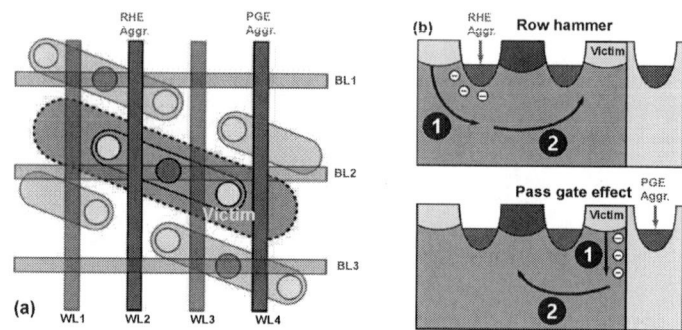

Fig. 1. (a) Schematic cross-sectional diagrams (x-axis and y-axis) of simulated DRAM BCAT cells. (b) Structure parameters table (heights, lengths, and thicknesses) of the simulated DRAM BCAT cells, respectively.

Fig. 2. (a) A diagram of DRAM cell array for the row-hammer (RHE) and passing gate effects (PGE), respectively. (b) A mechanism of RHE and PGE in a victim cell caused by applying aggressive signals to neighboring and pass gates, respectively.

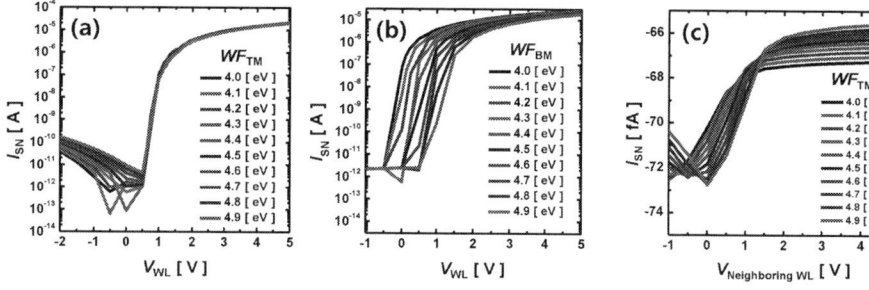

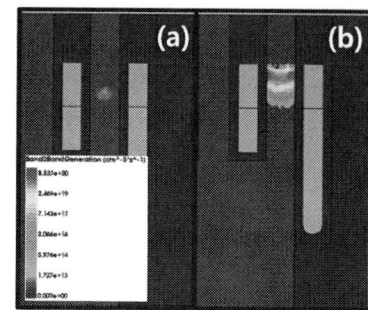

Fig. 3. (a) Simulated I_{SN}-V_{WL} curves of DRAM cells as a parameter of work-functions in a top-metal gate (WF_{TM}). (b) Simulated I_{SN}-V_{WL} curves of DRAM cells as a parameter of work-functions in a bottom-metal gate (WF_{BM}). (c) I_{SN}-$V_{Neighboring-WL}$ curves of the DRAM victim cell as a parameter of WF_{TMS}. As WF_{TM} of the increases, leakage current increase due to an increase of GIDL current.

Fig. 4. Simulated cross-sectional diagrams with the band-to-band tunneling (BTBT) generation as a parameter of WF_{TMS}. (a) $WF_{TM} = 4.0$ eV and (b) $WF_{TM} = 4.9$ eV.

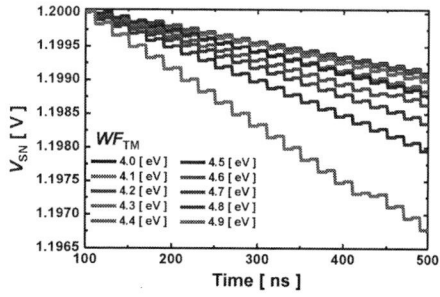

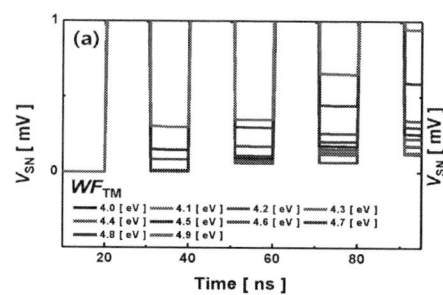

Fig. 5. Simulated V_{SN}-time plots of DRAM cells as a parameter of WF_{TMS} when aggressive signals are applied to the neighboring WL. As WF_{TM} of the increases, $|\triangle V_{SN}|$ increases in the DRAM victim cell due to an increase of I_{GIDL}.

Fig. 6. Simulated V_{SN}-time plots of DRAM cells as parameters of (a) WF_{TMS} and (b) WF_{BMS} when aggressive signals are applied to the passing WL, respectively. As the number of aggressive signals increase, V_{SN} increases in the DRAM victim due to positive charges accumulated in the storage node.

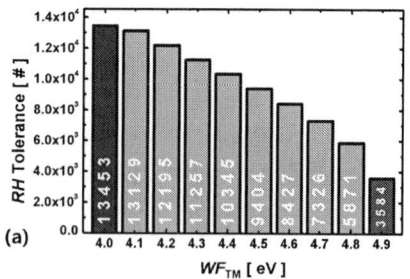

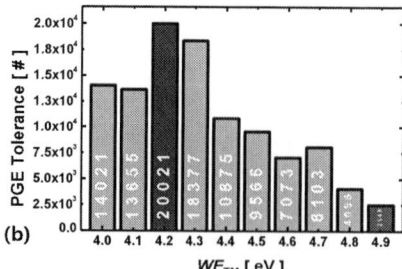

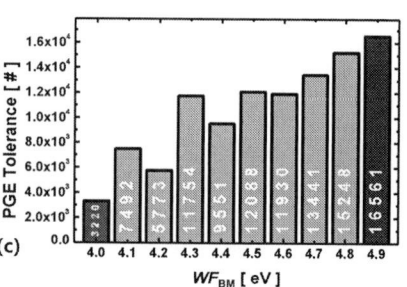

Fig. 7. (a) RH tolerance and (b) PGE tolerance of the DRAM victim cell as a parameter of WF_{TMS} and (c) PGE tolerance for WF_{BMS}. Since RHE is greatly affected by the I_{GIDL}, as WF_{TM} decreases, I_{GIDL} decreases, leading to increase the number of aggressive signals for bit flops. Since PGE is also affected by I_{GIDL}, as the WF_{TM} increases, the number of aggressive signals for bit-flops decreases, but below a certain WF_{TM} (~4.3 eV) the impact is greatly reduced.

979-8-3503-9164-0/24 $31.00 © 2024 IEEE

Analysis of Row Hammer and Passing Gate Effect in DRAM Cells by BCAT Structural Design

Jisung Im[1], Hansol Kim[1], Jinsu Kim[1], Seungmin Woo[1], Taeseong Kwon[1],
Young Jun Yoon[2], Jong-Ho Bae[3], Sung Yun Woo[1*]

[1]School of ECE, Kyungpook National University, Korea, [2]Dept. of EE, Andong National University, Korea, [3]School of EE,
Kookmin University, Korea
Email: sywoo@knu.ac.kr

Abstract — **In this paper, we investigate the impact of structural parameters such as gate angles, fin height through the control of gate overlaps and the distance from fin to source/drain on the row hammer effect (RHE) and passing gate effect (PGE) of DRAM cells. In a DRAM cell, the larger the angle of the gate profile and the lower the overlap height between the fin and gate of the cell transistor, the lower the RHE and PGE. Thus, the influence of adjacent and passing gates on the DRAM cell body potential is a key factor in RHE and PGE, and through optimization, the performance and reliability of DRAM technology can be improved.**

Keywords: DRAM, BCAT, TCAD simulation, row hammer effect(RHE), passing gate effect(PGE).

I. INTRODUCTION

In the field of Dynamic Random Access Memory (DRAM), the miniaturization of components has markedly improved performance and increased density. However, this progress introduces significant reliability challenges, particularly the row hammer effect (RHE) and passing gate effect (PGE), which severely compromise device integrity and functionality [1]-[2]. This research focuses on the structural parameters of DRAM, specifically variations in gate angles (θ_{angle}s), the reduction of fin height (H_{fin}) through decreased gate overlap and decreasing the inherent height of the fin to increase the distance from fin to source/drain ($L_{S/D}$), and their impact on these adverse effects. In this study, the systematic analysis of the correlation between the geometric modification of the DRAM cell structure and the RH/PG effects will contribute to the development of robust DRAM architectures.

II. SIMULATION RESULTS AND DISCUSSION

Fig. 1(a) shows the structure of the saddle-FinFET device in DRAM cells, implemented using Sentaurus TCAD 3D simulation. Additionally, lines in three directions, C1-C1', C2-C2' (Fig. 1(b)), and C3-C3' (Fig. 1(c)), are marked, and the device parameters are displayed on the right, with their values provided in Table 1. Fig. 2(a) illustrates a diagram of the DRAM cell array depicting the RHE and PGE. Fig. 2(b) shows the mechanism of RHE, where data changes from 1 to 0 due to the neighboring gate influence, while Fig. 2(c) shows the mechanism of PGE, causing data to shift from 0 to 1 under pass gate influence, explaining both effects. At first, to investigate the electrical characteristics with geometric modifications of the DRAM cells, the gate angle of the DRAM cells was adjusted from 0 degrees to 3 degrees in 1-degree steps, as shown in Fig. 3(a). The left image displays the DRAM cell with SiO_2 removed, while the right image shows a cross-section along the C2-C2' line. Fig. 3 (b) shows simulated I_{SN}-V_{WL} curves of DRAM cells as a parameter of θ_{angle}s. The DRAM cell with the θ_{angle} of 3 degrees exhibits the highest on-current at V_{WL} of 3 V, attributed to the reduction in channel length. Fig. 3(c) shows that as the θ_{angle} increases, the impact of the RHE is reduced. Similarly, Fig. 3(d) shows that the influence of the PGE decreases with an increase in θ_{angle}. This trend is due to the decreasing influence of the gate on the channel/body of the DRAM cell as the θ_{angle} increases. Fig. 4(a) shows the geometric modification of a DRAM cell with the gate overlap length adjusted. The fin height ($H_{fin\_overlap}$), which is the area covered by the gate above the fin, changes from 12.5nm to 50nm in 12.5nm steps. Fig. 4(b) demonstrates that as the $H_{fin\_overlap}$ increases, the on-current also increases, indicating improved gate controllability with larger $H_{fin\_overlap}$. Fig. 4(c) and 4(d) show the vulnerability of DRAM cells to RHE and PGE at the highest $H_{fin\_overlap}$, following the same trend where increased gate overlap leads to greater gate controllability and influence over adjacent gates. Fig. 5 examines the adjustment of the U-shaped fin's length to control the H_{fin} and also shows $L_{S/D}$. Same as Fig. 4, the H_{fin} simulated in 12.5 nm to 50 nm and observed along the C1 and C3 lines. Fig. 5(b), (c), and (d) exhibit the same trend as observed in Fig. 4, where a decrease in H_{fin} leads to an increase in channel length and a reduction in the area covered by the gate over the fin, resulting in a lower on-current. Furthermore, the increase in $L_{S/D}$ causes a reduction in RHE and PGE.

III. Conclusion

As dimensions of DRAM cells continue to decrease, these systems become increasingly susceptible to adverse effects such as the RHE and PGE. To address these challenges, this paper analyzes the impact of manipulating four parameters: θ_{angle}, H_{fin}, $L_{S/D}$ and gate overlap. Through this analysis, the study presents potential strategies to minimize these negative impacts, thereby enhancing device reliability and performance, such as increasing θ_{angle} or decreasing H_{fin}.

ACKNOWLEDGMENTS

The EDA tool was provided by the IC Design Education Center (IDEC), Korea.

REFERENCES

[1] H. Nam, et al. , *IEEE CAL*, vol. 22, no. 2, 2023.
[2] M. Suh, et al. , *IEEE TED*, vol.71, no. 4, 2022

P1-12

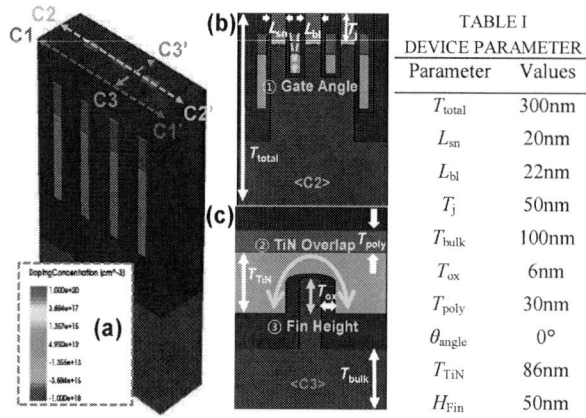

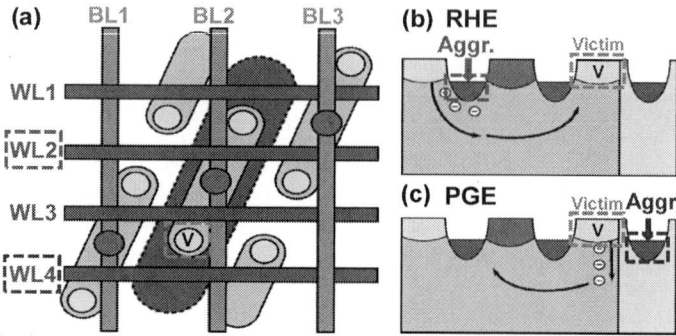

TABLE I	
DEVICE PARAMETER	
Parameter	Values
T_{total}	300nm
L_{sn}	20nm
L_{bl}	22nm
T_j	50nm
T_{bulk}	100nm
T_{ox}	6nm
T_{poly}	30nm
θ_{angle}	0°
T_{TiN}	86nm
H_{Fin}	50nm

Fig. 1. (a) 3D schematic diagrams, cross sectional views along (b) C2–C2' and (c) C3–C3' of simulated DRAM BCAT cells.

Fig. 2. (a) A diagram of DRAM cell array for the row-hammer (RHE) and passing gate effects (PGE), respectively. (b) A mechanism of RHE in a victim cell caused by applying aggressive signals to neighboring gates. (c) A mechanism of PGE in a victim cell caused by applying aggressive signals to pass gates.

① Gate Angle

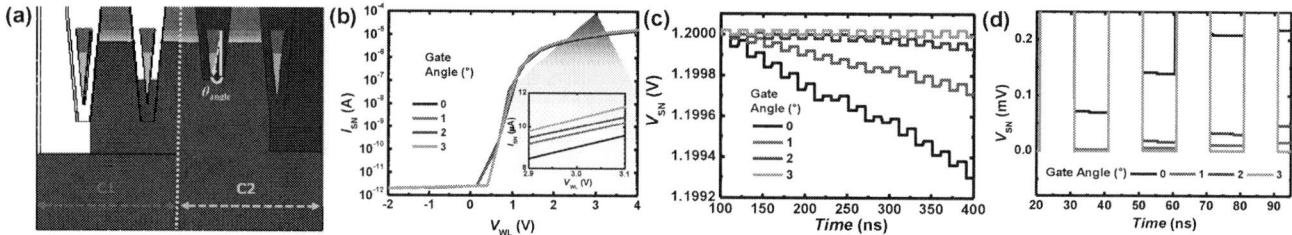

Fig. 3. (a) Cross-sectional views along C1–C1' and C2–C2' of simulated DRAM BCAT cells with a $\Delta\theta_{angle}$ of 3 degrees in gate profiles. (b) Simulated I_{SN}-V_{WL} curves of DRAM cells as a parameter of θ_{angle}s. (c) Simulated V_{SN}-$time$ plots of DRAM cells as a parameter of θ_{angle}s when aggressive signals applied to the neighboring WL. (d) Simulated V_{SN}-$time$ plots of DRAM cells as parameters of θ_{angle}s when aggressive signals applied to the passing WL.

② Gate Overlap (+ Fin Height)

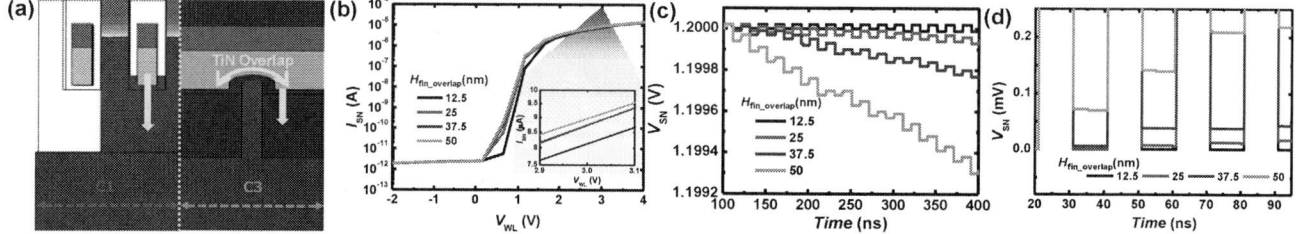

Fig. 4. (a) Cross-sectional views along C1–C1' and C3–C3' of simulated DRAM BCAT cells with TiN overlap, where $H_{fin\_overlap}$ = 12.5nm. (b) Simulated I_{SN}-V_{WL} curves of DRAM cells as a parameter of $H_{fin\_overlap}$ s. (c) Simulated V_{SN}-$time$ plots of DRAM cells as a parameter of $H_{fin\_overlap}$ s when aggressive signals applied to the neighboring WL. (d) Simulated V_{SN}-$time$ plots of DRAM cells as parameters of $H_{fin\_overlap}$ s when aggressive signals applied to the passing WL. As the length of the TiN overlap decrease, the RH and PG effects decrease because the influence of neighbor and pass WLs on the victim DRAM cell decreases.

③ Fin to S/D Distance (+ Fin Height)

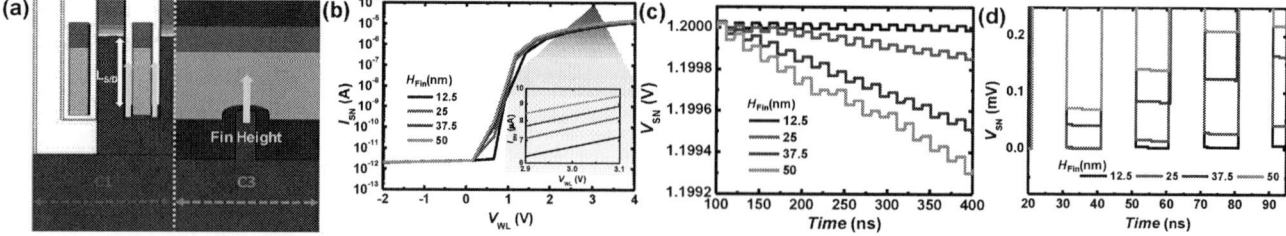

Fig. 5. (a) Cross-sectional views along C1–C1' and C3–C3' of simulated DRAM BCAT cells with fin distances from S/D regions to top of fins ($L_{S/DS}$), where H_{fin} = 12.5nm. (b) Simulated I_{SN}-V_{WL} curves of DRAM cells as a parameter of H_{fin}s. (c) Simulated V_{SN}-$time$ plots of DRAM cells as a parameter of H_{fin}s when aggressive signals are applied to the neighboring WL. (d) Simulated V_{SN}-$time$ plots of DRAM cells as parameters of $L_{S/DS}$ when aggressive signals are applied to the passing WL. As $L_{S/DS}$ decrease, the RH and PG effects decrease because the influence of neighbor and pass WLs on the victim DRAM cell decreases.

979-8-3503-9164-0/24 $31.00 © 2024 IEEE

Controllable Polarization Switching of Hafnia-Based Ferroelectric Bilayers

Jiae Jeong, Hyoungjin Park, and Jiyong Woo

School of Electronic and Electrical Engineering, Kyungpook National University, South Korea

Email: jiyong.woo@knu.ac.kr

Abstract — **For neuromorphic applications, achieving a precisely controllable remnant polarization (P_r) as well as coercive voltage (V_C) is crucial to achieve a multilevel polarization in the ferroelectric layer (FL). Therefore, we introduce bilayered ferroelectric thin films with Zr-doped HfO_2 (HZO) on top of Al-doped HfO_2 (HAO), allowing the adjustment of P_r over a broad voltage range relative to the ramping voltage. In the bilayer ferroelectric HZO/HAO, polarization switching behavior occurs sequentially through each FL, which results in tunable P_r characteristics.**

I. INTRODUCTION

The controllable polarization characteristics such as remnant polarization (P_r) and coercive voltage (V_C) in hafnia-based ferroelectric layer (FL) are necessary in ferroelectric field-effect transistor for neuromorphic synaptic devices [1]. However, the Zr-doped HfO_2 (HZO) or Al-doped HfO_2 (HAO) FLs, which have been studied extensively to date, encountered the limitation that the P_r remains unchanged even at higher voltages, lacking partial switching properties. Therefore, our study showed that stacking two different FLs allows polarization switching to be modulated precisely by the ramping voltage, resulting in P_r tunability.

II. EXPERIMENT

The single HZO, HAO and HZO/HAO stacks deposited by atomic layer deposition were encapsulated by W bottom and top electrodes deposited by DC sputtering. Finally, annealing for single stacks and ferroelectric bilayer was executed in a furnace in air ambient. Each fabricated stack was confirmed by cross-sectional transmission electron microscopy (TEM) images and depth profiles (Fig. 1).

III. RESULT AND DISCUSSION

The single HZO and HAO stacks were annealed at 400 °C and 600 °C for 3 min each, respectively. Additionally, the HZO/HAO stack was annealed at 400 °C for 4 min. We initially investigated the electrical behavior of the HZO and HAO FLs, and HZO/HAO stack. In the current–voltage (I–V) measurements of the fabricated devices, single stacks exhibited only one peak in each polarity, whereas two peaks occurred in the I–V curve of the HZO/HAO stack (Fig. 2). This implies that the polarization switching occurred sequentially in each FL of the HZO/HAO stack. In the polarization–voltage (P–V) curve of the HZO FL, a P_r value of 29.2 $\mu C/cm^2$ was exhibited with abrupt polarization change at V_C of ~ 1.6 V (Fig. 3). In contrast, a relatively gradual switching at a V_C value of ~1 was observed with a P_r value of 6.7 $\mu C/cm^2$. The stronger ferroelectricity with an abrupt transition at V_C and higher P_r

in the HZO film indicates that more orthorhombic (o) phases were formed and the dipoles in the phase can thus rapidly align as a response to the voltage [2]. Notably, unlike other single stacks, where rare change in P_r, the consistent change in P_r was observed in the P–V curve of the HZO/HAO stack, with a step voltage of 1 V from 4 V to 8 V. The value of P_r in the HZO/HAO stack decreased as the annealing temperature increased from 400 °C to 600°C (Fig. 4). Further, the P_r value enhanced until the annealing time reached 4 min. However, longer annealing time exceeding 4 min rather lowered P_r. In comparison to the single HZO and HAO, where P_r increment (ΔP_r) reached a maximum of 125 % and 210 % respectively, the ΔP_r for the HZO/HAO stack was 889 %, indicating an enhanced P_r controllability within a broad voltage range (Fig. 5). The reliable polarization switching was observed up to 10^5 cycles even for the bilayer HZO/HAO stack in cycling endurance measurements conducted at a frequency of 1 kHz (Fig. 6). The confirmation of o-phases in the HZO/HAO stack was achieved through grazing incidence X-ray diffraction (GIXRD) analysis (Fig. 7). The diffraction peaks at approximately 28.5 ° and 30.5 ° indicate the monoclinic and mixed phase of the o, tetragonal, and cubic phases, respectively.

A plausible mechanism for the HZO/HAO stack is described by two-step polarization switching process (Fig. 8). Initially, the more o-phases were formed in the HZO FL, resulting in a higher current level in I–V curve compared to that in the HAO FL. Therefore, the dipoles in the relatively insulating HAO responded first in the HZO/HAO stack. A smaller number of dipoles in HAO started to align gradually in low-voltage regime. Higher voltages were necessary for polarization switching to occur in the HZO FL. The switching of increased number of dipoles with fast response resulted in a larger P_r value in switching process under high-voltage regime in HZO.

IV. CONCLUSIONS

In this study, bilayered ferroelectric HZO/HAO structure enabling controllable polarization switching was designed for neuromorphic applications. In the HZO FL, a larger P_r value was exhibited; in contrast, a relatively smaller P_r was achieved in the HAO FL. Stacking two FLs resulted in a progressive increase in P_r due to their synergetic effects.

ACKNOWLEDGMENTS

This research was supported by National R&D Program through the National Research Foundation of Korea (NRF) funded by Ministry of Science and ICT (RS-2023-00258227).

REFERENCES

[1] K. Lee et al., ACS Applied Materials & Interfaces, vol. 11, no. 42, pp. 38929–38936, October 2019.

[2] B. Buyantogtokh et al., Journal of Applied Physics, vol. 129, no. 11, June 2021.

P1-13

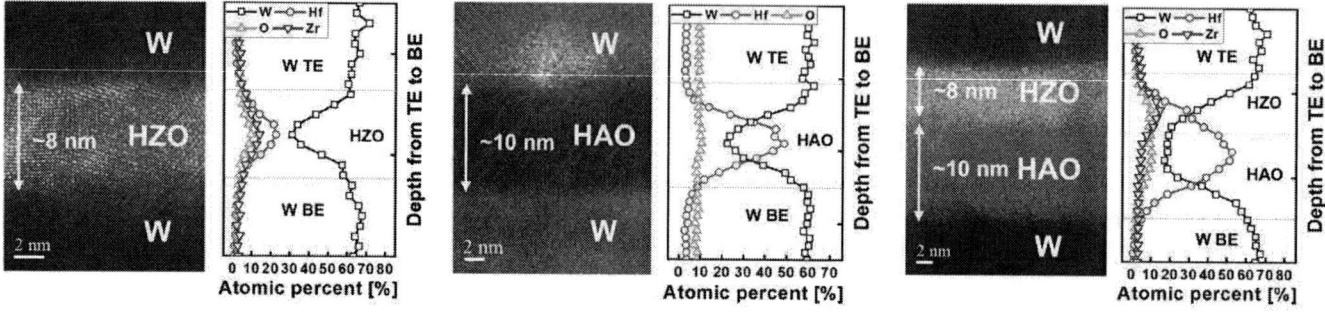

Fig. 1 TEM images and depth profiles from top electrode (TE) to bottom electrode (BE) of the single HZO and HAO FLs, and HZO/HAO bilayer.

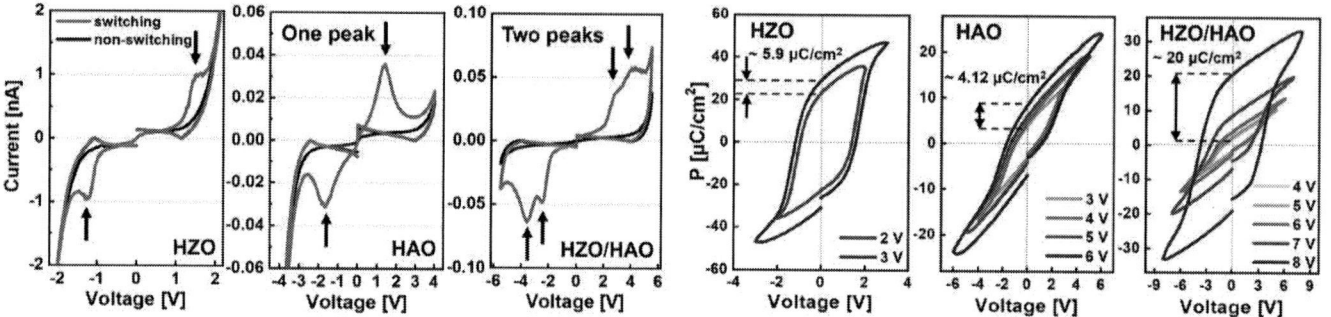

Fig. 2 I–V curves of single HZO, HAO FLs and HZO/HAO stack.

Fig. 3 P–V hysteresis loops of single HZO, HAO FLs and HZO/HAO stack.

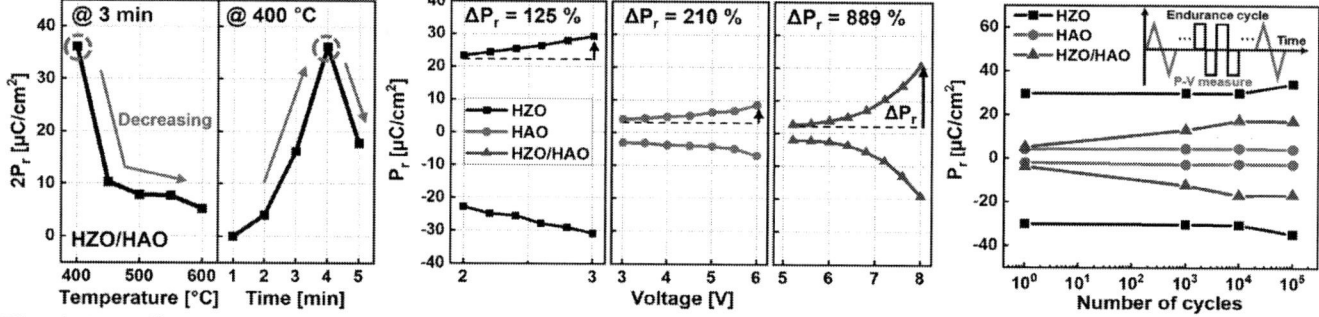

Fig. 4 Annealing temperature and time dependence of HZO/HAO stack in P–V loop.

Fig. 5 Comparison on the controllability of P_r values of the single HZO, HAO FLs and HZO/HAO stack.

Fig. 6 Endurance characteristics of the single and bilayer stack and pulse scheme for endurance test.

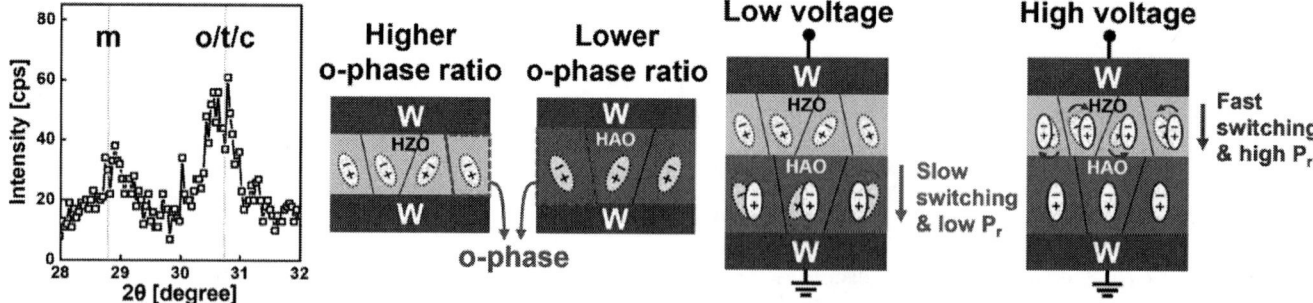

Fig. 7 GIXRD pattern of the HZO/HAO stack.

Fig. 8 Plausible mechanism of two step alignment process in HZO/HAO stack.

979-8-3503-9164-0/24 $31.00 © 2024 IEEE   94

Short-term and long-term T-O phase transition responsible for two stages of wake-up process in ferroelectric Hf$_{0.5}$Zr$_{0.5}$O$_2$ film

Danyang Chen[1], Qiang Gao[1], Yuyan Fan[1], Zikang Yao[1], Jingquan Liu[1], Mengwei Si[2] and Xiuyan Li[1]*

[1]Department of Nano/Microelectronics, [2]Department of Electronic Engineering, Shanghai Jiao Tong University, Shanghai, China
*Email: xiuyanli@sjtu.edu.cn

Abstract — **We design and carry out an *in-situ* grazing incidence X-ray diffraction (GIXRD) investigation in the phase transition behaviors during wake-up of the ferroelectric Hf$_{0.5}$Zr$_{0.5}$O$_2$ (HZO) films. Based on results, we demonstrate that short-term and long-term tetragonal (T)-orthorhombic (O) phase transitions are responsible for two stages of wake-up processes for polarization obtaining and increase in HZO, respectively.**

Keywords: Ferroelectric HZO, wake-up, *in-situ* GIXRD, phase transition

I. INTRODUCTION

Wake-up effect in ferroelectric (FE) Hf$_x$Zr$_{1-x}$O$_2$, which affects the endurance of devices, is one of major concerns on the reliability of memory devices for practical application. Understanding of physical origin is the key to engineer it and enhance the devices performance. Several viewpoints have been proposed so far, such as defects redistribution, phase transition, depolarization induced by interface layer [1-6]. Direct evidence, however, is still absent. In this work, by carefully designing a series of *in-situ* grazing incidence X-ray diffraction (GIXRD) measurements on FE Hf$_{0.5}$Zr$_{0.5}$O$_2$ (HZO) films during the wake-up process, we demonstrate that two kinds of short-term and long-term T-O phase transition occur in two stages of wake-up in FE HZO. Furthermore, the issue, how two stages of T-O phase transition are driven, is discussed based on the results.

II. EXPERIMENTAL METHODS

Fig. 1 shows the TEM image and fabrication process of the samples with TiN/HZO/TiN/W structures. The top electrode area is set as large as to 200×4000 μm^2 for *in-situ* GIXRD investigation. Three post-metallization annealing (PMA) conditions, 550°C 30 s, 450°C 30 s, and 350°C 3 h, were varied for crystallization of HZO. The schematic of *in-situ* GIXRD measurements in the wake-up of the HZO samples is shown in **Fig. 2**. ±2/4 V triangle waveforms were applied on the sample for polarization switching cycling. Along with this, polarization-voltage (P-V) characteristics were measured. Just after each set of polarization switching cycling, GIXRD measurement was carried out *in-situ* on the samples. The area of X-ray spot is controlled to be comparable with that of the electrode to get the information of the films after training waveform.

III. RESULTS AND DISCUSSIONS

A. Short-term T-O phase transition

The GIXRD spectrums of HZO samples after the first couple of polarization switching cycles along with corresponding P-V characteristics in them are shown in **Fig. 3**. It can be found that O (111)/T (101) peaks at 2θ of ~30.2° of all the samples with different PMA conditions show obvious shift from the higher 2θ to the lower one after the cycles. Namely, T-O phase transition occurs in the polarization switching even at the beginning. The increase in

peak intensity at 2θ of ~34.8° also indicates that the O (002) grains, which contributes to the out-of-plane polarization mainly, increase after the cycles. Note that the peaks change in the first half cycle, corresponding to the positive polarization increase from 0 as shown in **Fig. 3(c)**, is much more significant than the others. This suggests that the remnant polarization (P$_r$) obtaining at the beginning is a result of T-O phase transition rather than random polarization direction pooling. Here we defined it as a short-term wake-up process. Interestingly, the changes after the first half cycle are relatively slight and highly depend on the PMA condition. Namely, the XRD spectrums basically remain the same after the first cycle in sample with 550°C 30 s PMA, while the T-O phase transition lasts longer in the sample with lower PMA temperature.

B. Long term T-O phase transition

The T-O phase transition in long-term of polarization switching cycling is also investigated on 350°C PMA 3 h sample, as shown in **Fig. 4**. Here we successively applied 10^3, 10^4, 10^5 cycles under ±4 V and carried out *in-situ* GIXRD measurements between these electric waveforms. It can be found that the peak position shift at 2θ of ~30.2° and intensity increase at 2θ of ~34.8° corresponding to T-O phase transition still continue along with P$_r$ increase from 7 to 14 μC/cm^2. These results indicate two stages of wake-up process, corresponding to short-term and long-term, occur in the HZO samples. Note that long-term T-O phase transition seems continuing even under ±2 V pulse without polarization switching after 10^5 cycling, implying that the it is induced by the electric stress itself rather than polarization switching cycling.

C. Physical understanding of two stage T-O phase transition

We finally discuss how two stages of T-O phase transition occur in wake-up of FE HZO films. The T-rich phenomenon in the pristine state of HZO films indicates that T-phase is an energy favourable phase, namely it is with energy well in energy landscape. The electric stress helps it to overcome the energy barrier between T- and O-phase. Then, such a transition is consistent with the first-order ferroelectric phase transition which could be described by Landau-Ginzburg-Devonshire (LGD) model as reported [7-10]. In the transition, some part of T-phase grains may become unstable under electric field because the energy barrier between T- and O-phase vanish. In this case, short-term T-O phase transition and wake-up occurs and HZO films change from non-polarized to polarized phase (**Fig. 5(a)**). Meanwhile, other grains with relatively lower energy in T-phase remains stable. The relative energy between O- and T-phase in these grains is possibly changed by the defects redistribution or surface energy reduction induced by electric stress. As T-phase instability increases, these domains will undergo the same phase transition as short-term wake-up (**Fig. 5(b)**). In this case, long-term T-O phase transition and P$_r$ increasing occur.

979-8-3503-9164-0/24 $31.00 © 2024 IEEE

IV. CONCLUSION

In conclusion, short-term and long-term T-O phase transition corresponding to two stages of wake-up process for P_r obtaining and increasing in FE-HZO is clearly exhibited through *in-situ* GIXRD measurements. By combing the results and LGD model, it has been understood that short-term phase transition is induced the energy barrier disappearance under an electric field, while the long-term one is done on the energy landscape change with electric stress.

Acknowledgments: This work was supported by National Key R&D Program of China (2022YFB3608400), National Natural Science Foundation of China (62274109, 62111540163).

Reference: [1] P. Jiang *et al.*, Adv. Elec. Mater. **7**(1), (2021). [2] M. Lederer *et al.*, Phys. Status Solidi (RRL) **15**, (2021). [3] T. Schenk *et al.*, ACS Appl. Mater. Interfaces **7**(36), (2015). [4] Y. Cheng *et al.*, Nat. Commun. **13**(1), 645 (2022). [5] B. Saini *et al.*, Adv. Electron. Mater. **9**(6), (2023). [6] D. Chen *et al.*, Appl. Phys. Lett. **122**(21), (2023). [7] L. Onsager, Phys. Rev. **65**(3-4), 117 (1944). [8] A. F. Devonshire, Philos. Mag. **42**(327), 1065-1079 (1951). [9] U. Schroeder *et al.*, Adv. Electron. Mater. **8**(9), (2022). [10] D. H. Lee *et al.*, IEEE Electron Device Lett., (2023).

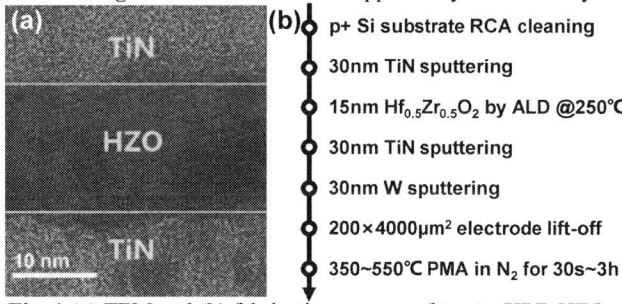

Fig. 1 (a) TEM and **(b)** fabrication process of *in-situ* XRD HZO samples. The size of top electrode was set to 200×4000 μm² to match the X-ray spot size for *in-situ* GIXRD investigation.

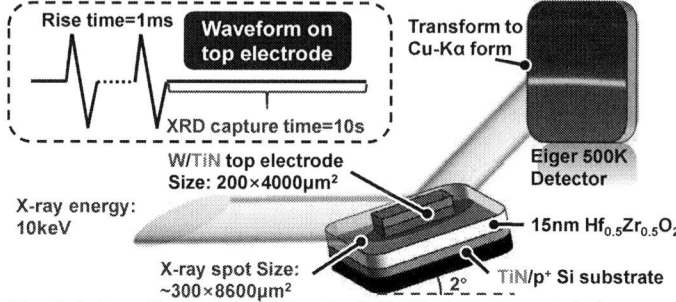

Fig. 2 Schematic diagram of *in-situ* GIXRD. P-V characteristics were measured on the HZO sample under ±2/4 V triangle waveforms. The GIXRD spectrums were transform into Cu-Kα form.

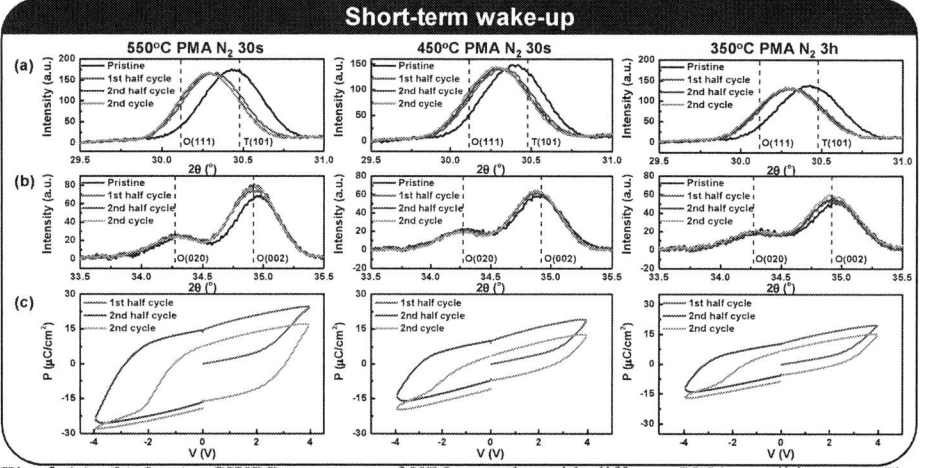

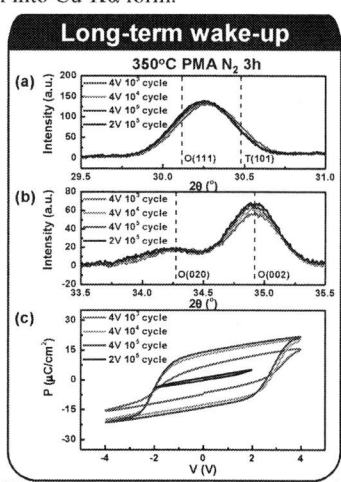

Fig. 3 (a), (b) *In-situ* GIXRD spectrum of HZO samples with different PMA conditions. **(c)** The corresponding P-V of above samples measured in the interval of GIXRD measurements. Two-stage T-O transition is clearly shown in the XRD peak shifting at ~30.2° and peak intensity increasing at ~34.8°. Here we only consider the O (002) grains contribution to the ~34.8° XRD peak.

Fig. 4 (a), (b) *In-situ* GIXRD and **(c)** corresponding P-V results after different cycles in 350°C 3 h sample. XRD peaks changes still continue.

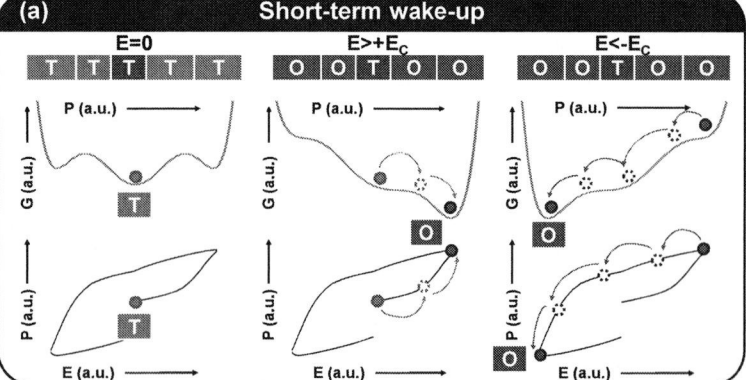

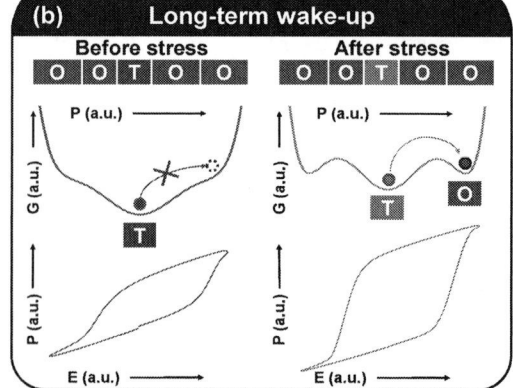

Fig. 5 Schematic diagram of physical understanding of **(a)** short-term and **(b)** long-term T-O phase transition in ferroelectric HZO film. For some parts of the T-phase grains, they become unstable when an electric field over coercive field applied, which is the origin of short-term T-O phase transition and P_r obtaining in HZO films. For those grains with more stable T-phase, the possible V_O redistribution or surface energy reduction induced by electric stress will change its energy landscape. And the T-phase in these grains finally transit into the former case, corresponding to long-term T-O phase transition process and P_r increasing.

979-8-3503-9164-0/24 $31.00 © 2024 IEEE

Gap in pagination due to unavailable papers.

Pages 97-100

Investigation on the Local Polarization Scheme in Ferroelectric Tunnel Field-Effect Transistors for Ternary Content-Addressable Memory Applications

Minjeong Ryu, Jae Seung Woo, and Woo Young Choi[*]

Department of Electrical and Computer Engineering and Inter-university Semiconductor Research Center (ISRC),
Seoul National University, Gwanak-gu, Seoul 08826, Republic of Korea
[*]Email: wooyoung@snu.ac.kr

Abstract — **This study investigates the influence of gate length scaling on ambipolar ferroelectric tunnel field-effect transistors in the context of implementing one-transistor (1T) ternary content-addressable memory (TCAM) based on local polarization. The analysis results demonstrate that the scalability of the local polarization-based 1T TCAM is on par with the previously widely researched uniform polarization-based counterpart. Consequently, the advantages of local polarization-based 1T TCAMs remain effective even with scaled ferroelectric transistors for data-intensive search applications.**

I. INTRODUCTION

Ferroelectric ternary content-addressable memories (TCAMs) have drawn significant interest due to the capability for data-intensive in-memory searching [1], [2]. Recently, we introduced a novel approach for one-transistor (1T) TCAMs, employing localized polarization switching in ambipolar ferroelectric tunnel field-effect transistors (FeTFETs), as depicted in Fig. 1 [3]. This strategy improves the accuracy of classification tasks in massive memory arrays by maximizing dynamic range and effectively reducing cell leakage currents compared to the previous scheme using uniform polarization. However, this approach imposes a previously unseen scaling constraint, as it necessitates the presence of opposing polar states separately within the ferroelectric layer at source- and drain-side to represent the don't care state 'X'. In this paper, the effect of gate length (L_G) downscaling on the performance of local polarization-based 1T FeTFET TCAM is discussed.

II. RESULTS AND DISCUSSION

We first fabricated a FeTFET with metal-ferroelectric-insulator-semiconductor (MFIS) structure using the CMOS gate-first process [3], [4]. Fig. 2a shows the cross-sectional transmission electron microscope image of the FeTFET device. Fig. 2b illustrates the measured ambipolar transfer curves wherein the 'X' state exhibits a programmed n-mode and an erased p-mode. The implementation of ternary states is achieved with the write bias schemes in Fig. 2c. To write bit 'X' from bit '1' in the 1T FeTFET TCAM cell involves the local switching of ferroelectric polarization at the source side from downward to upward. It selectively programs the n-mode from low-V_{th} to high-V_{th} state by increasing the tunnel barrier exclusively for electrons and not holes (Fig. 2d).

Regarding the local polarization-based 1T FeTFET TCAM cells, three mechanisms could potentially restrict scalability. Two of them are conventional ones associated with scaling ferroelectric transistors. One constraint relates to the sufficiency of ferroelectric domains for storing distinct memory states, while the other relates to the short-channel effect of transistors themselves [5], [6]. The L_G limits imposed by these two mechanisms are reported to be 30 nm and 20 nm, respectively [5], [7]. Furthermore, an additional limitation emerges concerning the lateral space required to achieve the necessary electric field distribution in the ferroelectric and the resultant segregation of locally upward and locally downward polarization near the source and drain, respectively.

The L_G limit imposed by this third mechanism is 20 nm, as depicted in Fig. 3. These are Sentaurus TCAD simulation results based on device parameters experimentally calibrated referring to the measured FeTFET characteristics. Down to L_G of 20 nm, the dynamic ranges are maximized by local polarization at the search voltages for bit '1' and bit '0' (-0.25 V and 2.0 V, respectively). A further decrease in L_G to 15 nm leads to dynamic range degradation and an increase in undesirable cell leakage currents. This is caused by the drain-induced barrier thinning [8] rather than insufficient formation of local polarization. As a result, utilizing local polarization-based 1T TCAM does not impose a stricter scaling limit in comparison to the uniform polarization-based one.

To validate the search operation of 1T TCAM array with scaled FeTFET cells, HSPICE simulations were performed using the compact model described in our previous work [3]. Fig. 4 shows the configuration of the simulated TCAM array, where input signals are applied to search lines (SLs), and the shorter falling delay of match line (ML) voltage indicates a higher degree of mismatch (DOM). Fig. 5 shows the L_G-dependent transient response of the ML sense amplifier outputs during the 1T TCAM search operations for a specific query. The last figure presents the ratio of search delay in the match state to that in the mismatch states. With L_G = 30, 20, and 15 nm, the MLs discharged more rapidly with increasing DOMs. However, L_G of 15 nm is impractical due to an insufficient sensing margin and high static power dissipation. It is confirmed that the FeTFET TCAM cell is scalable down to L_G of 20 nm. Therefore, the scalability of the local polarization-based 1T FeTFET TCAM is comparable to the previous uniform polarization-based 1T FeTFET TCAM.

III. SUMMARY

In this manuscript, the influence of L_G scaling of ambipolar FeTFETs on the performance of local polarization-based 1T TCAM was investigated. The additional scaling limitation introduced by the local polarization scheme is less dominant than the common limitations encountered in conventional uniform polarization-based TCAM approaches. In consequence, the scalability of the local polarization-based 1T FeTFET TCAM is comparable to that of uniform polarization-based TCAM, verifying the continued superiority of the former even with scaled devices.

ACKNOWLEDGMENTS

This work was supported by Samsung Research Funding & Incubation Center of Samsung Electronics under Project Number SRFC-TA2103-01.

REFERENCES

[1] S. Dutta *et al.*, in *IEDM Tech. Dig.*, 2021. [2] X. Yin *et al.*, *IEEE Trans. Circuits Syst. II, Exp. Briefs*, 2019. [3] M. Ryu *et al.*, *IEEE Electron Device Lett.*, 2024. [4] J. S. Woo *et al.*, *Adv. Intell. Syst.*, 2023. [5] H. Mulaosmanovic *et al.*, in *Proc. 4th IEEE Electron Devices Technol. Manuf. Conf.*, 2020. [6] E. Yurchuk *et al.*, *IEEE Trans. Electron Devices*, 2014. [7] A. M. Ionescu and H. Riel, *Nature*, 2011. [8] J. Appenzeller *et al.*, *IEEE Trans. Electron Devices*, 2008.

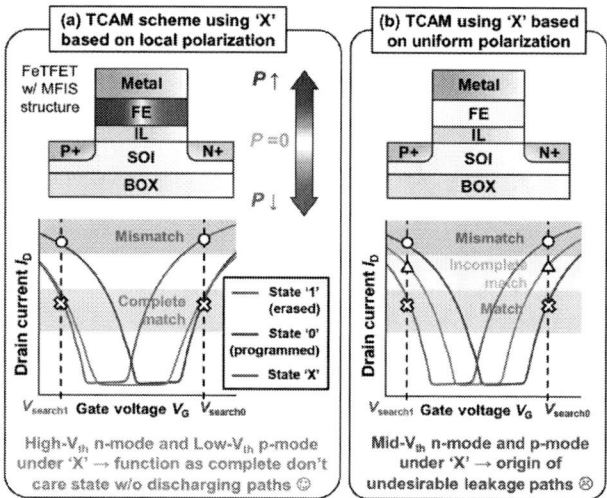

Fig. 1. Comparison of (a) local polarization- and (b) uniform polarization-based 1T FeTFET TCAM schemes. The former performs search operations with significantly higher accuracy.

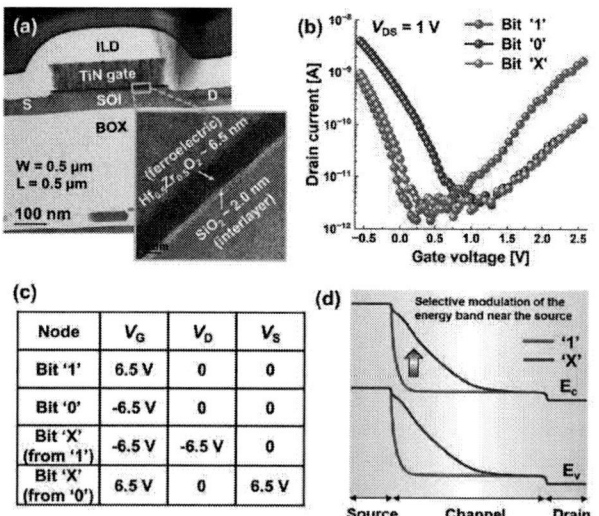

Fig. 2. (a) Cross-sectional view of the fabricated FeTFET device. (b) Measured I_D-V_G curves of the fabricated device corresponding to the ternary states. (c) Table on the write bias schemes for ternary states of local polarization-based 1T FeTFET TCAM cell. (d) Comparison of schematic lateral energy band diagrams of the FeTFET device under the '1' state and the 'X' state.

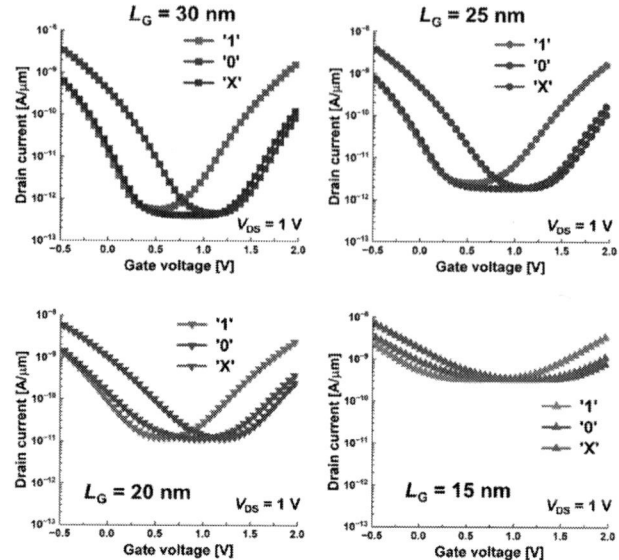

Fig. 3. Simulated I_D-V_G curves of FeTFETs each corresponding to L_G = 30 nm, L_G = 25 nm, L_G = 20 nm, and L_G = 15nm.

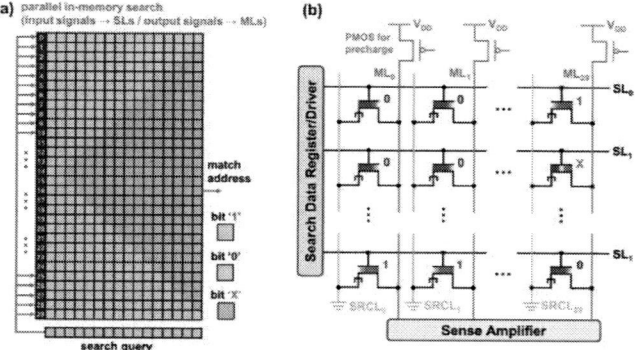

Fig. 4. (a) Simulated routing table represented with 30 × 16 TCAM array. (b) Structure of FeTFET-based 1T TCAM array for search including AND array and part of peripheral circuits.

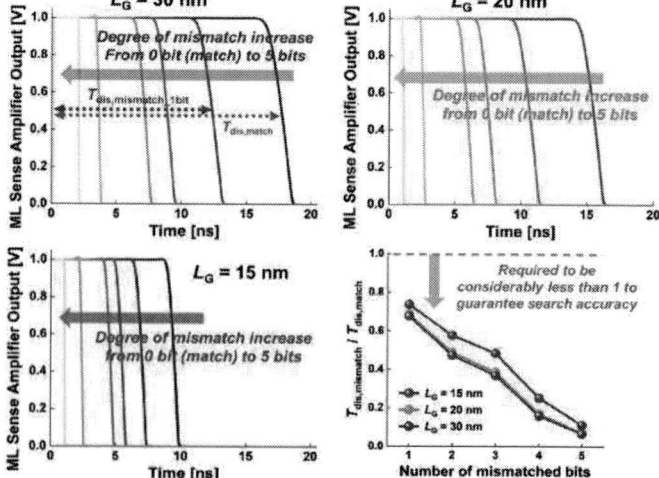

Fig. 5. Simulated transient response of ML voltage in the 1T FeTFET TCAM array corresponding to the L_G = 30 nm, L_G = 20 nm, and L_G = 15nm. The bottom-right graph shows T_{dis} ratio between the match and mismatch cases according to the degree of mismatch.

Study of Spacing-induced Fringing Effects on Emerging CFET Technology Nodes

You-Zheng Chen[1], Narasimhulu Thoti[2], and Ying-Tsan Tang[1]

[1]Department of Electrical Engineering, National Central University, Taoyuan, Taiwan, [2]Department of Microelectronics
Research Institute, University of Oulu, Finland
Email: narasimhulu.thoti@gmail.com, yttang@ee.ncu.edu.tw

Abstract — This paper simulates CFET characteristics and its inverter performance at A^0 technology nodes using TCAD simulation. We demonstrate CFET inverter's suitability for very low power consumption (0.2 V) and investigate reliability issues caused by fringing effects that retard corner currents through SRH recombination and current density simulations. This study investigates the separation thickness (D_{pn}) of p- and n-CFETs as well as the distance between nanosheets (D_{ns}) to explore the fringing effects, offering valuable design guidelines for CFET technology.
Keywords: CFET, nanosheet, footprint, inverter.

I. INTRODUCTION

CFET is a feasible architecture aimed at maximizing effective channel width by vertically stacking n-type and p-type devices [1]. This stacking preserves the channel width and reduces the number of tracks. CFET reduces the SRAM bit cell area and enables SRAM cross-coupled connections [2]. However, as nanosheets or metal tracks approach, fringing effects affect SRH recombination and slow current at the corners. Therefore, studying spacing between the p- and n-CFETs (D_{pn}) and the nanosheets (D_{ns}), respectively, impact on current density and SRH recombination is crucial for CFET performance. In this study, precise calibrations with proper physical models have been made under this demonstration. The device characteristics independently and inverter performance have been explored and demonstrated as suitable for very low power consumption.

II. DEVICE DESIGN AND MODELLING

The proposed design of CFET is depicted in Fig. 1 by vertically stacking n- on top of the p-CFET and with the combinations of gate-all-around together with nanosheet geometries. The key specifications referring to the regions and parameters that are shown on the right of Fig. 1(a) play a crucial role in deciding the footprint (FP) of CFET. Keeping typical effective oxide thickness (t_{ox}) and a few atomic layers of metal thickness (t_m) constant, the rest of the device specifications and their impact are delivered in this work. CFET is an inverter that saves a FP~50% than the conventional designs are delivered in Fig. 1(b) with the help of circuit symbol and layouts. The proposed CFET design is modelled through Sentaurus TCAD as a commercial tool. Primarily, the calibrations are performed with experimental data (CFET as design) to demonstrate the precise simulations, as shown in Fig. 2(a) [3]. In brief, the calibrations with experimental CFET design are demonstrated by tuning all the physical and carrier transport models, respectively. The tunneling models are also used at necessary regions for proper estimation of leakage current.

II. RESULTS AND DISCUSSIONS

The parameter that estimates the performance of a CFET is demonstrated and depicted in Figs. 2(b) and (d). Here, the performance of p- and n-CFETs are shown independently (at low (= |0.05| V) and high (= |0.5| V) gate bias (V_{GS}) to test its ability in terms of on-current (I_{on}) and drive current (I_{DS}) at constant drain bias (V_{DS}). The key point is a selection of 3-channel p- and 2-channel n-CFET devices to maintain similar I_{on} capabilities. Additionally, the energy band diagram indicating drain-induced barrier lowering (DIBL) of CFET for n-device is depicted in Fig. 2(c), which is identified as a very good figure. CFET is an inverter that benefits low power consumption because of the saved FP and reduced metal tracks with power rails (V_{DD}). Such inverter performance of CFET is depicted in Figs. 2(e) and (f) with the characteristics of voltage transfer (VTC) and pulsed input-output. It is seen from Fig. 2(e) that 0.2 V of V_{DD} saves an ample amount of power ($\propto V_{DD}$) effectively.

The CFETs scalability is also demonstrated with the physical significance, i.e., change in Shockley-Read-Hall (SRH) recombination and electron/hole current density terms at varied D_{pn} and D_{ns} from 20 nm down to 5 nm, respectively (see Fig. 3). The observations are, a profound reduction in D_{ns} (~5 nm) slightly pronounced to increase SRH recombination (Fig. 3(a)). Similarly, SRH is affected at a minimum D_{pn} of 5 nm (Fig. 3(b)). Both are caused by the fringing effect that we see as the two plates (here, nanosheets or metal tracks) approach closer [4, 5]. Nevertheless, the average contribution by either heavily scaled D_{pn} or D_{ns} is marginal. Hence, the electron/hole current density effect during D_{pn} is identified as marginal (see Figs. 3(c) and (d)).

II. CONCLUSIONS

The CFET design that is suitable for low power and high scalability has been demonstrated through TCAD simulations. Here, the precise calibrations with proper physical models have been made under this demonstration. The device characteristics independently and inverter performance have been analyzed and demonstrated as suitable candidate for very low power consumption (~ 0.2 V). The significance of D_{pn} and D_{ns} that define the overall FP reduction has been evaluated with its physical terms. Hence, the analyses are most useful under CFET technology.

ACKNOWLEDGMENTS

This work was supported by the National Science and Technology Council (NSTC) of Taiwan (No. 112-2218-EA49-013-MBK, 112-2622-8-A49-013-SB, NSTC 113-2119-M-A49-007) and Taiwan Semiconductor Research Institute (No. JDP113-Y1-035).

REFERENCES

[1] Chang S W et al 2019 IEDM, pp. 11.7.1-11.7.4.
[2] H. -H. Liu et al., in *IEEE TED* **70**, pp. 883-890
[3] C. -Y. Huang, et al., 2020 IEDM, pp. 20.6.1-20.6.4.
[4] Jung S.G., et al., *IEEE Access* **10** 41112–8, 2022.
[5] N. Thoti and Y. Li 2023 Nanotechnology **34** 505208.

979-8-3503-9164-0/24 $31.00 © 2024 IEEE

P1-18

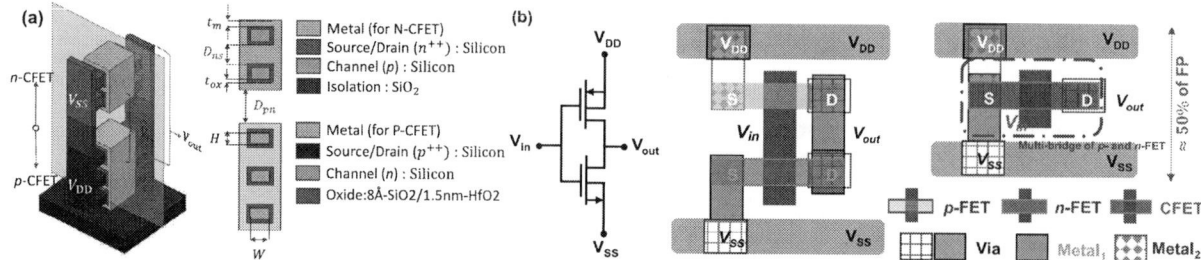

Fig. 1: (a) Design of CFET along with its region specifications. (b) Symbol and layout of conventional and CFET-inverter, respectively.

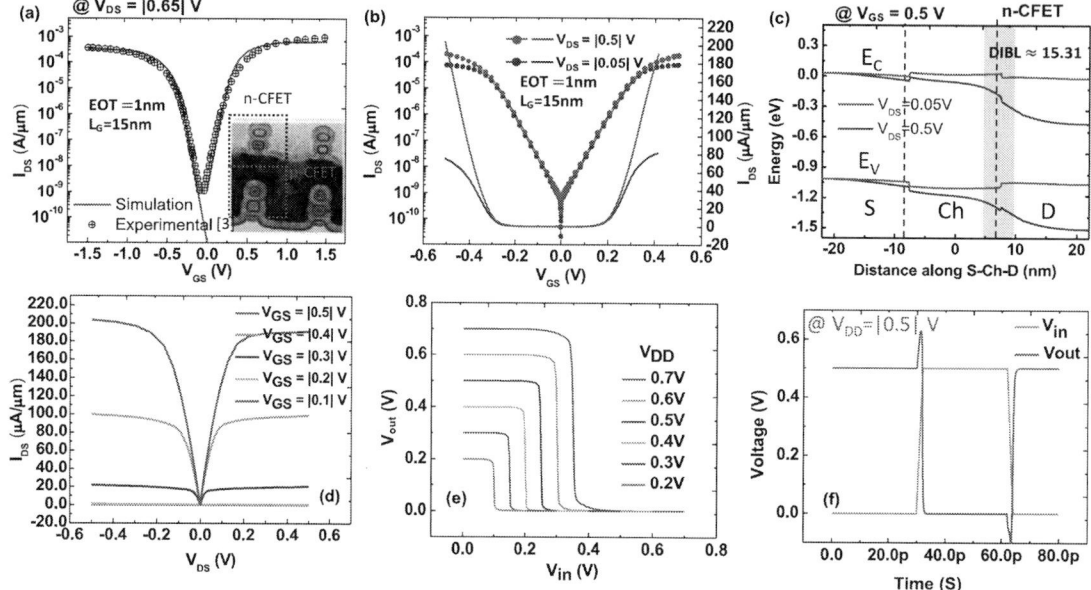

Fig. 2: (a) Calibrated results [3], (b) I_{DS}-V_{GS}, (c) energy-band diagram, (d) I_{DS}-V_{DS}, (e) VTC, and (f) pulse characteristics of CFET.

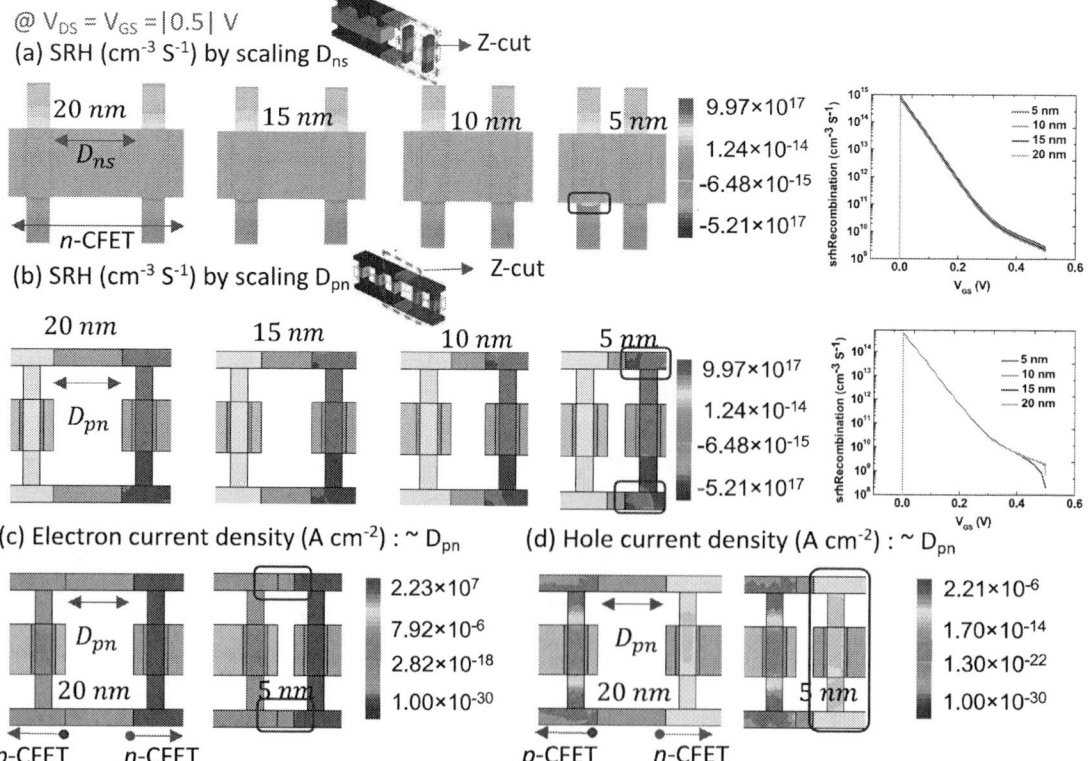

Fig. 3: D_{ns} and D_{pn} scaling of CFET with physical terms, i.e., SRH recombination terms.

979-8-3503-9164-0/24 $31.00 © 2024 IEEE   104

Compact Model of Feedback Field-Effect Transistor Using Artificial Neural Network

Seung Su Jeong[1], Jong Hyeok Oh[2], and <u>Yun Seop Yu</u>[3]

[1]Major of ICT & Robotics Engineering, Hankyong National University, Republic of Korea

Email: ysyu@hknu.ac.kr

Abstract — A compact model of feedback field-effect transistor (FBFET) using artificial neural network (ANN) for circuit simulation is introduced. Validation of this model is verified from comparing the current-voltage (I-V) and capacitanc-voltage (C-V) characteristics obtained from TCAD simulation of FBFET.

Keywords: FBFET, compact modelling, artificial neural network, TCAD

I. INTRODUCTION

Various devices are introduced to overcome the limitations of scale-down of MOSFETs, one of which is a feedback field effect transistor (FBFET) with advantages of extremely low subthreshold swing (SS), high on/off current ratio, and hysteretic behaviors [1]. The FBFFETs can be applied to memory devices for high-performance, high-density, and low-power memory applications [2], logic-in-memory [3], and spike neural network (SNN) [4]. Circuit simulations, including FBFETs, require compact models of FBFETs, which are critical for effective design and analysis. However, a simple analytical model of FBFET suitable for circuit design has yet to be developed. Existing compact models of FBFET have included procedures to solve numerically to couple different analytical and physical-based equations [5]. Recently, a compact model of machine learning (ML)-based FBFETs has been proposed to predict their hysteresis current-voltage (I-V) characteristics, and it predicts well latch-up voltage, latch-down voltages, saturation current, and memory window, but some regions of the I-V characteristics did not match the TCAD results [6]. Therefore, it is necessary to develop a deep learning-based FBFET compact model that can measure more accurate hysteresis I-V characteristics.

In this study, we introduce an artificial neural network (ANN) model of FBFET that can predict nonlinear models based on hysteresis I-V characteristics obtained from TCAD simulations for circuit simulations.

II. MODEL

Fig. 1 shows a schematic diagram of N-type FBFET, used in this study [2]. Fig. 2 shows an ANN architecture applied commonly for DC drain-source current (I_{ds}), total gate capacitance (C_{gg}), gate-source capacitance (C_{gs}), and gate-drain capacitance (C_{gd}). This ANN consists of one input nodes including two input nodes, two hidden layers, and one output layer, and each hidden layer consists of 16 nodes. Fig. 3 shows procedures to train pre-processing data of I-V characteristics with the ANN architecture as shown in Fig. 2 and to post-process them after inferring them by applying the trained weights and bias to the ANN architecture. Fig. 4 shows a flow-chart for predicting I-V and C-V characteristics

of FBFET in forward and reverse voltage sweeps to reflect its hysterestic characteristics. To implemnet the proposed model into HSPICE, Verilog-A [8] was used.

III. RESULTS

Figs. 5(a) and (b) show I_{ds}-V_{gs} characteristics of N-type FBFET in forward and reverse V_{gs} sweeps at different V_{ds}, respectively. Figs. 6(a) and (b) show $|I_{ds}|$-V_{gs} characteristics of P-type FBFET in forward and reverse V_{gs} sweeps at different V_{ds}, respectively. Figs. 7(a) and (b) show C_{gg}-V_{gs} characteristics of N-type FBFET in forward and reverse V_{gs} sweeps at different V_{ds}, respectively. Figs. 8(a) and (b) show C_{gg}-V_{gs} characteristics of P-type FBFET in forward and reverse V_{gs} sweeps at different V_{ds}, respectively. Figs. 9(a) and (b) show voltage transfer characteristics and transient analysis of monolithic 3D inverter (M3D-inverter) stacking vertically of N and P-type FBFETs, respectively. The HSPICE simulation results of the proposed ANN model (lines) agree considerably well with TCAD simulation results (symbols).

Figs. 10(a) and (b) shows a circuit diagram and timing diagram of memory operation of 2×2 cell of the monolithic 3-dimensional static random access memory with FBFET (M3D-SRAM-FBFET) [2], respectively. A cell consists of N-type FBFET and N-type MOSFET in series. Fig. 10(b) shows that the M3D-SRAM-FBFET provides a reliable RAM function as a non-destructive reading operation.

IV. CONCLUSIONS

A compact model was proposed to apply the hysteresis I-V characteristics of FBFETs to circuit simulations using ANN. The validation of the proposed ANN model was verified from comparing the HSPICE simulation results using the model with TCAD simulation results. Inverter and SRAM characteristics were confirmed through DC and transient analysis of M3D-inverter and M3D-SRAM-FBFET.

ACKNOWLEDGMENTS

This research was supported by the Basic Science Research Program through NRF of Korea funded by the Ministry of Education (NRF-2019R1A2C1085295).

REFERENCES

[1] A. Padilla *et. al.*, in *IEDM Tech. Dig.*, pp. 1–4, Dec. 2008.
[2] J. Oh *et. al.*, *Micromachines*, 13, p. 1625, 2022.
[3] Y. Yangm *et. al.*, *Scientific Reports*, 12, p. 3643, 2022.
[4] S. Woo *et. al.*, *IEEE TED*, 67, p. 2995-3000, 2020.
[5] Y. Taur *et. al.*, *Solid-State Electronics*, 134, p.1-8, 2017.
[6] S. Woo *et. al.*, *Micromachines*, 14, p.504, 2023.
[7] Silvaco Int. *ATLAS Ver. 5. 32. 1. R Manual*. 2022.
[8] HSPICE Reference Manual, Mountain View, CA, USA:Synopsys, 2022.

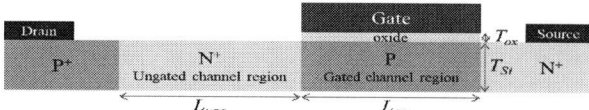

Fig. 1. Schematic view of n-type FBFET.

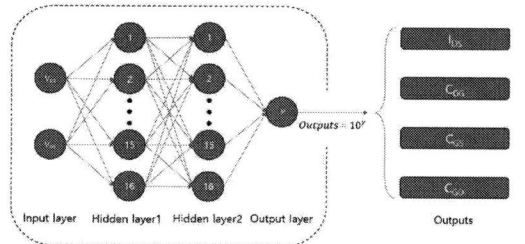

Fig. 2. Proposed ANN.

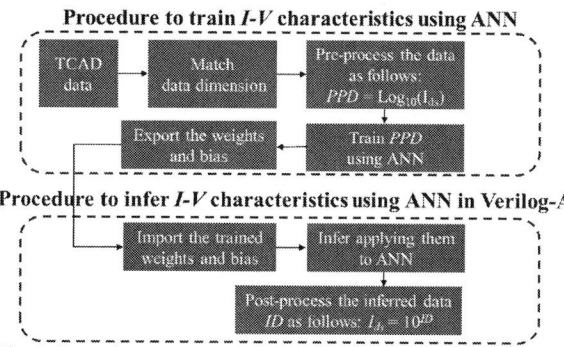

Fig. 3. Procedures pre-processing and post-procerssing data of *I-V* characteristics using ANN.

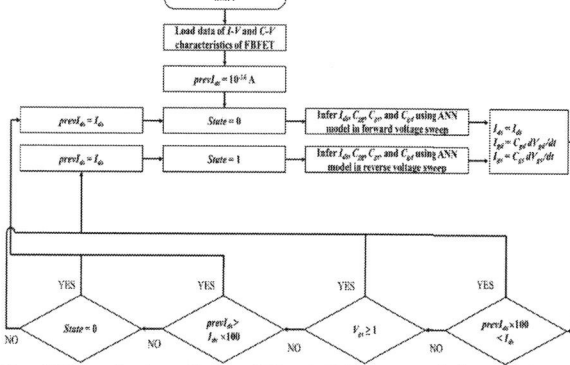

Fig. 4 Flow-chart predicting *I-V* and *C-V* characteristics of FBFET in forward and reverse voltage sweeps.

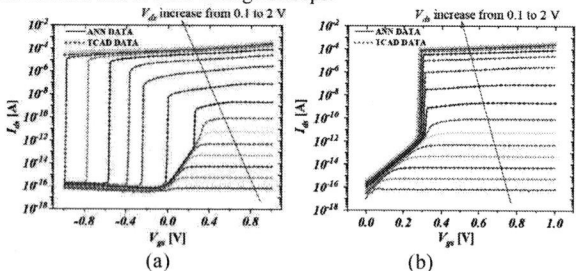

Fig. 5. I_{ds}-V_{gs} characteristics of N-type FBFET. (a) Forward sweep and (b) reverse sweep.

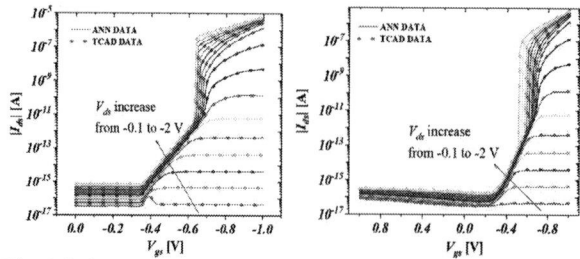

Fig. 6. $|I_{ds}|$-V_{gs} characteristics of P-type FBFET. (a) Forward sweep and (b) reverse sweep.

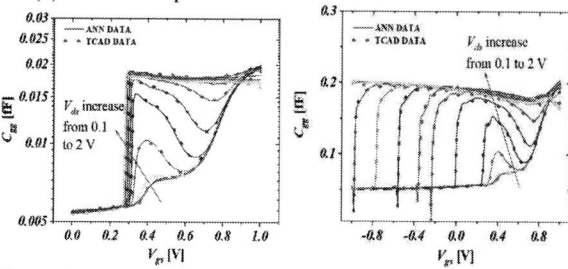

Fig. 7. C_{gg}-V_{gs} characteristics of N-type FBFET. (a) Forward sweep and (b) reverse sweep.

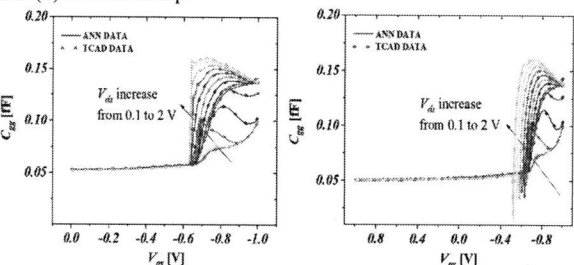

Fig. 8. C_{gg}-V_{gs} characteristics of P-type FBFET. (a) Forward sweep and (b) reverse sweep.

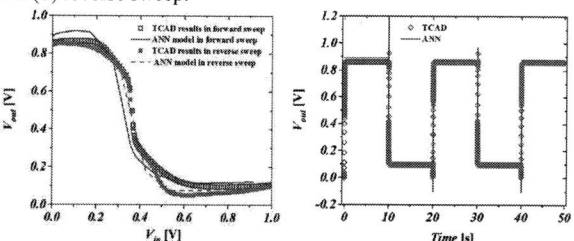

Fig. 9. Inverter characteristics. (a) Voltge transfer characteristics and (b) transient analysis.

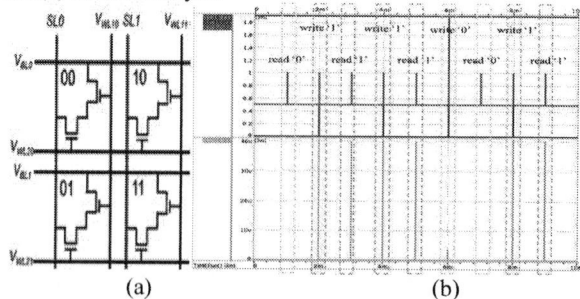

Fig. 10. M3D-FBFET-SRAM [2] 2×2 cell. (a) circuit diagram and (b) timing diagram of memory operation simulated by HSPICE icluding Verilog-A.

Neural Encoder Using "PN-Body Tied SOI-FET"

Masaki Kobayashi, Jiro Ida, Takayuki Mori

Kanazawa Institute of Technology, Japan, Email: c6300881@st.kanazawa-it.ac.jp

Abstract— **This study reports the potential of PN-Body tied silicon on insulator field-effect transistor (PNBT SOI-FET) as a neural encoder. When a suitable resistor was connected to the body terminal of the PNBT SOI-FET, the generation of spike signals was confirmed. It was also found that the spike frequency varied by changing the values of the body, gate, and drain voltages. This indicates that the spike frequency contains information on the change in voltage values, and the neural encoding can be reproduced.**

I. INTRODUCTION

Research in neuromorphic chips is currently active. Some researchers focus on spiking neural networks (SNN), which use spike signals to process information. To realize SNN, an encoding technology is required to convert analog signals into spike signals. In a previous study, an encoding method using a single device was proposed [1]. The flow of the neural encoder in SNN is shown in Fig. 1. The advantages of a single-device neural encoder include low power consumption and low hardware cost [2]. In this study, a neural encoder was reproduced using PN-Body tied silicon on insulator field-effect transistor (PNBT SOI-FET). As a type of encoding, we aimed to reproduce rate encoding, which handles information with spike frequencies.

II. DEVICE STRUCTURE AND OPERATION PRINCIPLE

Fig. 2 shows the structure of the PNBT SOI-FET. The PNBT SOI-FET has a PN junction attached to the body-tied structure of the SOI-FET. The characteristic of the body current I_B versus the body voltage V_B is shown in Fig. 3. The device shows steep characteristics due to the floating body effect and positive feedback by the pnpn structure, and can amplify the I_D [3]. In this study, a resistor was connected to the body terminal to reproduce a neural encoder. The relationship between V_B and the steep rise and fall of the drain current is shown by the following conditional equation [4].

$$\frac{R_{\text{body-down}}}{R_{\text{body-down}} + R_{\text{ext}}} \times V_B \geq V_{\text{latch-up}} \quad (1)$$

$$\frac{R_{\text{body-up}}}{R_{\text{body-up}} + R_{\text{ext}}} \times V_B \leq V_{\text{latch-down}} \quad (2)$$

$R_{\text{body-down}}$ is the body resistance just before the device steeply turns on, and $R_{\text{body-up}}$ is the body resistance just before it steeply turns off. Equation (1) indicates condition that when the device is OFF and the body terminal voltage is greater than $V_{\text{latch-up}}$, the steep switching occurs. Equation (2) means that when the device is ON and the body terminal voltage is less than $V_{\text{latch-down}}$, the device turns off abruptly. By repeating this operation with the PNBT SOI-FET, we aimed to convert the signal into a spike signal. These phenomena occur because the voltage divider ratio changes between changes as the PNBT SOI-FET is turned on and off, causing the voltage across the body terminal to switch. We extracted each parameter used in the conditional equation from Fig. 3 and calculated R_{ext} to induce spike operation. Table 1 summarizes these parameters. In this measurement, we used $R_{\text{ext}} = 1 \text{ G}\Omega$.

III. MEASUREMENT RESULTS

Figs. 4–9 show the measurement results of input current I_B and output current I_D at varying voltage values. The results of each measurement confirmed that the I_B was amplified.

Fig. 4 shows the drain current I_D when $V_G = 0.22$ V, $V_D = 1$ V, and V_B is varied. From this result, the input signal (V_B) is converted to a spike wave output, with the spike frequency changing as V_B change, as shown in Fig. 5. V_B changes the carrier charging rate. It can be said that rate encoding is reproduced.

Fig. 6 shows the drain current when $V_B = 2.0$ V, $V_D = 1$ V, and V_G is varied. Similar to the results in Fig. 4, the spike frequency varies with the change in V_G. However, the spike frequency changes with very small alterations in V_G, compared with the change observed with V_B. The relationship between spike frequency and V_G is shown in Fig. 7. It is considered that the V_G changes the barrier height between the source and body, and affects the positive feedback.

Fig. 8 shows the drain currents for $V_B = 0$–2.0 V, $V_G = 0.22$ V, and varying V_D. The relationship between spike frequency and V_D is shown in Fig. 9. The spike frequency is unchanged from $V_D = 0.1$ V to $V_D = 1$ V. V_D means the supply voltage and the frequency should not vary with supply voltage. However, over $V_D = 1$V, the spike frequency increases. The high V_D affects the width of the neutral region under the gate. It may affect the floating body effect and positive feedback.

IV. CONCLUSION

We demonstrated the potential of PNBT SOI-FET as a neural encoder. It was found that spikes were generated on PNBT SOI-FET by inserting a resistor, using the same concept of Ref [4]. The possibility of rate-encoding one PNBT SOI-FET at the input level of V_B was confirmed, and the frequency does not change until $V_D = 1$ V.

ACKNOWLEDGMENTS

This work is the result of collaboration with the High Energy Accelerator Research Organization (KEK) and LAPIS Semiconductor Co., Ltd. This work was supported by JSPS KAKENHI Grant Number JP21K14216 and JST-CREST Grant Number JPMJCR20Q1. This work was also supported through the activities of VDEC, the University of Tokyo, in collaboration with Cadence Design Systems and Siemens Electronic Design Automation, Japan K. K.

REFERENCES

[1] S. S. Radhakrishnan *et al.*, Nat. Commun., vol. 12, Apr. 2021.
[2] J. K. Han *et al.*, Adv. Sci., vol. 9, no. 2106017, Apr. 2022.
[3] T. Mori *et al.*, IEEE J-EDS, vol. 6, Oct. 2018.
[4] D. Lim *et al.*, IEEE EDL, vol. 42, no. 5, May 2021.

979-8-3503-9164-0/24 $31.00 © 2024 IEEE

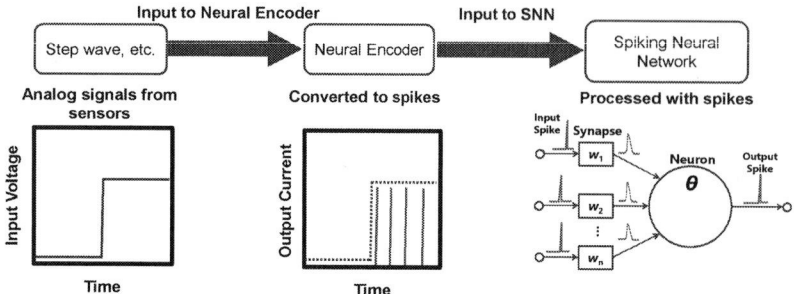

Fig. 1 Signal processing flow in spiking neural network

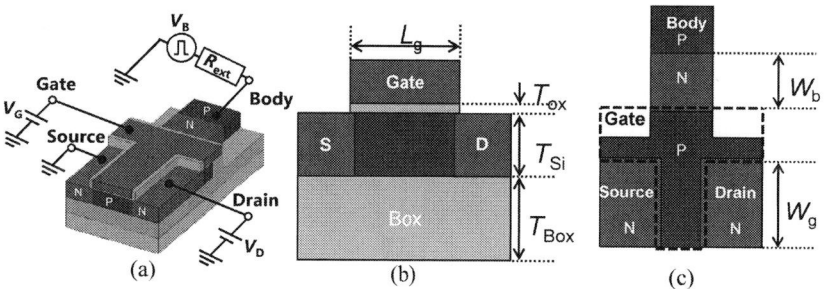

Fig. 2 Structure of PNBT SOI-FET.
(a) Bird's-eye view with resistor, (b) front view, (c) top view.

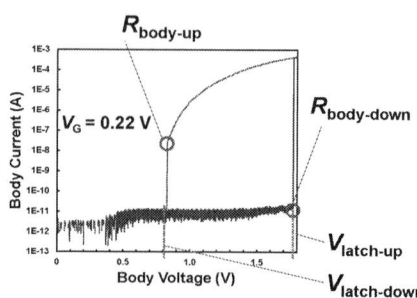

Fig. 3 I_B–V_B characteristics of PNBT SOI-FET.

Table. 1 Each parameter of Equations (1) and (2) extracted from Fig. 3.

V_B	2 V
$V_{latch-up}$	1.610 V
$V_{latch-down}$	0.842 V
$R_{body-up}$	3.197×10^7 Ω
$R_{body-down}$	1.091×10^{11} Ω
$\mathbf{4.393 \times 10^7 \leq R_{ext} \leq 2.640 \times 10^{10}}$	

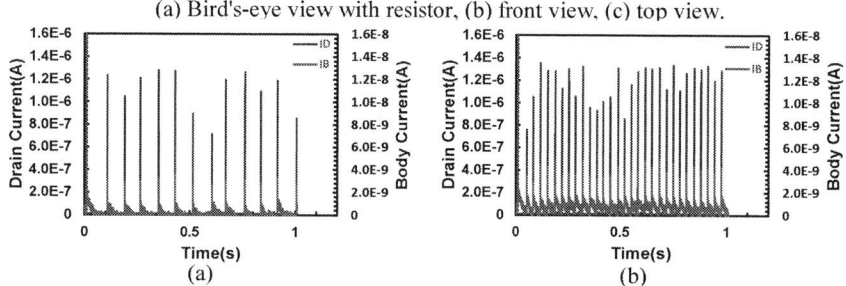

Fig. 4 Measurement results of PN-Body Tied SOI-FET at
(a)V_B = 0–2.0 V and (b)V_B = 0–2.5 V.

Fig. 5 Frequency-V_B characteristics.

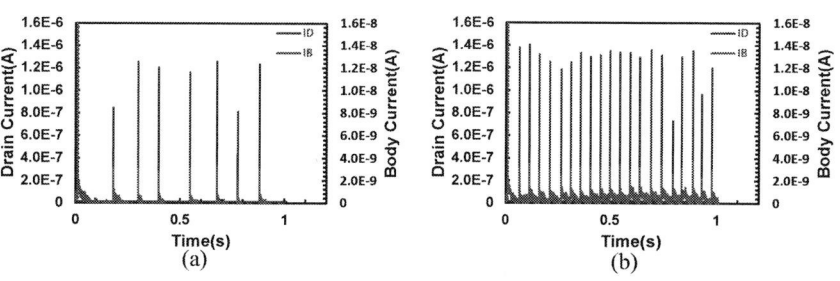

Fig. 6 Measurement results of PN-Body Tied SOI-FET at
(a) V_G = 0.217 V and (b)V_G = 0.225 V.

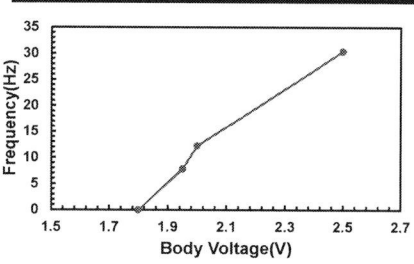

Fig. 7 Frequency-V_G characteristics.

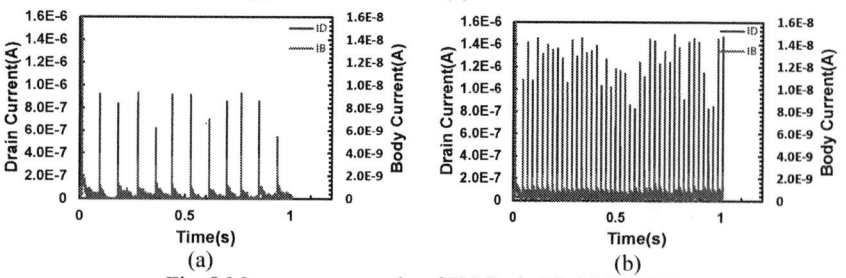

Fig. 8 Measurement results of PN-Body Tied SOI-FET at
(a) V_D = 0.1 V and (b)V_D = 1.5 V.

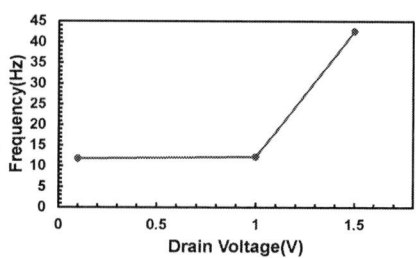

Fig. 9 Frequency-V_D characteristics.

979-8-3503-9164-0/24 $31.00 © 2024 IEEE

Engineering Strategies in HfO$_x$ RRAM-Based Analog Synapses Toward Linear Weight Update for Neuromorphic Hardware Accelerators

Yunsur Kim, Hyeonsik Choi, and Jiyong Woo

School of Electronic and Electrical Engineering, Kyungpook National University, South Korea

Email: jiyong.woo@knu.ac.kr

Abstract — This study introduces two distinct engineering approaches for HfO$_x$-based resistive memories (RRAMs) that can be implemented in various manners for neuromorphic hardware platforms. For highly dense cross-point synaptic arrays, we show that a gradual weight update can be achieved even in selector-free RRAM by integrating a thin Al$_2$O$_3$ nonlinear barrier. Meanwhile, in configurations that use conventional transistors as selectors, the weight update as quick as possible plays an important role in accelerating training process. We thus reveal that the abundant oxygen vacancies in sputtered HfO$_x$ enables nanosecond weight modulation driven by identical pulses.

I. INTRODUCTION

Resistive memory (RRAM) has been extensively studied as a synaptic device for neuromorphic applications [1]. For large-scale integration, selector-free RRAM (1R) arrays with inherent nonlinear current-voltage (I-V) behavior [2] have been proposed to minimize sneak path currents. At the expense of low synaptic density in the array, conventional one transistor-one RRAM (1T-1R) configurations can be alternative to implement robust neuromorphic hardware (Fig. 1). Therefore, we presented how HfO$_x$-based RRAMs can be appropriately designed for these representative two cases through stack engineering.

II. EXPERIMENT

A 6-nm-thick HfO$_2$ switching layer was formed through atomic layer deposition (ALD), which was defined as ALD-HfO$_2$ RRAM. For selector-free Al$_2$O$_3$/HfO$_2$ RRAM, a 1.2-nm-thick ALD deposited Al$_2$O$_3$ layer followed the deposition of the HfO$_2$ layer. Then, Ti and W were deposited by DC sputtering for scavenging layer and electrode, respectively.

III. RESULT AND DISCUSSION

A. Selector-free Al$_2$O$_3$/HfO$_2$ RRAM

To address the issue of abrupt set switching observed in the ALD-HfO$_2$ RRAM (Fig. 2a and Fig. 2b), we introduced a thin Al$_2$O$_3$ nonlinear barrier. This resulted in a nonlinearity of 17 [2] in the I-V response at the low resistance state (LRS) owing to the large bandgap of Al$_2$O$_3$, with uniform current distributions of both LRS and high resistance state (HRS) observed during repeated DC cycles (Fig. 3). The impact of the Al$_2$O$_3$ barrier on the switching behavior was further verified by the observation of two-step switching, rather than rapid switching, during real-time AC measurement (Fig. 4). The two-step transition facilitated analog weight updates with high linearity, defined as α [3], of -0.2 and -6.2 under pulses with widths of 5 μs and 50 μs, for potentiation and depression, respectively (Fig. 5).

B. Nanosecond synaptic operation in sputtered-HfO$_x$ RRAM

To date, the HfO$_2$ switching layer has been mostly deposited by ALD, making it difficult for weight update to be rapidly tuned by sub-microsecond pulses. To promote ion movement, we utilized RF sputtering for non-stoichiometric HfO$_x$ layer (Fig. 6). The sputtered-HfO$_x$ RRAM continued to exhibit gradual set switching. This can be explained by the increased amount of oxygen vacancies (V$_0$) in the HfO$_x$ layer (Fig. 7), which eventually enabled a more linear updates of the current state with pulse widths of less than 150 ns (Fig. 8).

The pattern recognition accuracies of fabricated Al$_2$O$_3$/HfO$_2$ and sputtered-HfO$_x$ RRAMs were evaluated with near-ideal accuracy of 83 % and 90 %, respectively (Fig. 9). Furthermore, we suggest a plausible conduction mechanism of the set-switching process in each device, which was directly related to the synaptic linearity (Fig. 10). While the formation of conductive filament (CF) led to an abrupt switching in the ALD-HfO$_2$ RRAM, the Al$_2$O$_3$ barrier induced gradual increase of current states with the nonlinear characteristic that effectively suppressed the LRS current. In the sputtered-HfO$_x$ RRAM, abundant V$_0$ prompted the formation of multiple weak CFs, enabling fast operating speed as well as linear weight updates.

IV. CONCLUSIONS

We showed stack engineering strategies in HfO$_x$ RRAM designed for neuromorphic synaptic devices. For high-density synaptic arrays, selector-free Al$_2$O$_3$/HfO$_2$ RRAM not only allowed nonlinear I-V behavior but also linearly modulated the current state. When the need for a compact two-terminal selector is alleviated by adopting the 1T-1R configuration, a fast update of weights of RRAM synapses may be preferred. Considering this scenario, we revealed that V$_0$-rich sputtered HfO$_x$ layer allows linear weight updates with sub-150 ns operation.

ACKNOWLEDGEMENT

This work was supported by the Technology Innovation Program (or Industrial Strategic Technology Development Program) (RS-2023-00231956 & 00236772, Customizable Low-power AI Accelerator and Software Based on Resistive ACiM for Edge Devices) funded By the Ministry of Trade, Industry & Energy (MOTIE, Korea) (1415187475).

REFERENCES

[1] Z. Li et al., TED, vol. 64, no. 6, pp. 2721-2727, 2017. [2] S. Lee et al., EDL, vol. 35, no. 10, 2014. [3] J. -W. Jang et al., IEEE ISCS, pp. 1054-1057, 2014.

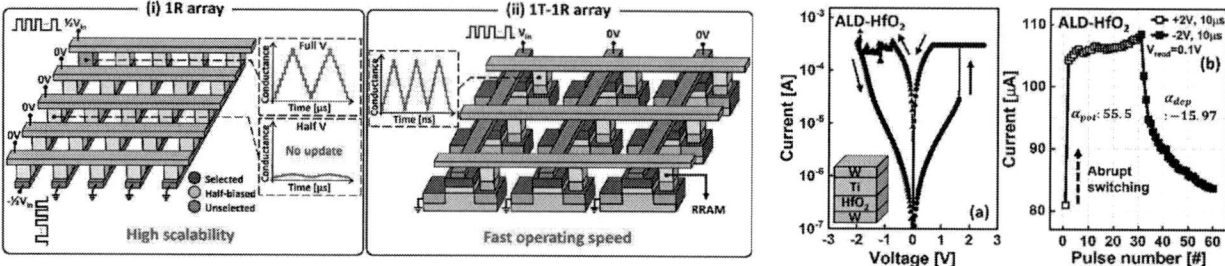

Fig. 1. Two scenarios for implementing neuromorphic hardware systems with cross-point arrays using RRAM-based synaptic devices.

Fig. 2. (a) DC I-V and (b) synaptic behavior of ALD-HfO$_2$ RRAM.

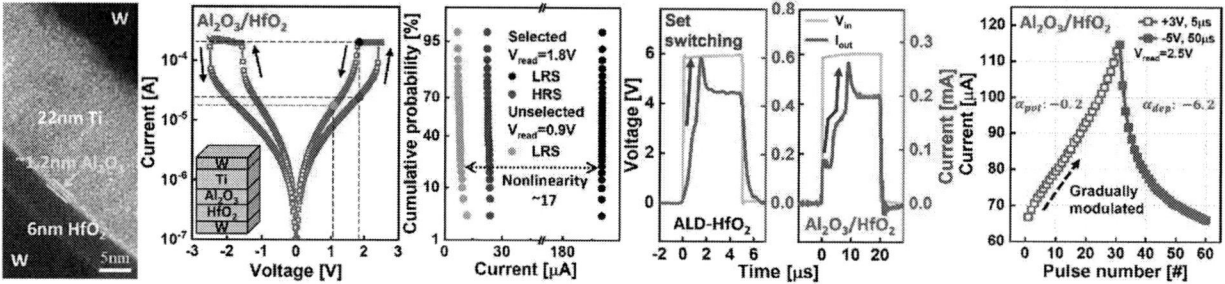

Fig. 3. Cross-sectional transmission electron microscopy (TEM) image and DC I-V response of Al$_2$O$_3$/HfO$_2$ RRAM with high uniformity of current levels with nonlinearity.

Fig. 4. Set switching analysis of ALD-HfO$_2$ and Al$_2$O$_3$/HfO$_2$ RRAMs.

Fig. 5. Gradual current update of Al$_2$O$_3$/HfO$_2$ RRAM.

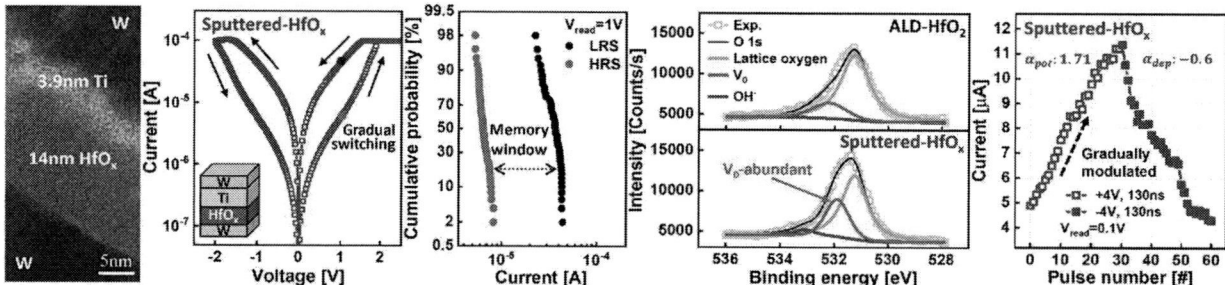

Fig. 6. Cross-sectional TEM image and gradual set switching observed at DC I-V response of sputtered-HfO$_x$ RRAM with uniform current distribution at repeated DC cycles.

Fig. 7. X-ray photoelectron spectroscopy analysis of sputtered-HfO$_x$ RRAM.

Fig. 8. Analog synaptic behavior of sputtered-HfO$_x$ RRAM.

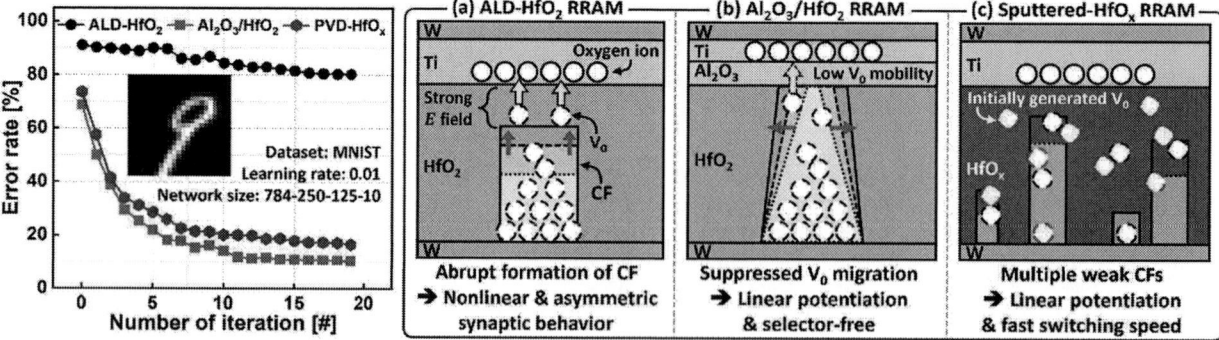

Fig. 9. Comparison of recognition accuracies of three RRAMs.

Fig. 10. Plausible conduction mechanism of (a) typical ALD-HfO$_2$ RRAM, (b) Al$_2$O$_3$/HfO$_2$ selector-free RRAM, and (c) V$_0$-abundant sputtered-HfO$_x$ RRAM.

979-8-3503-9164-0/24 $31.00 © 2024 IEEE

Enhancing Single-Electron Reservoir Computing Performance with Delay Function and Multiple-layer Reservoir Circuits

Shunya Watanabe, Takahide Oya

Graduate School of Engineering Science, Yokohama National University

Tokiwadai 79-5, Hodogaya-ku, Yokohama, 240-8501, Japan

Phone: +81-45-339-4125, E-mail: watanabe-shunya-cb@ynu.jp

Abstract —**We have previously designed single-electron reservoir computing (SERC) circuit and have confirmed its competence for waveform prediction. In this study, we observed that the virtual increase in connections from the reservoir to the output by implementing delay functions led to an improvement in the performance for waveform prediction. Furthermore, the SERC circuit with multiple-layer reservoir possessed the capability to predict complex waveforms, which our previous circuit could not deal with.**
Keywords: nanodevice, single electron circuit, reservoir computing, physical reservoir computing (PRC)

I. INTRODUCTION

We enhanced the function of the previously proposed single-electron reservoir computing (SERC) circuit[1] by implementing delay functions[2] and virtually increasing the connections from the reservoir to the output. This improvement resulted in better performance in waveform prediction. Additionally, we designed SERC circuit with multiple-layer reservoir, enabling prediction of complex waveforms that were not achievable with the previous SERC circuit design.

The reservoir computing (RC) is a type of recurrent neural network (RNN), which is a network of artificial neurons and includes feedback connections. The RC consists of three layers: the input, the reservoir, and the output. Unlike other typical RNN models, random network of neurons is in the reservoir, which corresponds to middle layer of RNN, and the connections between neurons in the reservoir have fixed weights. Only the weights of the neurons connecting to the output layer are trained. Nonlinear physical systems are allowed to implement the reservoir, such systems being referred to as physical reservoir computing (PRC). Several PRCs, utilizing various physical systems such as optical circuits[3], spin devices[4], etc., have been reported.

Our previous experiments have shown that a component called a single-electron oscillator (SEO), which includes a tunnel junction, exhibits behavior similar to biological neurons[5]. This led us to the idea that the single-electron circuit composed of an array of SEOs could be applied to RC, i.e., the SERC circuit.

II. CIRCUIT DESIGN AND SIMULATION

The reservoir in SERC circuit consists of honeycomb-arrayed multiple-tunnel-junction SEOs (MJSEOs) (Fig. 1), which has a smaller effect due to stochastic behavior by electron tunneling (Fig. 2). As a demonstration of learning, waveform prediction was conducted on the SERC circuit. We used normalized root-mean-square error (NRMSE) as the evaluation metric.

The increase in the number of data obtained from the reservoir, i.e., the increase in the number of connections between the reservoir and the output, is expected to improve performance of SERC. To increase the connections, a delay function was adopted to SERC (Fig. 3). The delay function, implemented in software, records the state of the original reservoir for a certain period and creates a virtual reservoir state that is delayed by a certain time. Additionally, it is possible to have several virtual reservoirs delayed at different intervals. Data from the original reservoir and all delayed virtual reservoirs are combined and sent to the output. Based on this data, learning is ultimately conducted. In other words, learning is conducted using a mixture of current and past stat of reservoir. This allows for an increase in the types of data obtained from the reservoir without increasing the size of the reservoir circuit. It was confirmed that the increase in the number of connections by the implementation of the delay function resulted in improved performance for sin-wave prediction (Fig. 4). It was also verified that its capability to predict other types of waveforms: square- and triangle-waveforms (fig. 5).

Furthermore, we designed a multiple-layer SERC circuit (Fig. 6). Its reservoir consists of stacked layers of honeycomb circuits with varying parameters of resistors of MJSEOs, interconnected to each other. This design aims to extract more complex data from the reservoir. It was confirmed that our multiple-layer SERC circuit represented smaller NRMSE, which was 0.071, for complex waveform prediction than NRMSE of our previous SERC with delay function, which was 0.133, i.e., the proposed multiple-layer SERC circuit showed better performance (Fig. 7).

ACKNOWLEDGMENTS

This work was partly supported by JSPS KAKENHI Grant No. JP23H00169, and Kayamori Foundation of Information Science Advancement.

REFERENCES

[1] S. Watanabe and T. Oya, SNW 2023, P01, 2023.
[2] Y. Sakemi et. al., Scientific Reports, 10, 21794, 2020.
[3] M. Nakajima et. al., Communications Physics, 4(1), 2021.
[4] R. Nakane et. al., IEEE Access, 6, pp. 4462-4469, 2018.
[5] T. Oya, et al, Int'l Journal of Unconventional Computing, 1, pp. 177-194, 2005.

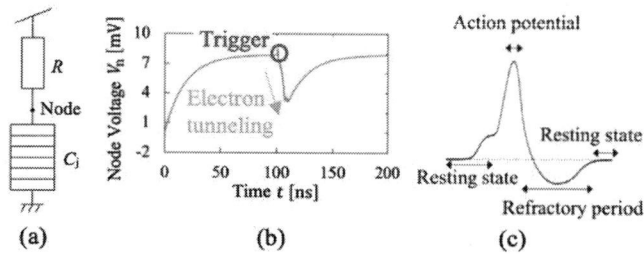

(a) **(b)** **(c)**

Fig. 1 **(a)** Schematic of MJSEO. Resistor and multiple-tunnel-junction are connected in series. **(b)** Voltage variation of MJSEO's node. Trigger voltage being applied to its node, sudden voltage change occurs due to electron tunneling. **(c)** Membrane voltage variation of spiking neuron.

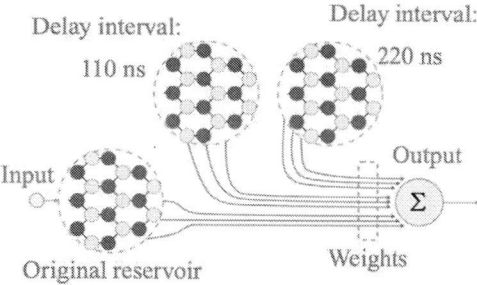

Fig. 3 Schematic of SERC circuit with delay function. Two virtual reservoirs are prepared by the delay function. Delay intervals are 110 ns and 220 ns, respectively. Each reservoir: original reservoir and two delayed reservoirs, has 16 connections to output; totally, 48 connections are joined to output.

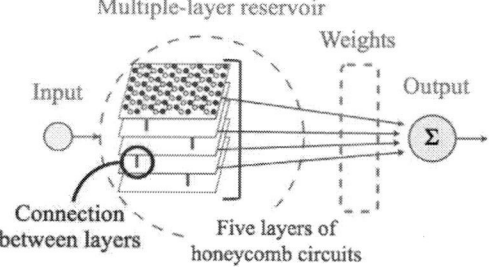

Fig. 6 Schematic of multiple-layer SERC circuit. Reservoir has 5 layers of honeycomb circuits of MJSEO, interconnected with each other through capacitor. Each layer consists of different parameters from other layers. The main layer where input is received has 16 connections to the output, and other layers have 8 connections, totaling 48 connections.

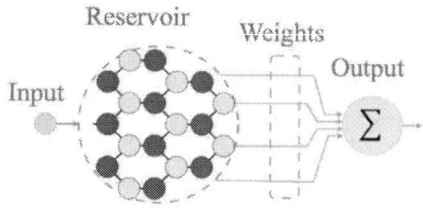

Fig. 2 Schematic of SERC. Reservoir contains honeycomb-arrayed MJSEOs, where MJSEOs are connected through capacitors. Input is trigger voltage applied to nodes of MJSEO, and output takes sum of voltage changes from several nodes of MJSEO in reservoir and weight for each voltage. Learning process is conducted by repeated adjustment of weights.

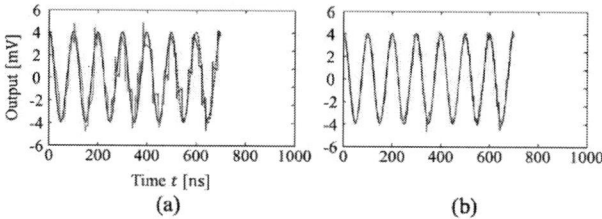

(a) **(b)**

Fig. 4 Simulation result for sine-waveform prediction. Orange-colored line describes under leaning, green does prediction, blue does expected output. **(a)** Result of original SERC, i.e., 16 connections: NRMSE is 0.134. **(b)** Result of SERC with delay function, i.e., 48 connections: NRMSE is 0.052.

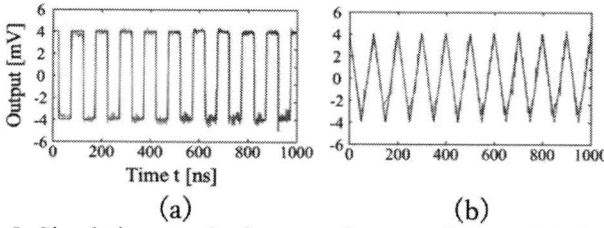

(a) **(b)**

Fig. 5 Simulation result for waveform prediction. **(a)** Square-waveform prediction: NRMSE is 0.118. **(b)** Triangle-waveform prediction: NRMSE is 0.059.

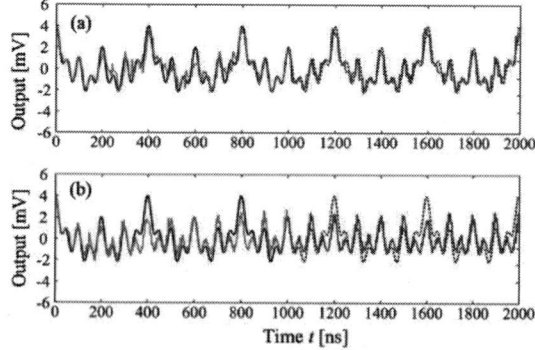

Fig. 7 Simulation result for complex waveform prediction. **(a)** Result of multiple-layer SERC: NRMSE is 0.071. **(b)** Result of SERC with delay function: NRMSE is 0.133.

Design of Single-Electron Circuit Representing Brownian Motion of Particles to Implement Circuit Capable of Computing Diffusion Limited Aggregation Model

Ryoga Miyakoshi, Takahide Oya

Graduate School of Engineering Science, Yokohama National University

Tokiwadai 79-5, Hodogaya-ku, Yokohama, 240-8501, Japan

Phone: +81-45-339-4125, Email: miyakoshi-ryoga-wg@ynu.jp

Abstract — We propose a new single-electron circuit that can calculate Diffusion Limited Aggregation (DLA) simulation model. DLA can be observed in nature, such as the growth of metal dendrite. To calculate DLA model, representation of particles' Brownian motion as a circuit operation is important. For this, the Brownian motion was represented on the single-electron circuit by node voltage changes of arrayed single-electron oscillators in it as the first step of this study. As the results, the obtained behavior was close to Brownian motion as we desired.

Keywords: single-electron circuit, Diffusion Limited Aggregation model, Brownian motion

I. INTRODUCTION

We propose a new single-electron circuit (SE circuit) that can calculate Diffusion Limited Aggregation (DLA) simulation model.

The SE circuit can control individual electrons by utilizing quantum effects, i.e., electron tunnelling and Coulomb blockade. As its features, it is known to have stochastic and parallel operation. However, an optimal information processing methods have not been established.

A phenomenon called DLA can be observed in various places in nature, such as the growth of metal dendrite, snow crystal and lightning bolts. In the DLA algorithm, if particles in Brownian motion adjacent to a nucleus aggregate, it become part of the aggregate. Gradually, it forms a dendritic structure. A common way to analyse this phenomenon using a computer is to perform random simulations for each particle[1]. But in this way, increasing the number of particles requires a huge amount of computation.

In this study, we aim to design a computationally efficient DLA simulation circuit that takes advantage of the stochastic and parallel operation which are features of SE circuits, and establish an optimal information processing methods with SE circuits.

II. CIRCUIT DESIGN

In this study, we use SE oscillators[2] (SEOs) (Fig.1). When the node voltage exceeds the threshold voltage, electron tunnelling occurs stochastically, and the node voltage drops suddenly. By connecting some SEOs with coupling capacitors and setting the bias voltages of them to positive and negative alternately (e.g., Fig. 2), the change of the node voltage due to electron tunnelling induces electron tunnelling at the adjacent oscillator. As a result, the behaviour like a wave propagation is obtained (Fig. 3).

In order to represent the DLA model in circuit operation, it is necessary to represent the Brownian motion of particles, as mentioned above. However, in the original circuit shown in Fig. 2, the Brownian motion of particles (wave propagation) in the reverse direction cannot be realized once the wave has travelled because of insufficient charging time for the node voltage. Therefore, we add "one-way" SE circuits and one-dimensional array of SEOs (Fig. 4). The one-way circuit consists of the branching structure of SEOs and one SEO which bias voltage set low. Due to this structure, it can limit the direction of signal propagation to one direction[3]. Also, one-dimensional array of SEOs makes the charging time of node voltage to continue propagation when propagation returns between SEOs as a coordinate (SEOs marked in red in Fig. 4).

III. SIMULATION RESULTS

We constructed the circuit in Fig. 4 in two-dimensions and named it Brownian motion circuits. The simulation results of Brownian motion circuits are shown in Fig. 5. And Fig. 6 visualizes the propagation process of Fig. 5 with arrows. As the results, the propagation in the reverse direction can be confirmed, and it seems to move with equal probability in the four directions (up, down, left, right). We think that we were able to represent the behaviour like Brownian motion with SE circuits. In conclusion, as a first step in pioneering a single-electron circuit capable of computing the DLA model, we have shown that the Brownian motion behaviour of particles can be expressed using our single-electron circuit.

In the near future, we plan to simulate multiple particles in parallel in the same two-dimensional Brownian motion circuits.

As a next step of this study, we also plan to design our SE-DLA circuit based on the Brownian motion circuits.

ACKNOWLEDGMENTS

This work was partly supported by JSPS KAKENHI Grant No. JP23H00169, and Kayamori Foundation of Information Science Advancement.

REFERENCES

[1] X. Guo, J.L. Wang, "Modeling of the fractal-like adsorption systems based on the diffusion limited aggregation model," Journal of Molecular Liquids, 324, 114692, (2021).

[2] T. Oya, T. Asai, T. Fukui, and Y. Amemiya, "Reaction-diffusion systems consisting of single-electron circuits," Int. Journal of Unconventional Computing, 1, pp. 177-194, (2005).

[3] T. Oya, A. Schmid, T. Asai, Y. Leblebici, and Y. Amemiya, "On the fault tolerance of a clustered single-electron neural network for differential enhancement," IEICE Electronics Express, 2, pp. 76-80, (2005).

P2-5

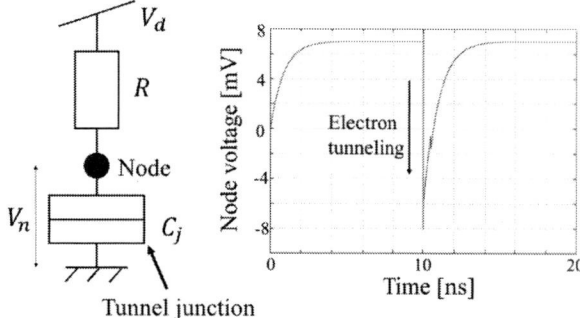

Fig. 1 Schematic of single-electron oscillator and its operation

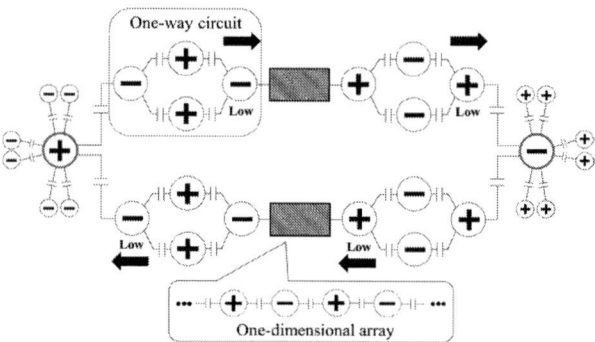

Fig. 4 Brownian motion circuit by single-electron oscillators

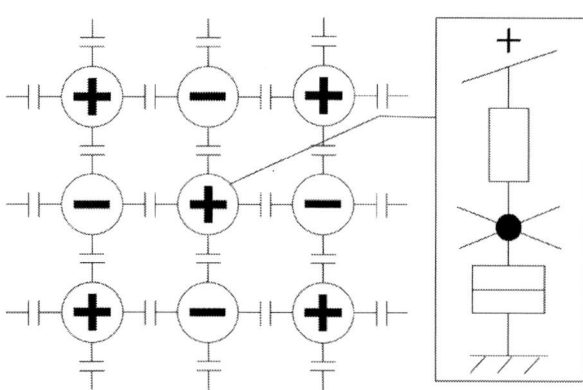

Fig. 2 Two-dimensional array of single-electron oscillators
(Configuration of oscillator is described in circle. "+" and "−" in circle indicates polarity of bias voltage for oscillators.)

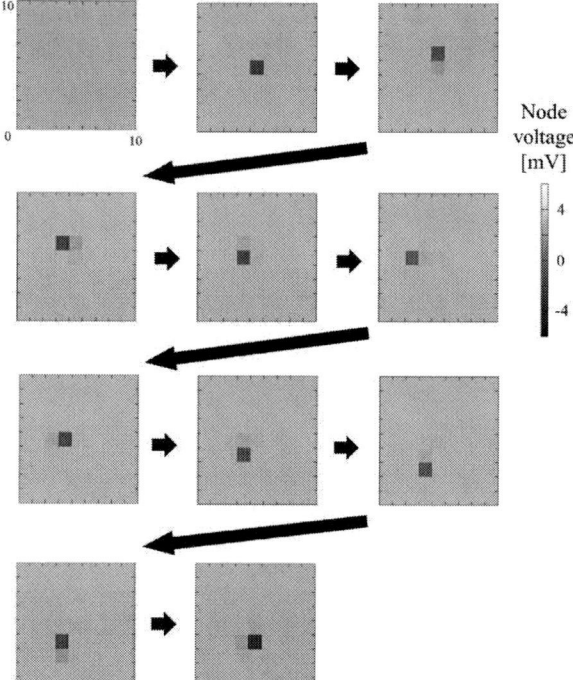

Fig. 5 Simulation result of two-dimensional Brownian motion circuit (10 × 10 red-circled oscillators shown in Fig. 4)

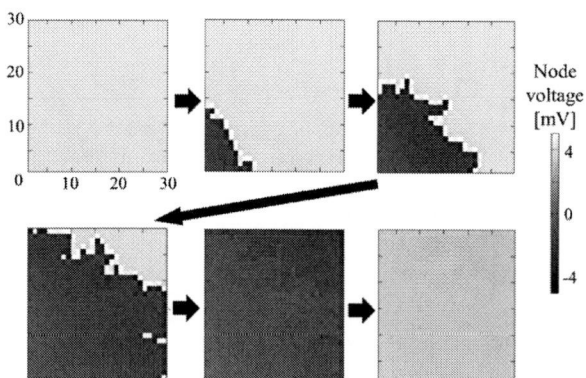

Fig. 3 Simulation result of two-dimensional array of single-electron oscillators (30 × 30 oscillators)
(White means high node voltage and black means low node voltage.)

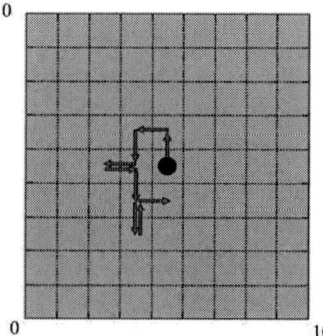

Fig. 6 Propagation transitions in Fig. 5 shown with arrows

979-8-3503-9164-0/24 $31.00 © 2024 IEEE   114

A Machine-Learning-based Model for Emerging Memories Featuring Multiple States

Zian Wang[1], Zhao Rong[2], Runsheng Wang[3], Mansun Chan[2], Lining Zhang[1]

[1]School of ECE, Peking University Shenzhen, China, [2]Department of ECE, Hong Kong University of Science and Technology, Hong Kong, China [3]School of Integrated Circuits, Peking University, Beijing, China

Email: eelnzhang@pku.edu.cn

Abstract — **Machine learning approaches have been studied for nanoelectronics device modeling, including emerging memories. In this work, a recurrent network is developed to cover the multi-state memory. A multi-state envelope hysteresis function (MSEHF) is constructed, for which a recurrent module is designed to reproduce. The sequential data for the network training is designated by combining different modes for all the state transitions. With a feedforward module for the nonlinear states, the obtained composite recurrent neural network is shown to capture key multi-state memory properties. While the proposed approach is also compatible with two-state memories, the machine learning model is applicable in circuit simulations with the desired high accuracy and correct functions.**

Keywords: Emerging memory, multi-state modeling, neural network.

I. INTRODUCTION

Machine learning methods are used to model emerging memories to accelerate the development cycle and improve the accuracy of their compact models [1], [2]. Nowadays, many kinds of emerging memories show the electrical characteristics of multi-value storage [3], [4], which means that different storage values can be identified in different resistance states under the same read voltage. Some previous works have proposed neural network based models for emerging memories[5], [6]. Up to now, a neural network based memory model that can support multi-value storage has not been proposed due to its complexity. In this work, we present a generic emerging memories model based on a non-linear auto-regressive with exogenous inputs (NARX) network, which enables multiple different storage states and their inter-switching. The model is data-driven with high accuracy and its capability of circuit simulations is also achieved.

II. DEVELOPING THE NEURAL NETWORK MODEL

The emerging memory is modeled using NARX because it needs to consider the impact of past historical information. The structure of the model based on the NARX network is shown in **Fig. 1**. R_{norm}^{t-i} represents the past output value at t−i moments, where the value of i is determined by hyperparameter optimization [7].

The training of the NARX-based neural network model has two main parts: recurrent module and feedforward module [13]. First, multi-state switching is trained using the recurrent NARX. As shown in **Fig. 2**, a multi-state envelope hysteresis function (MSEHF) is constructed according to the storage state of the memory. Read voltage (e.g., −1V in Fig. 2) is selected and resistance values of the memory are noted

under this voltage to be the value of MSEHF. Next, the training dataset of NARX is designed based on the state values of MSEHF, and the complete training dataset is shown in **Fig. 3(a)**. We need to design a mode for each case of resistive state switching. Examples of the design method are shown in **Fig. 3(b)(c)(d)(e)**. After NARX training is completed, the final model is generated by combining the experimental data through a feed-forward module to capture key multi-state memory properties.

III. RESULTS AND DISCUSSIONS

The proposed model in this work can be applied to memories with multiple storage states with high accuracy. **Fig. 4(a)** shows two-state RRAM model built with proposed method, which fits the experimental data well. **Fig. 4(b)** and **(c)** show the three-resistance state FTJ model based on the dataset of **Fig. 3(a)**. **Fig. 4(d)** shows the model of the four-resistance state FTJ corresponding to the MSEHF of **Fig. 2**.

The model presented in this work can be applied to circuit simulation. We implemented the model shown in **Fig. 4(b)** and **(c)** in Verilog-A and built a 2*2 1T1R array with it, as shown in **Fig. 5(a)**. **Figure 5 (b)** shows the operation of programming (set, reset) and reading in multi-resistor memory by applying the corresponding voltage signals to bit line, word line, and source line. The results show the model can correctly reflect the characteristics of the storage unit.

IV. CONCLUSION

A machine-learning-based approach is developed for modeling multi-state memories. The network structure and sequential dataset for training have been designed. Examples show the model's accuracy and functions, which is helpful for design technology co-optimizations of memories.

ACKNOWLEDGMENTS

This work was supported by the Natural Science Foundation of China under grants of 62074006, and 62125401, and by the ACCESS–AI Chip Center for Emerging Smart Systems under the InnoHK Funding, Hong Kong.

REFERENCES

[1] Wong, HS. Salahuddin, S., *Nature Nanotech* 10, 191, 2015.
[2] S. Yu and P. -Y. Chen, *IEEE Solid-State Circuits Magazine*, 8(2), 43, 2016.
[3] Andre Zeumault, et al. *J. Appl. Phys.*, 131(12), 2022.
[4] S. George et al. ISVLSI, 2016.
[5] A. S. Lin et al. *IEEE Access,* vol. 9, 3126, 2021.
[6] Z. Rong et al. *EDL*, 44(8), 1272, 2023.
[7] N. Mohajerin, S. L. Waslander, *TNNLS*, 30(11), 3370, 2019.

P2-6

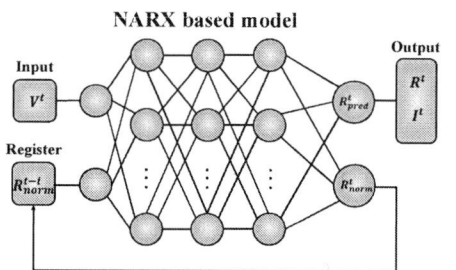

Fig. 1. Structure of a NARX-based neural network model. The output of the historical state is fed back to the inputs through registers, which together with the current inputs determine the state of the memory.

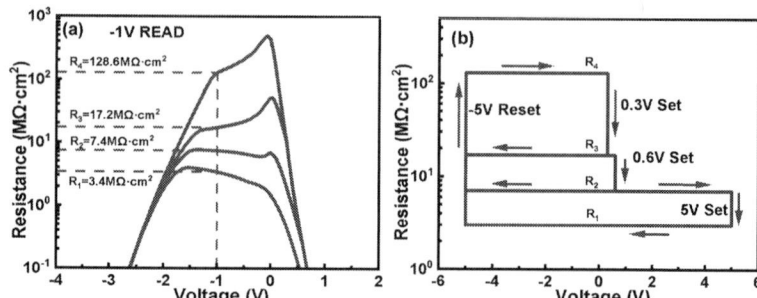

Fig. 2. (a) R-V characteristics and (b) MSEHF of memories, taking four-resistive FTJ as an example. The values of the resistance states should correspond to the resistance values in the experiment data or TCAD data.

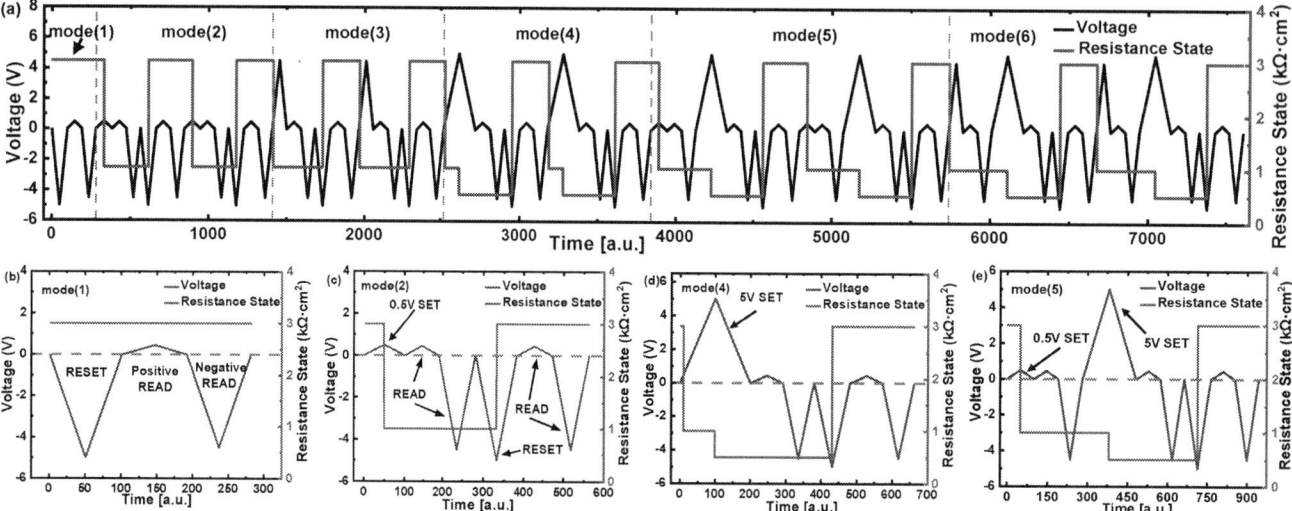

Fig. 3. (a) Training dataset of NARX for multi-state modeling, taking triple-resistive FTJ as an example. Initialisation is done first, and then each resistive state switching mode is repeated twice to improve training accuracy. (b) State switching mode 1. The device is set to high resistive state (RESET), and later read with positive and negative voltages, respectively. (c) State switching mode 2. The device is set to intermediate resistive state, then READ and RESET. (d) State switching mode 4. The device is set to low resistive state directly. (e) State switching Mode 5. The device is firstly set to the intermediate resistive state and then set to the low resistive state.

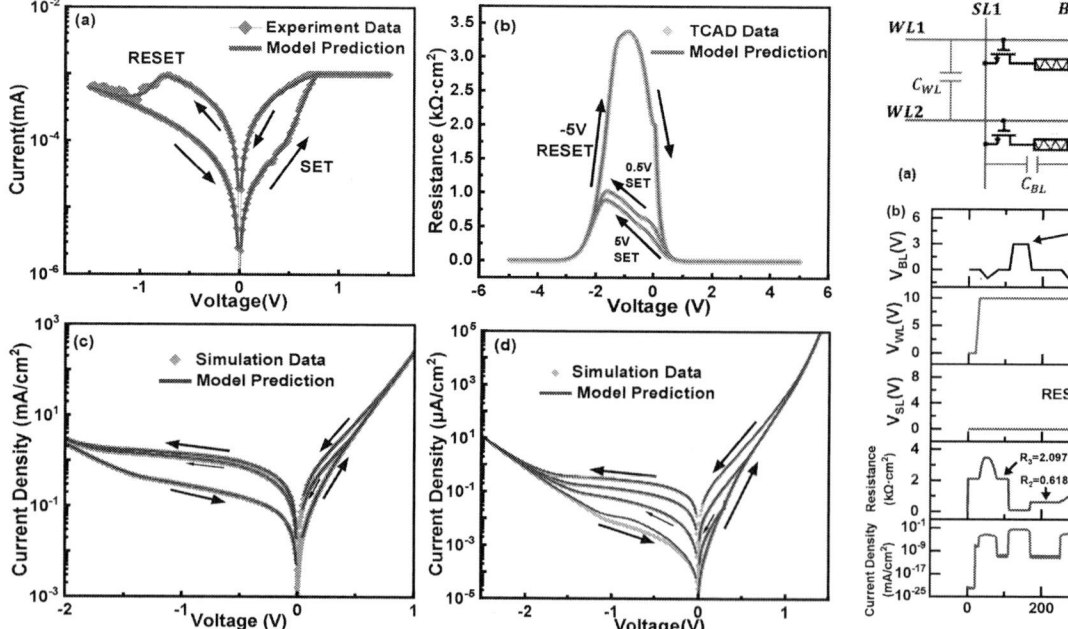

Fig. 4. Comparison of the test results of the model with the training data. (a) I-V characteristics of two-state RRAM. (b) R-V and (c) I-V characteristics of triple-resistance FTJ. (d) I-V characteristics of four-resistance FTJ.

Fig. 5. (a) A 2*2 1T1R array circuit schematic with BL, WL and SL. (b) Simulation results showing correct SET, RESET and READ operations.

979-8-3503-9164-0/24 $31.00 © 2024 IEEE   116

Automated Recipe Creation & Verification of Single Wafer Wet Etching: Ensemble Learning with Backcasting & Forecasting AIs using Scarce Data

Koki Shibata[1], Hiroshi Horiguchi[2], Chihiro Matsui[1], and Ken Takeuchi[1]

[1]Dept. of Electrical Engineering and Information Systems, The University of Tokyo, Tokyo, Japan
[2]Application Software Engineering Operations, SCREEN Semiconductor Solutions Co., Ltd., Japan
koki.shibata@co-design.t.u-tokyo.ac.jp

Abstract — Towards smart semiconductor fabrication, the automated process recipe determination with AI is required under scarce data (NOT big data) conditions. This paper proposes precise recipe determination by ensemble learning combining backcasting & forecasting AIs. Proposed backcasting AI predicts etching recipe to optimize the equipment arm motion, traditionally set manually, to achieve the desired etching amount. Automatically estimated etching recipe is optimized in real time and both throughput and yield improve. By means of incorporating additional features such as the differential and variance of etching results and employing ensemble learning with multiple neural networks, RMSE (Root Mean Squared Error) of the objective variable representing the periodic motion of the arm reduces by more than 40%. In addition, forecasting AI that predicts etching results from a recipe made by backcasting AI automatically validates the recipe.

Keywords: Process optimization, Wet etching, Ensemble learning, Scarce data, Backcasting AI, Forecasting AI

I. INTRODUCTION

With the scaling of semiconductor devices, real-time precise control of the etching is becoming increasingly important. Traditionally, the creation of recipes relied on the experience of experts, and recipes have been verified by obtaining actual etching results (Fig. 1). However, process optimization by machine learning is attracting attention [1, 2]. In this paper, backcasting AI (Fig. 2) and forecasting AI (Fig. 3) are combined to automatically create recipes from etching results and validate the estimated recipes. The backcasting (forecasting) AI are trained on a small number of actual recipe/result pairs, to provide recipes (etching results) from results (recipes) (Fig. 1). Combining backcasting & forecasting AIs in cyberspace can produce promising recipes at extremely low cost (Fig. 1). Proposed backcasting AI adopts jackknife-based ensemble learning to build a robust and reliable recipe from a small data set of 144 samples (Fig. 4).

II. RECIPE CREATION BY BACKCASTING AI

In the wet etching process, the etchant is sprayed onto the rotating wafer from the tip of a nozzle that moves in a cyclic manner (Fig. 5). The distance from the wafer center to the tip where the etchant is injected is referred to as the nozzle position. Fig. 6 shows dataset of measured etching volume and corresponding arm motion. In the backcasting AI, the periodic motion of the arm as the objective variable is represented by an 8-dimensional vector. The eight variables are the minimum and maximum points, the period T, and the arrival time T1-T5 of the arm at the point where the amplitude is divided into six equal parts (Fig. 7). Fig. 8 shows the difference and variance features extracted from the etching results. These features are combined with the raw etching results and input to a backcasting AI for training (i.e., feature engineering). Fig. 9 shows hyperparameters in the difference and variance features. These hyperparameters are tuned by grid search among [50, 150, 500] options, resulting in width of diff = 500, stride of diff = 50, width of var = 150, and stride of var = 150.

The proposed backcasting AI employs ensemble learning. Multiple (such as 50) models are operating in parallel, and the mean and median are calculated from a set of predictions to obtain the final prediction result. The ensemble learning is effective in dealing with diverse data patterns and reducing the effects of noise and outliers [3]. In this paper, the ensemble learning uses a jackknife method to select subsets of training data so that each subset is unique and free of duplicates. Each training dataset is trained on a single neural network. This single NN model has two fully-connected layers with 256 hidden nodes. Simple two-layer NN is preferred to prevent overfitting, when dealing with small datasets. If overfitting happens, NN model memorizes the training data and introduces noise and anomalies that degrades the inference accuracy of new unknown inference data. By operating multiple models in parallel, the impact of noise and anomalies are mitigated, leading to more reliable inference, especially with limited training data. Fig. 10 shows RMSE versus the number of models for ensemble learning. The finally converged RMSE is smaller when the median rather than the mean is employed for the objective variables because the median is more robust to outliers than the mean. As a result, RMSE of the objective variables improves by 0.029 compared to the case without feature engineering, and furthermore by 0.021 compared to the case without ensemble learning (Table I). The final estimated recipe is shown in Fig. 11.

III. RECIPE VERIFICATION BY FORECASTING AI

The conventional forecasting model transforms the nozzle arm scanning movement to the accumulated dispense time [4]. A challenge with the previous model [4] is that the accumulated dispense time exhibits extremely large values at the positions where the arm motion reaches its minimum and maximum points, leading to too wide dynamic range. To address this issue, this paper introduces a data pre-processing technique known as clipping, which aims to mitigate the impact of outliers [5]. In Lower Clipping, data pre-processing is applied to round values below the bottom 5% to 0. In Upper Clipping, data pre-processing is applied to round values above the top 5% (Fig. 12). When both Upper and Lower Clipping (Both Side Clipping) are applied, RMSE and MEP (Mean Error Percentage) decrease by 0.0022 and 0.57, respectively (Table II). Fig. 13 compares the etching result predicted by the forecasting AI with the original measured outcome. This comparison is made by converting the recipe predicted from the measured etching result, using backcasting AI. By using backcasting AI for the recipe creation and forecasting AI for the recipe verification, Both Side Clipping improves RMSE from 0.088 to 0.069, which is closer to the measured etching result. These results show that highly accurate NN models are obtained from a small amount of data.

IV. CONCLUSION

This paper proposes a wet etching optimization for creating etching recipes using backcasting AI and for verifying recipes using forecasting AI. After training from only 144 samples, the estimated etching results achieve RMSE of 0.069. Real time automatic recipe determination becomes feasible for wet etch production lines that have been relying on human hands.

REFERENCES

[1] D. M. Fried, *EDTM*, 2023, pp. 241-243. [2] C. C. Chu, *VLSI Symposium* 2022 Workshop. [3] M.A. Ganaie, et al., *EAAI*, 2022, vol.115, pp.105151. [4] S. Yoshikiyo, et al., *SNW*, 2023, pp. 2-3. [5] C. Sakr, et al., *ICML*, 2022.

979-8-3503-9164-0/24 $31.00 © 2024 IEEE

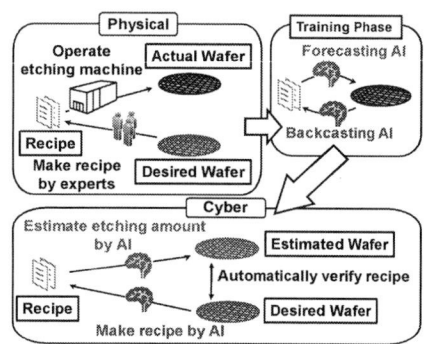

Fig. 1 Proposed backcasting & forecasting AIs. Real-world data trains backcasting AI models for etching recipe creation. Forecasting AI verifies etching recipe.

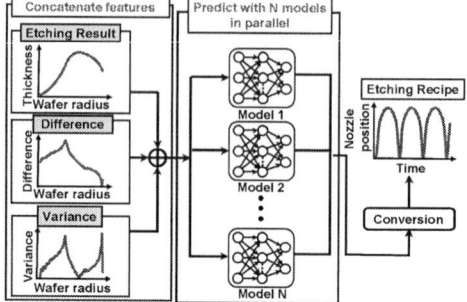

Fig. 2 Proposed ensemble backcasting AI. Features such as difference and variance are added for training. Final recipes are created by referring to mean or median (see Fig.10).

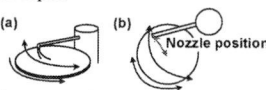

Fig. 3 Proposed clipped forecasting AI model. Arm movements are changed to accumulated dispense time and 5% clipping above and below is used as NN input.

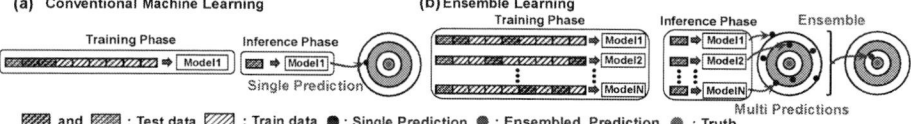

Fig. 4 Ensemble learning with scarce data. (a) Conventional machine learning with test/train split, providing a single prediction per data point (blue block). (b) Proposed ensemble learning, where multiple subsets of data are non-recursively extracted to train multiple models, resulting in numerous predictions for one data point (blue block).

Fig. 5 Nozzle arm scanning in wet etching. (a) Diagonal above view. (b) Direct overhead view.

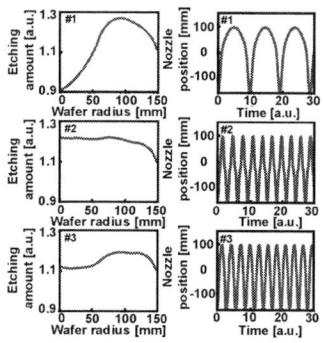

Fig. 6 Measured etching results and periodic nozzle arm scanning motion of #1, 2 and 3.

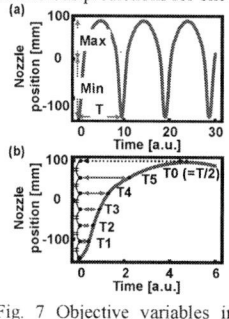

Fig. 7 Objective variables in backcasting AI. (a) Minimum and maximum positions and period T. (b) Time at 6th quartile point of amplitude.

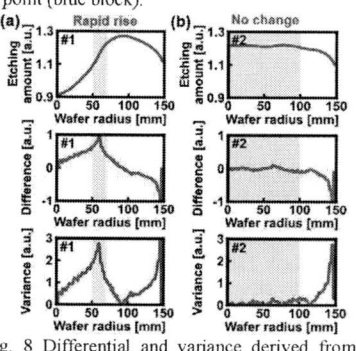

Fig. 8 Differential and variance derived from measured etching results. (a) Significant etching increases diff and var. (b) Unchanged measured etching results lead to minimal diff and var.

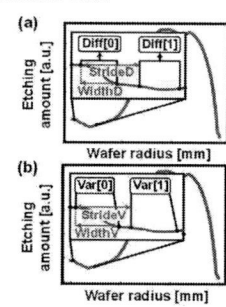

Fig. 9 Hyperparameters in explanatory variables in proposed backcasting AI. (a) Stride and width in difference features (StrideD and WidthD). (b) Stride and width in variance features (StrideV and WidthV).

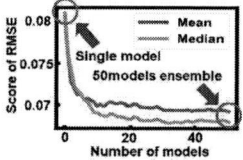

Fig. 10 Decreasing RMSE with increasing ensemble models. Comparison between adopting mean and median as the overall prediction from multiple sets of predictions. Using median achieves smaller RMSE.

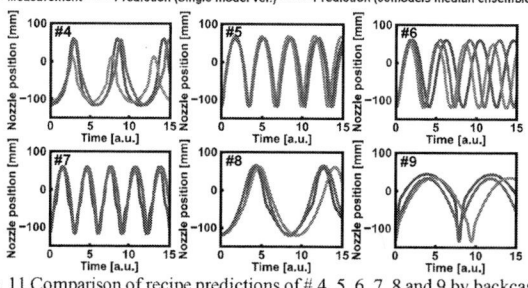

Fig. 11 Comparison of recipe predictions of # 4, 5, 6, 7, 8 and 9 by backcasting AI with and without ensembles. 50 models with median ensemble well reproduces the real measured recipe.

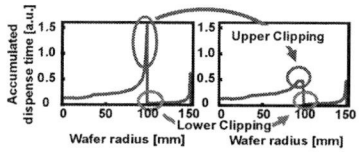

Fig. 12 Upper Clipping and Lower Clipping. Values above the upper 5% are rounded (Upper Clipping) and below the lower 5% are rounded to 0 (Lower Clipping).

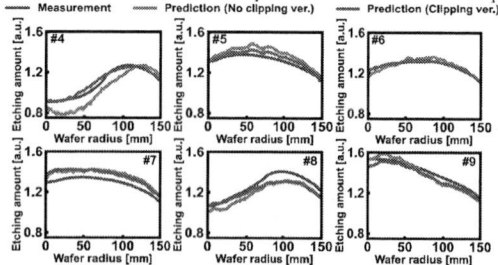

Fig. 13 Comparison of etching results verified by forecasting AI based on the recipe created by backcasting AI (Fig. 11). By using Both Side Clipping, predicted etching amount well reproduces measured results.

Table I Prediction accuracy of backcasting AI for etching recipe creation

Feature	Raw	Raw+Difference+Variance		
Type	Single model	Single model	50 models (mean)	50 models (median)
RMSE	0.1181	0.0890	0.0693	0.0680

Over 40% Decrease

$RMSE = \sqrt{(p - \hat{p})^2/8}$
p: True wave features
$\hat{p}$: Predicted wave features

Table II Prediction accuracy of forecasting AI for etching recipe verification

	Conv.	Lower Clip	Upper Clip	Both Side Clip
RMSE	0.0541	0.0551	0.0527	0.0519
Norm. RMSE	0.0306	0.0304	0.0297	0.0296
MEP [%]	11.39	11.2	10.83	10.82
Norm. MEP [%]	7.68	7.29	7.04	6.93

$RMSE = \sqrt{(E - \hat{E})^2/150}$
$MEP = \max(|E - \hat{E}|/E \times 100)$
E: Measured etching amount
$\hat{E}$: Predicted etching amount

979-8-3503-9164-0/24 $31.00 © 2024 IEEE

Design of the 2-nm Nanosheet NAND-type TCAM with High Speed and Compat Cell-size: 45% Layout-reduction of 3-nm TCAM

L.-A. Yu[1], C.-Y. Chou[1], C.-C. Lin[1], T.-Y. Tsai[1] and E Ray Hsieh[1,*]

[1]Dept. of Electrical Engineering, National Central University, Taoyuan, Taiwan

*E-mail: erayhsieh@cc.ncu.edu.tw

Abstract — For the first time, we design a high-performance NAND-type TCAM cells with low-power consumption and high operation speed. To realize these cells, important electrical characteristics I_dV_{gs}, I_dV_{ds}, I_{on}, S.S., and V_{th} of FinFETs and Nanosheet (NS) MOSFETs have been simulated. Furthermore, we have found that better performance of NS-inverter, compared to that of the FinFET inverter. Next, the RSNM and WNM of FinFET- and NS-based 6T SRAM cells have been investigated as well, and the design guideline of the 6T SRAM has been given; thus, the NAND-type TCAM with FinFET and NS devices have been constructed. We have found that the pull-up the pull-down and the pass-gate with the ratio of 1:3:3 is the optimized design for high-performance purpose. Then, we compare the operation speed and the power of the FinFET and NS TCAM cells. The results show that the layout of the NS TCAM cells can save 45% layout area, compared to that of the FinFET TCAM.

Index Terms — Nanosheet, FinFETs, SRAM, 2-nm CMOS Devices

I. INTRODUCTION

As the CMOS devices rapidly shrink with the Moore's law scaling, to develop the next in-memory-searching technologies and improves the performance for high-speed and low-latency 6G communication, in this paper, we design the FinFET [1] and NS MOSFET [2] using 2-nm CMOS technology and the 6T-SRAM[3] cell, as well. Then, we firstly propose the NAND-type TCAM [4] based on the 2-nm CMOS devices. Furthermore, we evaluate the transfer curve of FinFET- and NS- based inverter, SRAM and TCAM. We then optimize the ratio of Pull-up (PU): Pull-down (PD): Pass-gate (PG) as 1:3:3 to enhance the RSNM and WNM. Most importantly, the operation speed and the power consumption of the both TCAM designs have been evaluated, and the NS-one is superior to that of the FinFET-one. Finally, the layouts of the TCAM unit cells have been constructed, . These design guide-lines should be very useful for the future high-speed in-memory-searching to meet the need of the 6G communication.

II. SPEC OF THE FINFET AND NANOSHEET MOSFET

Table. (1) shows the physic parameters of FinFETs and Nanosheet (NS) MOSFETs, **Fig. 1 (a)** to **(b)** shows the structure of FinFET and Nanosheet MOSFET with the Si-channel and the SiO2 (IL)/ HfO2 (Gate-dielectric) Oxide layers, whose E.O.T is 1-nm.

Structural Parameters	FinFETs	Nanosheet MOSFET
S/D Doping (/cm^3)	8.00E+19	
SDC Doping(/cm^3)	8.00E+19	
Channel Doping (/cm^3)	1.00E+17	
Body Doping (cm^3)	5.00E+18	--------
Fin_Width (nm)	5	--------
Fin_Hight (nm)	42	--------
Sheet_Width (nm)	--------	24
Sheet_Hight (nm)	--------	3.3(Ptype)&3.5(Ntype)
EOT (nm)	1	0.75

Table. (1) the physical parameter data of the two structures

III. RESULTS AND DISCUSSIONS

A. Electrical Characteristic of the 2-nm Nanosheet Devices

Fig. 2 to **Fig. 5** are the electrical characteristics of the I_d-V_{gs} and the I_d-V_{ds} for the FinFETs and Nanosheet MOSFETs, respectively. **Fig. 6** to **Fig. 9** shows the performance, including the S.S., Ion, V_{th}-roll-off, and the DIBL of the NS devices is better than those of the FinFET since the gate-all-around structure of Nanosheet improves the gate-to-channel

controllability than the gate-covered Fin-channel on 3-side-walls in the FinFETs.

B. The Design of the 2-nm 6T SRAM Cells

Fig. 10 shows the electrical characterizes of the inverter transfer curves of the FinFETs and NS-MOSFETs. The performance of Nanosheet is superior to that of the FinFET thanks to the higher I_d-on/off ratio. **Fig. 11(b)** shows the RSNM of the NS-MOSFET is larger than that of the FinFET according to a stronger pull-down capability provide by a higher I_{on} of the n-type NS-MOSFETs. **Fig. 12** and **Fig. 13** show the comparison of the RSNM and WNM with different operating-voltages (V_{dd}). To further improve the RSNM and WNM, we design different size-compositions of the PG, PD, PU devices (2:2:1 ;2:2:2 ; 3:3:1; 3:3:2). The PG/PD ratio = 1, so the RSNM remains stably; meanwhile, by increase in the PG and PU size, the WNM is improved. **Fig. 14** also confirmed this trend. Therefore, we chose the ratio of 3:3:2 to construct the NAND-type TCAM cell.

C. The 2-nm NAND-type TCAM Design

Fig. 15 shows the circuitry of the NAND-type T-CAM cell. We first extract the read and write power of the TCAM cell in **Fig. 16**. The power of NS one is bigger than that of the FinFET because of its larger I_{on}. Next, we will study the correlation of the loading capacitance between the operation-speed, energy consumption. **Fig. 17** shows as the capacitance increases, the energy will not be affected sharply. On the other hand, **Fig. 18** shows the operation-speed will be retarded. As a result, it is found that the operation-speed of the FinFET TCAM will be much slower, affected by the loading capacitance than that of the NS-TCAM. **Fig. 19** shows the layout design of TCAM cell, composed of FinFETs. **Fig. 20** is the standard layout design of the T-CAM with NS devices. **Fig. 21** is another layout design of the NS- TCAM with the STI-saving design. The layout of the NS TCAM cells can save 45% layout area, compared to that of the FinFET TCAM.

IV. CONCLUSION

In conclusion, the high-speed and low power 2-nm NAND-type TCAM with the Nanosheet MOSFETs and FinFETs have been thoroughly designed, and the design issues have been carefully considered. Thanks to the better on/off ratio of the I_d, the Nanosheet TCAM has better performance than that of the FinFET TCAM in the 2-nm technology. The RSNM of the Nanosheet TCAM is also wider than that of the FinFET one. Furthermore, the operation speed of Nanosheet TCAM is faster, meanwhile, with a smaller layout-area (45% smaller) of the TCAM cell than that of the FinFET TCAM. As a result, the 2-nm Nanosheet NAND-type TCAM cell design will be a charming candidate of the in-memory-searching technology for the future low-latency communication in the 6G-era. (Table2)

ACKNOWLEDGEMENT

This work was supported by the National Science and Technology Council, Taiwan, under NSTC 112-2628-E-008-003.

REFERENCES

[1] Y. Junting , C. Shuming, C. Jianjun, H. Pengcheng. RADECS, 2016, pp. 1-4

[2] G. Rzepa, Krishna K. Bhuwalka, O. Baumgartner, D. Leonelli, Hui-Wen Karner, F. Schanovsky, C. Kernstock, Z. Stanojevic, Hao Wu, F. Benistant, C. Liu, M. Karner., IEDM, 2022, pp. 15.1.1-15.1.4

[3] Y. Ma, L. Zhang, J. Liu., ICICM, 2020, pp. 251-254

[4] Meng-Chou Chang, Kai-Lun He and Yu-Chieh Wang., GCCE, 2014, pp. 358-359

979-8-3503-9164-0/24 $31.00 © 2024 IEEE

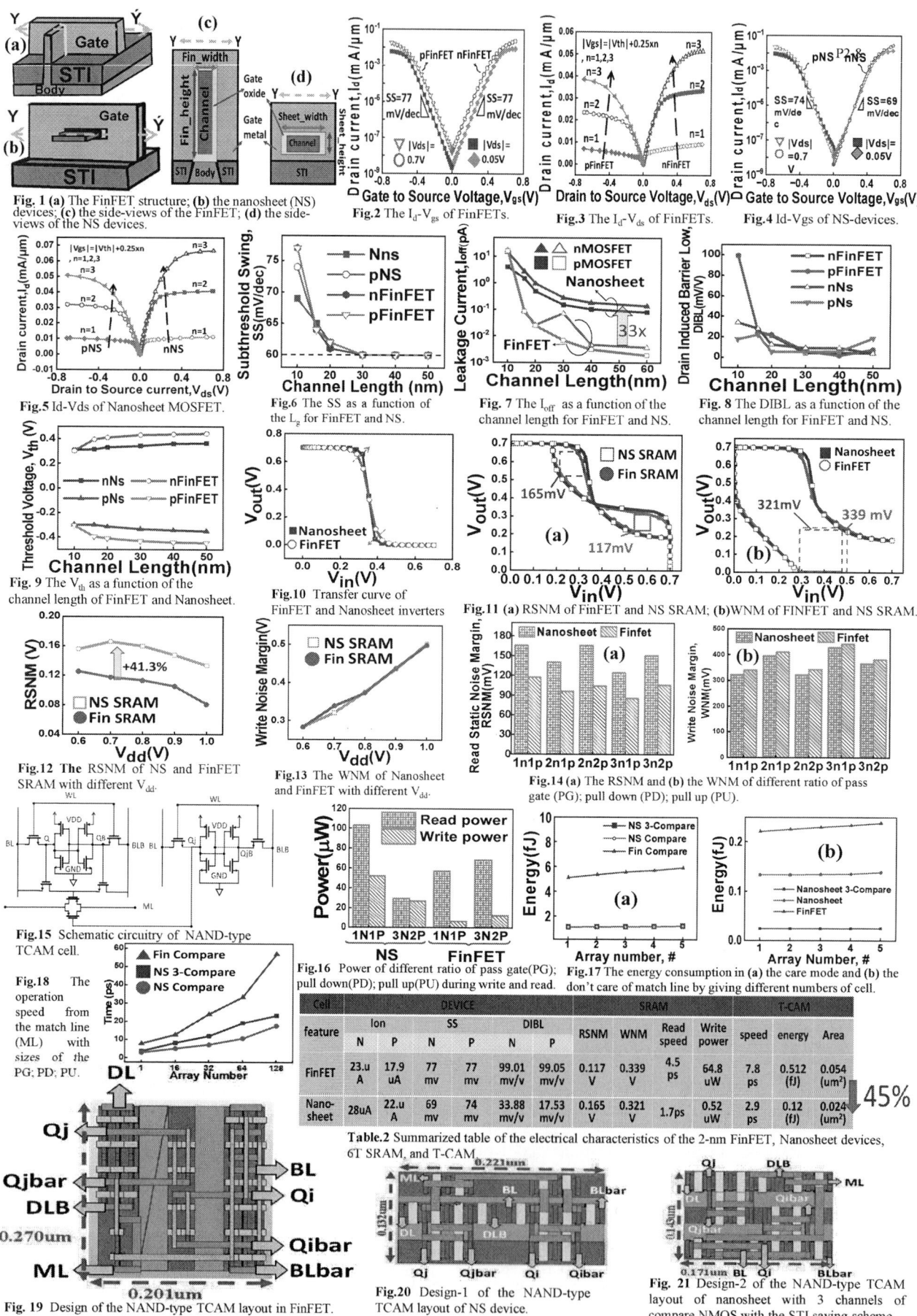

Design of single-electron information-processing circuit for particle computation

Soki Mizuno, Takahide Oya

Graduate School of Engineering Science, Yokohama National University
Tokiwadai 79-5, Hodogaya-ku, Yokohama, 240-8501, Japan
Phone: +81-45-339-4125, E-mail: soki-mizuno-gb@ynu.jp

Abstract — We propose a unique single-electron (SE) circuit to perform particle computation. The circuit is composed of "commanding direction for particles" circuits and "collision detection" circuits, both of which are composed of single electron oscillators (SEOs). In this study, we designed and simulated a 2-input particle computation logic gate with two particles and two commands. As results, we confirmed that our circuit operated correctly as desired.

I. INTRODUCTION

We propose a unique single-electron (SE) circuit to perform particle computation. SE circuits are those that can control electrons individually. Also, since they use tunneling effect, they have the characteristics of stochastic and nonlinear operation. However, the optimal information processing method for SE circuits has not yet been established. Therefore, we focused on particle computation [1], which is the information processing method using particles and obstacles.

In particle computation, a planar grid workspace containing a number of unit-size particles and some fixed unit-square obstacles. Each particle occupies one grid cell. All particles are commanded in unison. The commands are "Go Up" (u), "Go Right" (r), "Go Down" (d), and "Go Left" (l). The particles all move in the commanded direction until forward progress is blocked by a stationary obstacle or another blocked particle. Through this repetition, information processing such as logical computation is performed (Fig. 1). Particle computation has great potentiality because it can realize various types of information processing depending on the arrangement of particles and obstacles. We believe that proposed new SE information processing circuit with applicability can be realized by imitating particle computation. In this research, we aim to implement particle computation in the SE circuits. As a first step, we design an SE information processing circuit that performs logic computation.

II. CIRCUIT DESIGN AND SIMULATION

In this study, single electron oscillators (SEOs), which consist of a resistor R and a tunnel junction C_j connected in series (Fig. 2), are mainly used for the circuit design. It is known that, by connecting SEOs to each other, a change of node voltage V_n of each SEO due to electron tunneling occurs in a chain-reaction manner [2].

To represent particle computation, we designed "commanding direction for particles" circuits that move particles in the commanding direction and "collision detection" circuits that records the particle's stationary position. Commanding direction circuits can be constructed by connecting SEOs that represent the position of the particle in one direction that the propagation of the voltage change is in one direction and arranging them together (Fig. 3). The movement of one particle can be represented by the voltage change in one row of the commanding direction circuits. Obstacles can be represented by setting the bias voltage of the SEO representing the position of the particle to 0 mV. In this way, a maze of particle computation is configured in the commanding direction circuits. On the other hand, the collision detection circuits adjust the bias voltage of the SEO so that can record the position where the particles stop in the commanding direction circuits (Fig. 4). The locations where the voltage change occurs in the collision detection circuits show the locations where the particles stop. A part of commanding direction circuits and collision detection circuits can conduct one command.

In this study, we configured and simulated a 2-input particle computation logic gate with two particles on two commands, one in the up direction and the other in the right direction (Fig. 5). Since there are two instructions, two pairs of commanding direction circuits and collision detection circuits were prepared. This time, A, B, $\bar{A}$, and $\bar{B}$ were given as inputs to perform the logic operation. The results when A and B are given as inputs are shown in Fig. 6. The simulation results were as expected, and the logic computation using particle computation was able to be executed. In the future, we plan to design a circuit structure with an increased number of particles and instructions.

ACKOWLEDGMENTS

This work was partly supported by JSPS KAKENHI Grant No. JP23H00169, and Kayamori Foundation of Information Science Advancement.

REFERENCES

[1] Becker, A.T., et al., *Nat Comput* vol.18, pp.181-201, (2019).
[2] Oya, T., et. al., *Int'l J. Unconv. Comp.*, vol. 1, pp. 177-194, (2005).

Commands: < d, l, r, d >

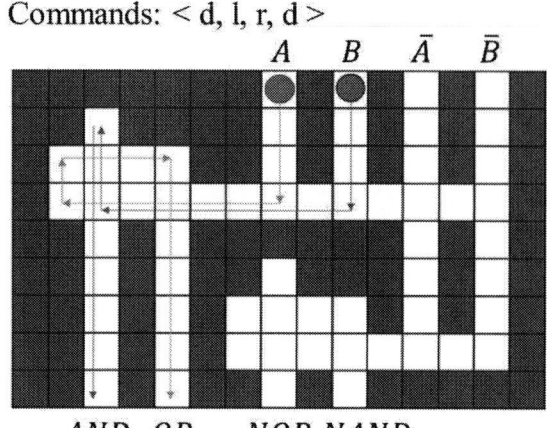

Fig. 1. Example of 2-iput logic gate using particle computation

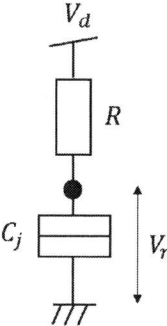

Fig. 2. Circuit configuration of single-electron oscillator

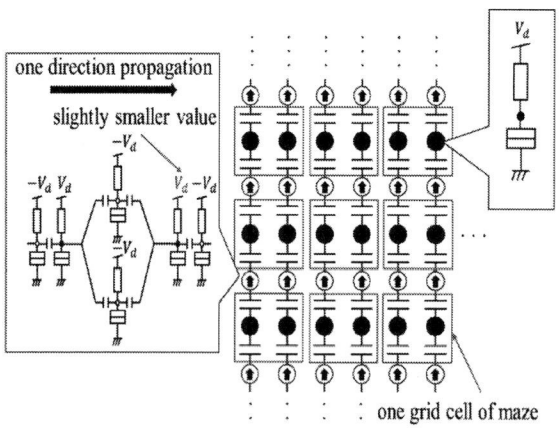

Fig. 3. Schematic of commanding direction circuits

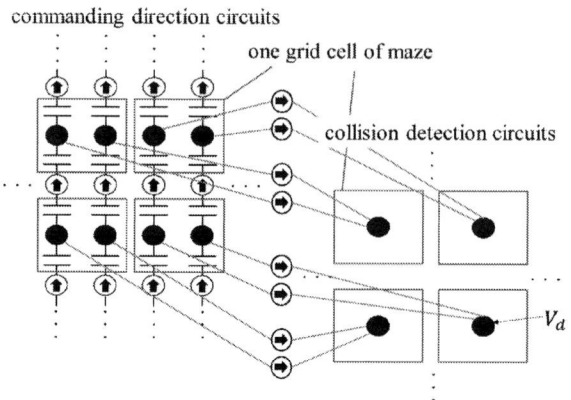

Fig. 4. Structure of collision detection circuits

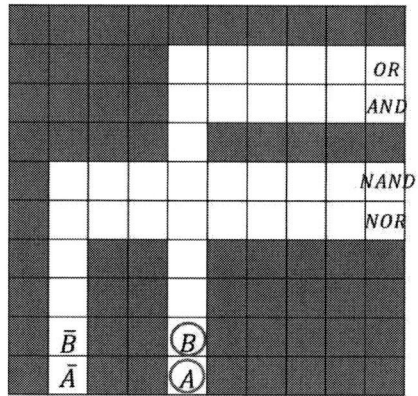

Fig. 5. The maze for 2-input particle computation logic gate with two particles and two commands (commands: u, r)

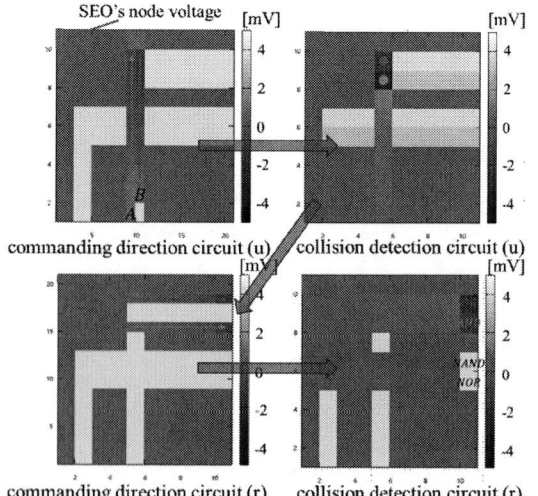

Fig. 6. Simulation results from designed circuit with input A and B.

979-8-3503-9164-0/24 $31.00 © 2024 IEEE   122

Predicting the Retention Property of Scaled Cylindrical IGZO 2T0C DRAM Cells

Sang-Mok Jeong[1] and Sung-Min Hong[1]

[1]School of Electrical Engineering and Computer Science, Gwangju Institute of Science and Technology, Republic of Korea
Email: smhong@gist.ac.kr

Abstract —A cylindrical IGZO 2T0C DRAM has attracted much interest due to its potential scalability and superior retention characteristics. However, the experimental devices reported up to now have relatively large areas. In this work, we predict the retention property of scaled cylindrical IGZO 2T0C DRAM cells.
Keywords: 2T0C DRAM, IGZO, retention, TCAD

I. INTRODUCTION

The Dynamic Random Access Memory (DRAM) has been the primary storage device for computing systems and digital devices since its invention in the 1970s [1]. However, the conventional one-transistor-one-capacitor (1T1C) DRAM has reached a critical scaling limit [2].

A cylindrical two-transistor (2T0C) indium-gallium-zinc-oxide (IGZO) DRAM is an attractive candidate for the next generation DRAM. **Fig. 1** compares the schematics of a conventional 1T1C DRAM and a 2T0C DRAM. Instead of a dedicated capacitor, the source of the write transistor (Tw) is used as a storage node (SN), which also acts as the gate of the read transistor (Tr). Due to its relatively small capacitance, the SN is vulnerable to the leakage current. Since IGZO has a wide bandgap, it is expected that the IGZO DRAM exhibits a much lower leakage current than its silicon counterpart. By stacking two transistors (Tw and Tr) vertically and adopting the channel-all-around (CAA) structure, the cell area can be potentially reduced down to $4F^2$ [3-5].

However, the experimental results reported up to now have been limited to large devices. Therefore, the prediction of the retention property of a scaled cell is urgently required. Also, the SN voltage suffers from the coupling effect, as shown in **Fig. 2**, which is difficult to estimate in an analytic way. In this work, by using an in-house device simulator, the electrical performance of a scaled cylindrical IGZO 2T0C DRAM cell is systematically investigated.

II. RESULT AND DISCUSSION

Important geometric parameters drawn in **Fig. 3** are considered in this work. Based upon the well-calibrated in-house device simulator (**Fig. 4**), impacts of various structure parameters on the retention property have been investigated. When a certain parameter is varied, all other five parameters are set to be their reference values. In the DC simulation, the drain voltage is 0.1 V. The off current has been extracted at $V_{gs} = -1.0$ V and $V_{ds} = 0.1$ V.

Impacts of the CD are discussed. As shown in **Fig. 5**, the drain current (normalized with the effective width) is heavily affected by the CD. Especially, a high off current is observed for a small CD value. It means that the CD scaling introduces detrimental effects. Increased off currents at small CD values can be easily understood from the internal physical quantity shown in **Fig. 6**. Although all other parameters are identical between two devices, the device with a CD of 50 nm exhibits much higher (undesirable) electron density even at $V_{gs} = -1.0$ V. In the cylindrical IGZO 2T0C DRAM considered in this work, due to the CAA structure, the gate controllability is degraded with a small CD, because the ratio of the IGZO channel volume to the gated interface area is increased. As an extreme case, when the CD becomes vanishingly small, the gate contact can hardly control the channel. Another quantity related with the write/read operations is the initial SN voltage right after the write phase. Due to the coupling effect depicted in **Fig. 2**, the gate voltage of the Tw affects the SN voltage. For various CD values, the SN voltage is calculated in **Fig. 7**. Since the coupling capacitance between the Tw gate and the SN is reduced with a small CD, basically the coupling effect is also reduced with a small CD. However, when the CD is further scaled, an excessive leakage current drastically reduces the initial SN voltage. Therefore, the CD scaling eventually introduces detrimental effects to the initial SN voltage as well as the leakage current. In **Fig. 8**, the retention time is drawn as a function of the CD. As a result, a rapid decrease of the retention time is observed at small CD values (50 nm ~ 100 nm).

IV. CONCLUSIONS

In summary, when the CD of the cylindrical IGZO 2T0C DRAM cell is scaled, the off current is significantly increased due to higher curvature. According to the simulation results obtained by changing other geometric parameters (not shown), we need a long channel (which increases the cell area), a thin gate oxide (which increases the capacitance between the SN and the Tw), and a thin IGZO channel (which suffers from a variability issue), in order to keep the off current reasonable low. Simultaneously, the initial SN voltage should be maintained in a reasonable range. All of these considerations introduce non-trivial challenges for the CD scaling.

ACKNOWLEDGMENTS

This research was supported by the NRF of Korea grant funded by the Korea government (NRF-2023R1A2C2007417).

REFERENCES

[1] R. H. Dennard, Nature Electronics, vol. 1, p. 372, 2018.
[2] S. Shiratake, in International Memory Workshop, pp. 1-3, 2020.
[3] X. Duan et al., IEEE Transactions on Electron Devices, vol.69, pp. 2196-2202, 2022.
[4] Q. Chen et al., IEEE Electron Device Letters, vol. 43, pp. 894-897, 2022.
[5] C. Chen et al., in International Electron Devices Meeting, pp. 615-618, 2022.

979-8-3503-9164-0/24 $31.00 © 2024 IEEE

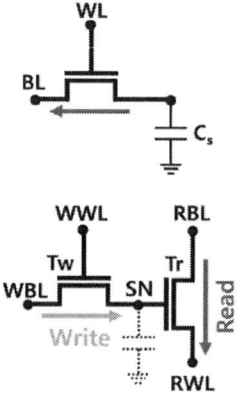

Fig. 1. 1T1C DRAM (top) and 2T0C DRAM (bottom).

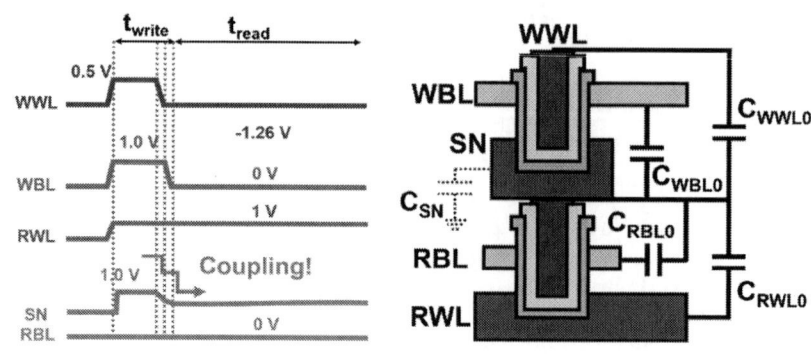

Fig. 2. SN voltage lowering due to the coupling effect (left) and parasitic capacitances between different electrodes (right).

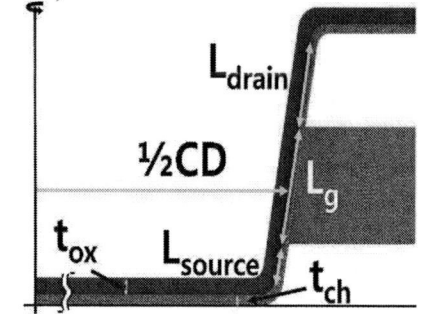

Fig. 3. Cylindrical 2T0C IGZO DRAM structure based on [5].

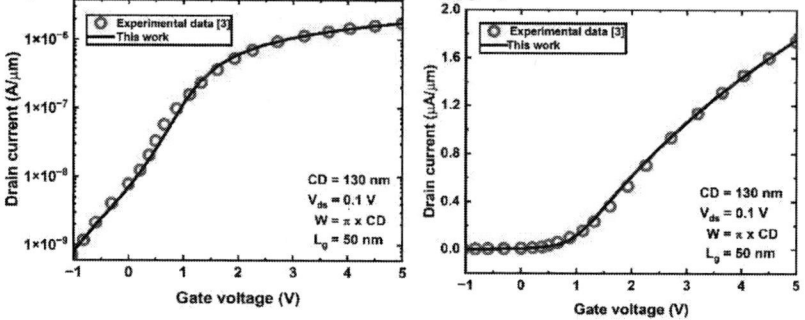

Fig. 4. Input characteristics of a device with CD = 130 nm in the semi-logarithmic scale (left) and the linear scale (right).

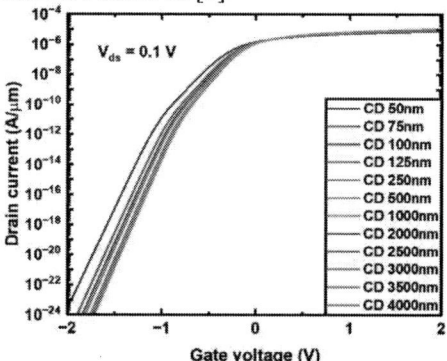

Fig. 5. Input characteristics of the IGZO DRAM for several CD values.

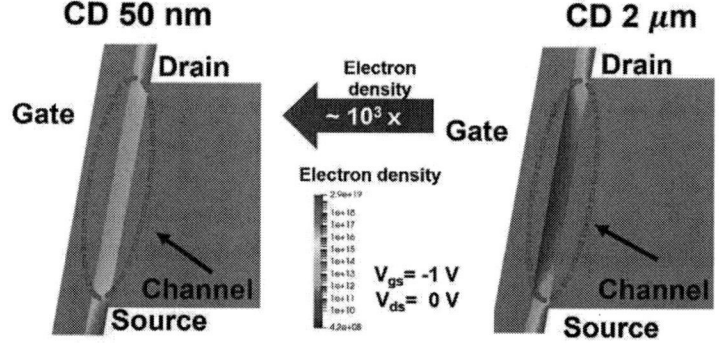

Fig. 6. Electron density profiles of two devices with CD = 2 μm and CD = 50 nm. Much higher electron density in the channel region is observed when CD = 50 nm.

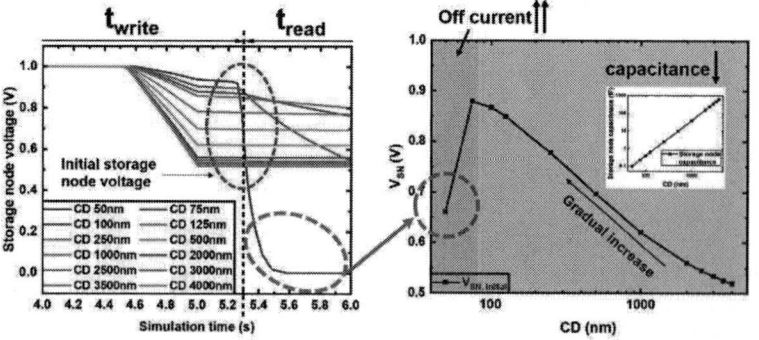

Fig. 7. Timing diagram of the SN voltage for several CD values (left) and the initial SN voltage as a function of the CD (right).

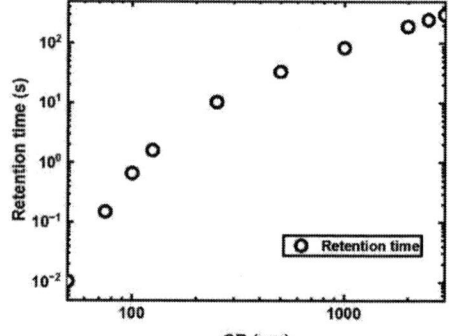

Fig. 8. Retention time as a function of the CD.

979-8-3503-9164-0/24 $31.00 © 2024 IEEE   124

Gap in pagination due to withheld paper.

Pages 125-126

Simulation of Trap-Induced Noise Characteristics in 3-nm Complementary FET

Jinghan Xu, Zheng Zhou, Fei Liu, Xiaoyan Liu

School of Integrated Circuits, Peking University, Beijing, 100871, China.

Email: liuxiaoyan@pku.edu.cn

Abstract — **This study investigates trap-induced noise characteristics in a 3-nm node complementary FET transistor (CFET) through TCAD simulations and theoretical calculations. A noise power spectral density (PSD) model tailored for the 3D gate-all-around structure is developed. Through theoretical calculations, the dependence of noise PSD on trap profiles and bias conditions is investigated.**

Keywords: CFET, TCAD, Power Spectral Density

I. INTRODUCTION

For sub-5-nm technology nodes, gate-all-around (GAA) FETs and nanosheet (NSH) FETs are seen as promising due to their superior gate controllability and scalability. To further optimize device layout area, the complementary field-effect transistor (CFET), which stacks n- and pFETs, has been proposed [1]. As device scale down, trap density within the gate dielectric (GD) increases, raising reliability concerns and making trap-induced noise increasingly significant. This study investigates CFET noise characteristics resulting from GD traps using a commercial TCAD tool [2]. We examine the voltage transfer characteristics (VTC) curve shifts and low-frequency noise (LFN) under different biases. We also developed a noise PSD calculation model for the 3D gate-all-around structure, which exhibits excellent agreement with TCAD simulations. Through theoretical calculations, we investigated the dependence of noise PSD on trap energy levels and positions across different bias conditions.

II. DEVICE STRUCTURE AND SIMULATION METHODS

Fig.1 shows the schematics and cross-sectional view of the 3-nm node CFET. The structural parameters and doping concentrations are detailed in Table I [3].

Fig.2 illustrates the workflow for TCAD simulations and theoretical calculations. The noise PSD is simulated using the impedance field method (IFM) [4]. The theoretical noise calculations incorporate electrical characteristics from TCAD simulations, including the GD band diagram and the transconductance (g_m). Traps are characterized by energy level E_t (relative to intrinsic level E_i in Si), position x_t (distance from the GD/channel interface) and density N_{bt}. The electron capture and emission of the traps within GD are modeled as the sum of inelastic phonon-assisted processes [5] and elastic processes [6]. The noise PSD (S_{Id}) is calculated based on Carrier Number Fluctuations (CNF) theory [7]:

$$S_{Id} \cong q^2 \cdot \left(\frac{g_m^2}{W_{g,eff} L_g \cdot C_{ox}^2} \right) \cdot \int N_{bt} \cdot \left(1 - \frac{x_t}{t_{ox}} \right)^2 \cdot S_{\Delta N} \cdot dx_t \quad (1)$$

where $W_{g,eff}$ represents the efficient width of the FET, C_{ox} symbolize the equivalent capacitances of the GD, t_{ox} is the EOT of GD and $S_{\Delta N}$ is the noise spectrum associated with a single trap.

Fig.3 shows the simulated Electrostatic Potential distribution in the nFET, revealing symmetry across the two nFET channels and the 4 directions of the surrounding GD

layer. For the studied 3D structure, a 1-dimensional cutline adequately approximates the band structure distribution throughout the entire GD region. $W_{g,eff}$ is computed by summing the widths of the surrounding GD layers. For example, in nFETs, it is expressed as:

$$W_{g,eff} \cong 2 * (W_{ch} + T_{ch}) * N_{nMOS} \quad (2)$$

III. RESULTS AND DISCUSSION

Fig.4 shows the trap-induced shifts on the VTC curve. With uniformly distributed traps in the GD, a trap density of $1e19 cm^{-3} eV^{-1}$ results in a VTC curve shift of 2mV, whereas a higher trap density of $1e20 cm^{-3} eV^{-1}$ caused a significant VTC curve shift of 20mV, showing direct correlation between increasing trap density and VTC curve shift.

Fig.5 depicts the dependence of S_{Id} on Vg. Across various bias conditions, the results of the theoretical calculations align closely with those from the TCAD simulations, demonstrating the efficacy of the calculation model.

By theoretical calculations, we studied the dependence of noise on trap profiles under various bias conditions. For traps at specific E_t and x_t, the noise PSD exhibits a characteristic Generation-Recombination (GR) noise profile. For traps at fixed x_t and varying E_t, the noise firstly increases before E_t becomes exactly the active E_t and then sharply decline, as shown in Fig.6 (a) and (b). For Vg=0.4V and 0.7V, traps at the respective active E_t are labeled Trap1 and Trap2, their positions are visualized in Fig.7. For traps located at different x_t, those further from the interface make major contribution to the lower frequency noise, as illustrated in Fig.6 (c) and (d).

Fig.7 presents the Local Noise Source (LNS) diagram, offering a visual representation of the GR noise generated by discrete traps. The conduction band energy of silicon at the channel/GD interface is denoted as $E_C(0)$. Traps with E_t above $E_C(0)$ are dominated by elastic model, characterized by a very small time constant and thus contribute negligibly to the LFN. Traps with E_t below the $E_C(0)$ are dominated by inelastic model, with those near the Fermi level E_f making a significant contribution to the noise.

III. CONCLUSION

In this work, we study the trap-induced noise characteristics of CFET through TCAD simulations and theoretical calculations, and visualize the noise source under various bias conditions. These findings provide valuable guidance for evaluating CFET noise performance and identifying specific traps.

REFERENCES

[1] S. Liao et al., IEDM., pp. 1-4. 2023. [2] Synopsys, I. "Sentaurus device user guide: Ver. O-2018.06," 2018. [3] S.-G. Jung et al., IEEE Access, vol. 10, pp. 41112–41118, 2022. [4] F. Bonani et al., IEEE TED, vol. 45, no. 1, 1998. [5] A. Palma et al., Physical Review B, vol. 56, no. 15, pp. 9565–9574, 1997. [6] F. Jiménez-Molinos et al., JAP, vol. 91, no. 8, pp. 5116–5124, 2002. [7] R. Asanovski et al., IEEE TED. vol, 68, no. 10, pp. 4826–4833, 2021.

P2-13

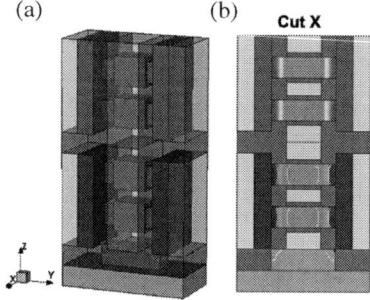

Fig1. (a) 3-D schematic of the CFET, (b) CFET's cross-sectional view through the channel.

Structural parameters	Value	
Physical gate length (L$_g$)	16nm	
Channel width (W$_{ch}$)	10nm	
Channel thickness (T$_{ch}$)	10nm	
Gate oxide thickness (T$_{ox}$)	2nm (HfO2)	
Number of stacked n-channel (N$_{nMOS}$)	2	
Number of stacked p-channel (N$_{pMOS}$)	2	
Doping concentrations	nMOS	pMOS
Source/drain doping (N$_{S/D}$)	1x10^{21} cm^{-3}	1x10^{21} cm^{-3}
Substrate doping (N$_{sub}$)	/	1x10^{19} cm^{-3}

Table I. Structural parameters and Doping concentrations used for 3-nm CFET.

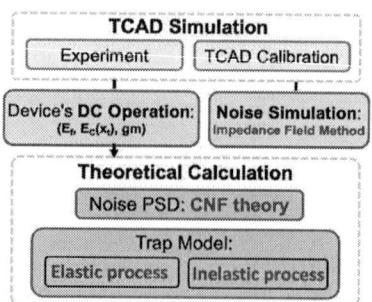

Fig2. Workflow of TCAD simulations and theoretical calculations of trap-induced noise.

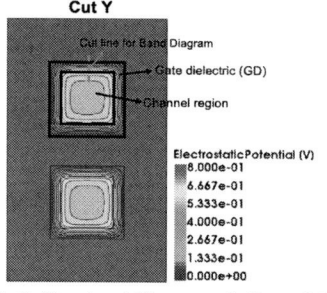

Fig3. Simulated Electrostatic Potential distribution of nFET. Symmetry is observed between the two nFET channels and the 4 directions of the surrounded GD layer.

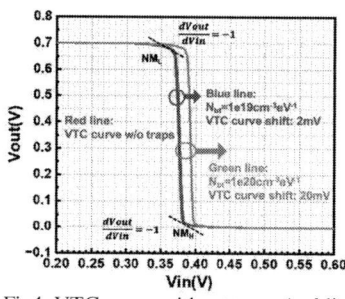

Fig4. VTC curve without traps (red line) and the shifted curve (2mV/20mV) with uniformly distributed traps in the GD, with N$_{bt}$=1e19cm^{-3}eV^{-1}(blue line) and 1e20cm^{-3}eV^{-1} (green line).

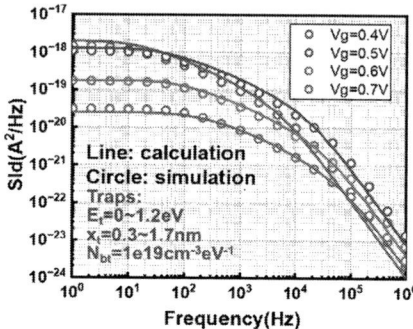

Fig5. Dependence of S_{I_d} on Vg. The theoretical calculation results fit well with the TCAD simulation results.

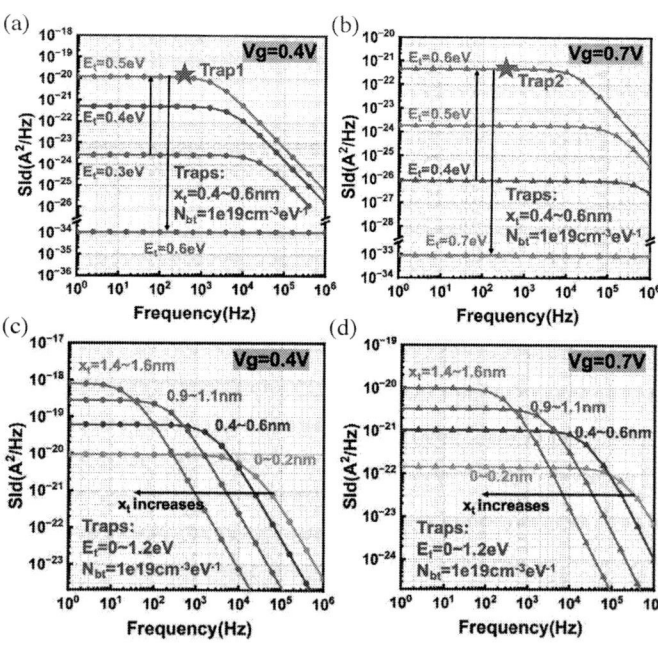

Fig6. Dependence of S_{I_d} on trap energy level E_t at (a) Vg=0.4V and (b) Vg=0.7V; dependence of S_{I_d} on trap position x_t at (c) Vg=0.4V and (d) Vg=0.7V.

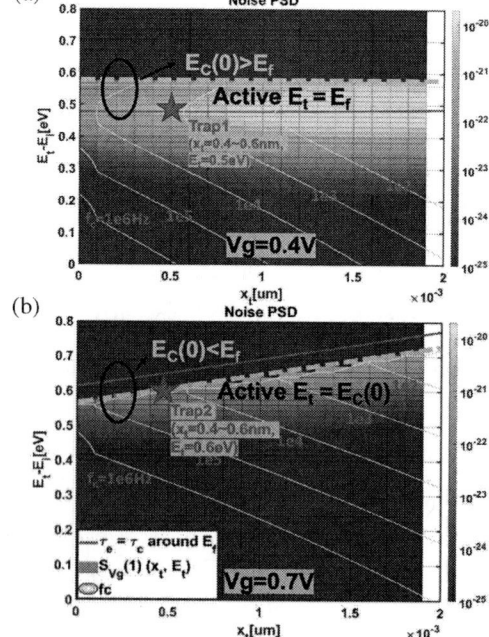

Fig7. Local Noise Source (LNS) figure, visualizing GR noise generated by traps at discrete x_t and E_t under different Vg.

979-8-3503-9164-0/24 $31.00 © 2024 IEEE

Gap in pagination due to unavailable papers.

Pages 129-132

Characteristics of Aluminum-based Oxide with ALD SiO₂ Interfacial Layer as the Gate Dielectric of the Silicon Carbide (SiC) MOS Capacitor with RTA Annealing

Cheng-Li Lin*, Bo-Xian Su and Yu-Lun Lee

Department of Electronic Engineering, Feng Chia University, No. 100, Wenhwa Rd., Seatwen, Taichung, Taiwan 407, R.O.C.

Phone: +886-4-24517250 Ext. 4969 Fax: +886-4-24510405 *E-mail: clilin@fcu.edu.tw

Abstract — **This work investigates the characteristics of aluminum-based oxide (Al_2O_3 and AlON) with ALD SiO_2 interfacial layer as the gate dielectric of the SiC MOS capacitor under various temperatures (400°C, 600°C, and 800°C) of RTA annealing. The experiment results show that the RTA at 400°C reveals the best performances – that is, larger voltage ramp dielectric breakdown (VRDB), larger accumulation capacitance, and lower interface state density (D_{it}). In summary, the Al_2O_3/SiO_2 gate stack on SiC substrate shows better characteristics than that of the AlON/SiO_2 gate stack.**

I. INTRODUCTION

Consequent to the fast development of electric vehicles (EVs), the power device requirement for EV applications has been increased. Silicon carbide (SiC) possesses superior properties, particularly wide bandgap, higher breakdown field, and high thermal conductivity [1, 2]. SiC has been considered the most important semiconductor material for power device applications. In particular, the MOSFET of SiC can be used for the logic circuit and can be fabricated with SiC power devices for monolithic SiC power device system applications.

Various crystal structures of SiC can be used as the substrate of field effect transistors (FETs). The 4H SiC reveals higher mobility than other crystal structures [1]. In addition, oxide growth (as a result of the thermal oxidation of SiC) reveals a poor interface state (D_{it}) due to the carbon accumulated at the interface of the SiO_2/SiC [3]. Thus, the gate oxide or high-k dielectric deposited by the CVD or ALD can be considered another method to improve interface states [4]. Notably, while Al_2O_3 possesses a higher energy bandgap, there have been few studies focusing on high-k dielectrics and the SiC MOSFET [5, 6]. This work investigates the characteristics of aluminum-based oxide with ALD SiO_2 interfacial layer as the gate dielectric of the SiC MOS capacitor.

II. EXPERIMENTAL

Fig. 1 shows the schematics of the TiN/Al_2O_3/IL-SiO_2/n-SiC and TiN/AlON/IL-SiO_2/n-SiC MOS capacitors. First, 4°-off axis (11-20) orientation n-type 4H-SiC wafers are substrates covered by an 11 μm-thick epitaxy layer and with a nitrogen doping concentration of 6×10^{15} cm⁻³. Each wafer is 4 inches in size and is cut to 14 mm ×14 mm with a resistivity of 0.015~0.028 Ω-cm. After RCA cleaning, an interfacial layer of SiO_2 (2 nm) is deposited by ALD. Two high-k dielectrics of Al_2O_3 (10 nm) and AlON (10 nm) are deposited by ALD separately. Subsequently, a TiN metal electrode is deposited by PVD and patterned by photolithography with a device area of 100μm × 100μm. Various RTA annealing with temperatures of 400°C, 600°C, and 800°C (30s) were conducted separately on capacitors to evaluate the gate stack quality and interface state of the capacitors. Finally, the Al film was evaporated by an E-Gun as a back electrode. The detailed device process is shown in **Fig. 2**. The electrical characteristics of I-V, C-V, and VRDB were measured to study the high-k gate stack characteristics.

III. RESULTS AND DISCUSSION

Fig. 3 shows the I_g-V_g leakage characteristics of two gate stack MOS capacitors under various RTA annealing. The as-deposited and RTA-treated (400°C) capacitors have lower leakage currents. Higher RTA temperatures of 600°C and 800°C show higher leakage currents. This finding indicates that higher

RTA annealing results in a high-k dielectric microstructure and/or interface state change, leading to a higher leakage current. In addition, the gate stack of Al_2O_3/SiO_2 in the RTA at 400°C reveals the lowest leakage current. **Fig. 4** shows the VRDBs at various temperatures of RTA annealing. Large VRDBs and hard breakdown behaviors are observed at as-deposited and RTA-treated (400°C and 600°C) Al_2O_3 and AlON gate stacks. An obvious soft breakdown is observed in the RTA (800°C) for two gate stacks. **Fig. 5** shows the box plots of VRDBs of the two gate stack capacitors under various temperatures of RTA annealing. The RTA at 400°C reveals the largest VRDB and the lowest variation for the Al_2O_3 and AlON gate stacks. In addition, the AlON gate stack reveals higher VRDB than that of the Al_2O_3 gate stack (11.26V vs. 10.83V) in the RTA at 400°C. **Fig. 6** and **Fig. 7** show the C-V characteristics of the Al_2O_3 and AlON gate stacks, respectively, under various temperatures of RTA annealing. For the Al_2O_3 gate stack, the RTA at 400°C shows a steep C-V curve and less hysteresis than other RTA temperatures. Additionally, the maximum capacitance (that is, the capacitance at accumulation) is also the largest in the RTA at 400°C. This finding indicates that the oxide trapping in the gate stack is less after the RTA annealing at 400°C, and the steep slope of the C-V curve shows the lowest interface state density [7]. Notably, similar results were observed for the AlON gate stack (as shown in **Fig. 7**). In the comparison between the maximum capacitances of the Al_2O_3 and AlON gate stacks in the RTA at 400°C, the Al_2O_3 gate stack reveals a larger capacitance (0.5 μF/cm² vs. 0.47 μF/cm²). The detail of the interface state density (D_{it}) at (E_c-E) energy is calculated using the high and low frequency method [8], as shown in **Fig. 8**. Two gate stacks show a lower D_{it} in the RTA at 400°C than other temperatures of RTA annealing. Additionally, the Al_2O_3/SiO_2 gate stack shows the lowest D_{it} at E_c-E=0.3eV (2.08×10^{12} cm⁻²eV⁻¹).

Table I shows the summary results of the electrical characteristics of the leakage current (J_g), C_{max}, VRDB, and the D_{it} of the Al_2O_3/SiO_2 and AlON/SiO_2 gate stacks. According to the experimental results, the Al_2O_3/SiO_2 gate stack has a better capacitance and a lower D_{it}. The VRDB is also approximate to the AlON gate stack. Thus, the Al_2O_3/SiO_2 gate stack on the SiC substrate shows better characteristics.

IV. CONCLUSION

This work investigates the characteristics of the Al_2O_3 and AlON high-k dielectrics with the ALD SiO_2 interfacial layer as the gate stack dielectric on SiC substrates under various RTA annealing. From the experiment results, the RTA at 400°C reveals the best performance, and the Al_2O_3/SiO_2 gate stack on the SiC substrate shows better characteristics than that of the AlON/SiO_2 gate stack.

ACKNOWLEDGMENTS

This work was financially supported by the National Science and Technology Council (NSTC), Taiwan, under Contract No. NSTC 112-2221-E-035-081 and NSTC 111-2221-E-035-074. We are thankful to the Taiwan Semiconductor Research Institute (TSRI) and Nano Facility Center (NFC) of the National Yang Ming Chiao Tung University (NYCU) of Taiwan for supporting the device fabrication process.

REFERENCES

[1] M. E. Levinshtein et al., *Properties of Advanced Semiconductor Materials GaN, AlN, InN, BN, SiC, SiGe.* New York, NY, USA: Wiley, **2001**, pp. 93-95.
[2] Chia-Lung Hung *et al.*, *IEEE TED*, 69(10), pp.5742-5748, **2022**.
[3] Menghua Wang *et al.*, *IEDM*, 69(10), **2021**, pp.765-768.
[4] Keita Tachiki *et al.*, *Applied Physics Express*, 13, 121002, **2020**.
[5] Bo-Xian Su, *Master Thesis*, Feng Chia University, Jan, **2024**.
[6] Yu-Lun Lee, *Master Thesis*, Feng Chia University, Jan, **2024**.
[7] Takuma Kobayashi *et al.*, *Applied Physics Express*, 13, 091003, **2020**.
[8] D. K. Schroder, *Semiconductor Material and Device Characterization*, 2nd, Chap 6, John Wiley & Sons, **1998**, pp.368-372.

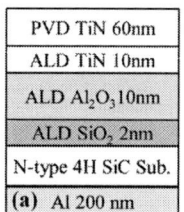

| PVD TiN 60nm |
| ALD TiN 10nm |
| ALD Al₂O₃ 10nm |
| ALD SiO₂ 2nm |
| N-type 4H SiC Sub. |
| **(a)** Al 200 nm |

| PVD TiN 60nm |
| ALD TiN 10nm |
| ALD AlON 10nm |
| ALD SiO₂ 2nm |
| N-type 4H SiC Sub. |
| **(b)** Al 200 nm |

Fig. 1 The schematics of AlOx based gate dielectric of (a) Al₂O₃/SiO₂/SiC and (b) AlON/SiO₂/SiC MOS capacitor structures.

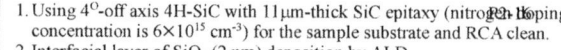

1. Using 4°-off axis 4H-SiC with 11μm-thick SiC epitaxy (nitrogen doping concentration is $6×10^{15}$ cm^{-3}) for the sample substrate and RCA clean.
2. Interfacial layer of SiO₂ (2 nm) deposition by ALD.
3. ALD Al₂O₃ and ALD AlON as gate dielectrics for Sample (a) and (b), respectively.
4. TiN (10 nm) top electrode deposited by ALD in the sample chamber of dielectric film.
5. PVD TiN 60 nm by sputtering as the top electrode.
6. Photolithography and patterning of the top electrode with a device area of 100μm × 100μm.
7. RTA annealing with 400°C, 600°C and 800°C (30s), separately.
8. Al film deposited by E-gun on back of wafer as back electrode

Fig. 2 Process steps of metal/oxide film/IL/SiC substrate (MOS) structures

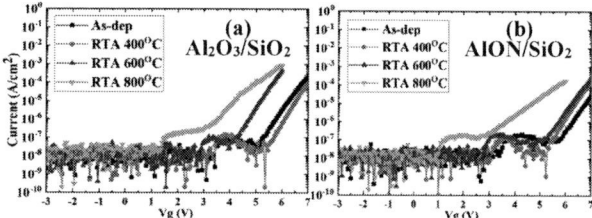

Fig. 3 Ig-Vg characteristics of (a) TiN/Al₂O₃/SiO₂/n-type SiC and (b) TiN/AlON/SiO₂/n-type SiC MOS capacitors with a device area is 100μm ×100μm.

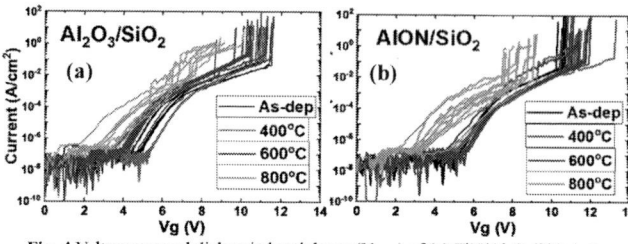

Fig. 4 Voltage ramped dielectric breakdown (V$_{RDB}$) of (a) TiN/Al₂O₃/SiO₂/n-type SiC and (b) TiN/AlON/SiO₂/n-type SiC MOS capacitors with a device area is 50μm ×50μm.

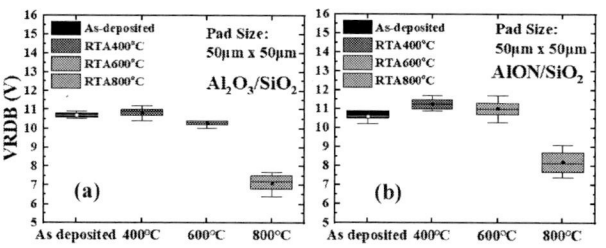

Fig. 5 Box Plots of the voltage ramped dielectric breakdown (V$_{RDB}$) of (a) TiN/Al₂O₃/SiO₂/n-type SiC and (b) TiN/AlON/SiO₂/n-type SiC MOS capacitors at different annealing temperatures with a device area is 50μm×50μm.

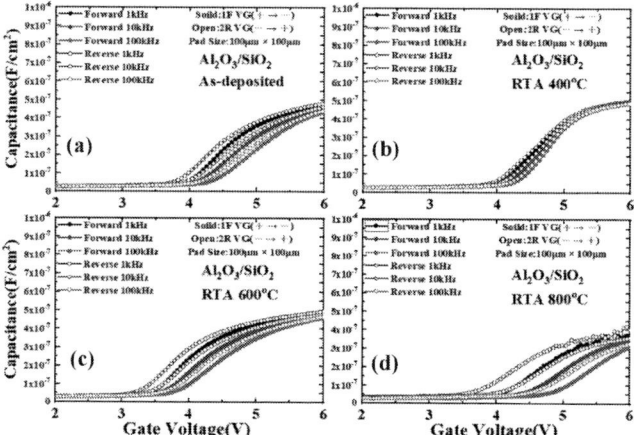

Fig. 6 Capacitance-voltage characteristics of (a) As-deposited, (b) RTA 400°C, (c) RTA 600°C, and (d) RTA 800°C of TiN/Al₂O₃/SiO₂/n-type SiC MOS capacitors under forward bias (+6 to 0V) and reverse bias (0V to +6V), respectively.

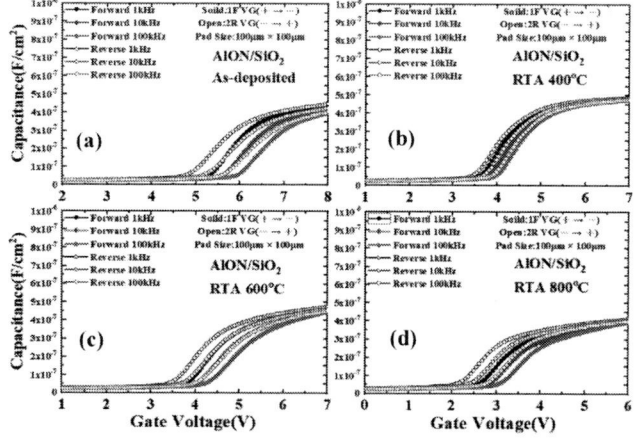

Fig. 7 Capacitance-voltage characteristics of (a) As-deposited, (b) RTA 400°C, (c) RTA 600°C, and (d) RTA 800°C of TiN/AlON/SiO₂/n-type SiC MOS capacitors under forward bias (+8V to 0V) and reverse bias (0V to +8V), respectively, with a device area is 100μm ×100μm.

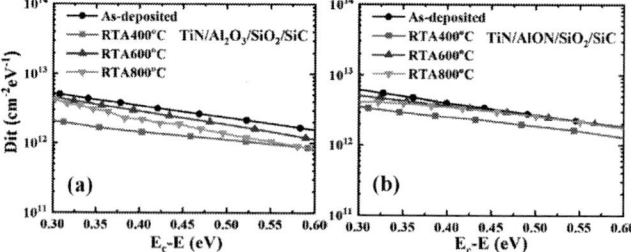

Fig. 8 Interface traps density state (Dit) of (a) TiN/Al₂O₃/SiO₂/n-type SiC and (b) TiN/AlON/SiO₂/n-type SiC MOS capacitors with a device area is 100μm ×100μm.

Table I Summary of the leakage current, VRDB, C$_{max}$ and Dit parameters of the (a) TiN/Al₂O₃/SiO₂/n-type SiC and (b) TiN/AlON/SiO₂/n-type SiC MOS capacitors under various RTA annealing.

Gate Stack Structure	RTA Anneal	Jg (A/cm²) at +6V	VRDB (V)	Cmax (×10⁻⁷ F/cm²) at 6V (1KHz)	Dit (×10¹²) at Ec-E= 0.3eV (cm⁻²ev⁻¹)
TiN/Al₂O₃/SiO₂/SiC	As-dep	3.66E-06	10.71	4.83	5.26
	400°C	6.22E-07	10.84	5.00	2.08
	600°C	4.61E-04	10.29	4.96	4.59
	800°C	8.12E-04	7.11	3.73	4.24
TiN/AlON/SiO₂/SiC	As-dep	3.14E-07	10.61	2.64	5.48
	400°C	1.56E-06	11.26	4.70	3.52
	600°C	4.15E-06	11.02	4.25	5.13
	800°C	7.39E-04	8.25	4.08	4.10

Electrical Characteristics of BEOL-Compatible Top-Gate In₂O₃ Transistors

Peiyan Hong, Jiayao Hao, and Xuefei Li*

Wuhan National High Magnetic Field Center and School of Integrated Circuits, Huazhong University of Science and Technology, Wuhan 430074, China. *Email: xfli@hust.edu.cn

Abstract — **In this work, high-performance top-gate indium oxide (In₂O₃) transistors with high-*k* dielectrics HfLaO were demonstrated. The top-gate In₂O₃ transistor with HfLaO gate dielectric at a channel length (L_{ch}) of 1 μm exhibits a high field-effect mobility of 40 cm²/V·s, a high current on/off ratio of ~10¹⁰ and enhanced-mode operation. The electrical performance of top-gate indium oxide (In₂O₃) transistors with different L_{ch} values from 2 μm to 0.2 μm was investigated systematically. This work provides technological opportunities for monolithic 3D integration.**

Keywords: Oxide transistor, In₂O₃, top-gate, scaling

I. INTRODUCTION

Amorphous oxide semiconductors are promising channel materials for logic and memory applications due to their high mobility, wafer-scale uniformity, and low thermal budget [1]-[4]. Among them, back-gate indium oxide (In₂O₃) transistors fabricated by atomic-layer deposition (ALD) have shown excellent performance with high output current [4] and high mobility, which makes In₂O₃ a competitive channel material for high-performance back-end-of-line (BEOL) compatible electronic devices for monolithic 3-D integration. To date, top-gate In₂O₃ transistors still suffer from a decrease in mobility, threshold voltage (V_{th}), and on/off ratio due to poor interface quality [5], [6]. To meet the need for monolithic 3D integration, high-quality top-gate dielectrics are essential and high-performance top-gate In₂O₃ transistors remain unexplored.

II. EXPERIMENTS

The top-gate In₂O₃ transistors were fabricated on a low-resistivity Si substrate with 8 nm HfO₂ as illustrated in Fig. 1. Fig. 2 illustrates the fabrication process flow. We deposited 8 nm HfO₂ as the back-gate dielectric by ALD. Then, a 2.6-nm thick In₂O₃ channel was grown by ALD at 225 °C and isolated by a diluted hydrochloric acid wet etching process, and a 6-nm thick HfLaO gate dielectric was formed by ALD at 300 °C. The top gate metal of 20 nm Pt was deposited finally followed by a rapid-thermal-annealing treatment at 300 °C in an O₂ ambient. The low temperature process ($\leq$ 300 °C) in this work shows BEOL-compatible potential in integrating with Si-based circuits.

III. RESULTS AND ANALYSIS

Fig. 3 shows the transfer characteristics of the In₂O₃ top-gate transistor with 6-nm HfLaO. The In₂O₃ transistor exhibits a high on/off of ~10¹⁰ at $V_{ds} = 0.1$ V. The field-effect mobility μ_{FE} extracted from the maximum transconductance (g_m) at a V_{ds} of 0.1 V reaches 40 cm²/V·s, as shown in Fig. 3. The V_{th} of the top-gate In₂O₃ transistors was defined as the top-gate voltage (V_{tg}) at $I_{ds} = (W_{ch}/L_{ch})$

×100 pA. A V_{th} of 0.12 V demonstrates enhanced-mode operation of the device and the current at $V_{tg} = 0$ V is close to 10^{-5} μA/μm, which is significantly lower than that in previous work (~1 μA/μm) [7], [8]. The excellent off-state characteristics could be attributed to a significant reduction in oxygen vacancies and hydrogen contamination as well as H doping during the ALD process using high-*k* HfLaO dielectrics [9], [10]. Fig. 4 shows the corresponding I_{ds}-V_{ds} characteristics of the same device, which exhibits a high maximum I_{ds} of 99.3 μA/μm even with an L_{ch} of 1 μm at $V_{tg} = 3$ V and $V_{ds} = 1$ V. Fig. 5 presents the I_{ds}-V_{gs} characteristics of an In₂O₃ top-gate transistor with an L_{ch} of 200 nm, exhibiting a V_{th} of -0.25 V at $V_{ds} = 0.1$ V, and the on/off ratio is as high as approximately 10^{10} at $V_{ds} = 0.5$ V. The 2.6-nm In₂O₃ tends to form a fully depleted channel, while the 6-nm HfLaO provides efficient electrostatic control. Fig. 6 shows the output characteristics of the same device as V_{gs} increases from -2 V to 3 V by 0.5 V. The maximum on-state current I_{ds} reaches 285 μA/μm under $V_{ds} = 1$ V and $V_{gs} = 3$ V. With the scaling of the channel length as shown in Fig. 7, g_m exhibits a monotonic increase, showing the channel length scaling benefit of the In₂O₃ top-gate transistors. The subthreshold slope (SS) and V_{th} of the top-gate In₂O₃ transistors with L_{ch} dependences are shown in Fig. 8 and Fig. 9, respectively. The SS of the In₂O₃ transistors is ~ 100 mV/dec, indicating a good interface quality. With decreasing L_{ch}, V_{th} has a slight negative shift.

IV. CONCLUSION

Ultrathin ALD HfLaO was developed to implement as the gate dielectric of In₂O₃ transistors, which possessed excellent electrical performance, including a decent mobility of 40 cm²/V·s, a high on/off ratio of ~ 10^{10}, and a small SS of ~100 mV/dec. This is mainly attributed to the reduction of donor traps such as oxygen vacancies and hydrogen states. Our approach provides a simple method to enable high-performance oxide top-gate transistors and can be compatible with CMOS technology.

ACKNOWLEDGMENTS

This work was supported by the National Science and Technology Major Project (Grant No. 2020AAA0109005).

REFERENCES

[1] H. Ye et al., IEDM. 28.3.1 (2020).
[2] K. Chen et al. VLSI. T2-4 (2022).
[3] S. Wahid et al., IEEE EDL. 44, 951 (2023).
[4] M. Si et al., Nat. Electron., 5, 164-170, 2022.
[5] P.-Y. Liao et al., IEEE TED, p. 2052, 2023.
[6] M. Si et al., IEEE TED, p. 6605, 2021.
[7] M. Si et al., IEEE TED, pp. 1075, 2021.
[8] A. Charnas et al., APL, pp. 2107.1. 2021
[9] Y. Nam et al., RCS Adv., p. 5622, 2018.
[10] C. R. Allemang et al., IEEE EDL, p. 1120, 2019.

P2-17

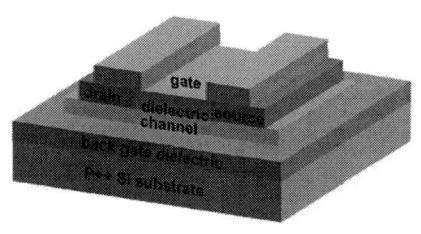

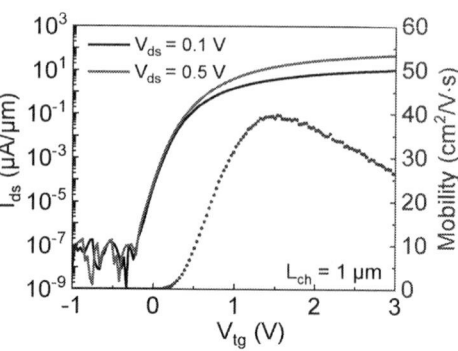

▢ P++Si substrate cleaning

▢ ALD 8 nm HfO$_2$

▢ ALD 2.6 nm In$_2$O$_3$ channel

▢ Isolation by HCl

▢ S/D contact, 20/20 nm Ni/Au

▢ ALD 6 nm HfLaO top gate

▢ Post anneal: 300 °C in O$_2$

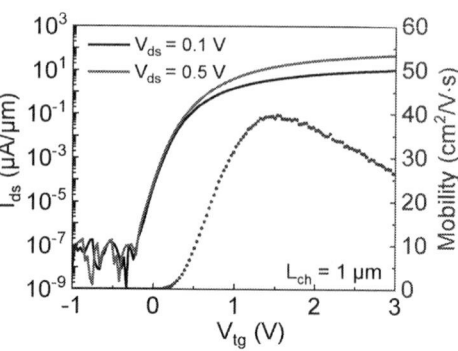

Fig. 1. Schematic diagram of a top-gate In$_2$O$_3$ transistor.

Fig. 2. Fabrication process flow of top-gate In$_2$O$_3$ transistors.

Fig. 3. I$_{ds}$-V$_{tg}$ characteristics of a top-gate In$_2$O$_3$ transistor with L$_{ch}$ of 1 μm. μ$_{FE}$ is extracted from the maximum g$_m$ at V$_{ds}$ of 0.1 V from the transfer curve.

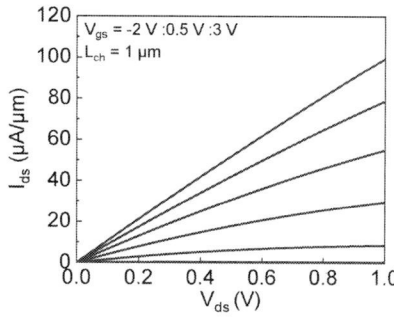

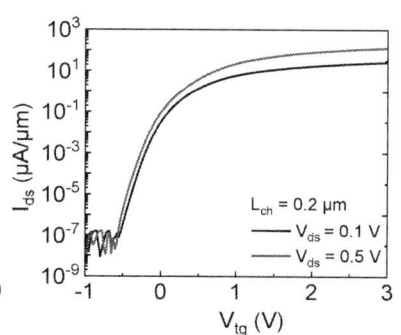

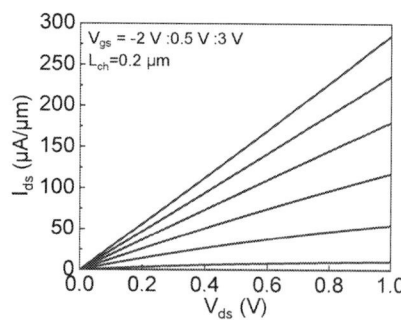

Fig. 4. I$_{ds}$-V$_{ds}$ characteristics of the top-gate In$_2$O$_3$ transistor with L$_{ch}$ of 1 μm in Fig. 3.

Fig. 5. I$_{ds}$-V$_{tg}$ characteristics of a top-gate In$_2$O$_3$ transistor with L$_{ch}$ of 0.2 μm.

Fig. 6. I$_{ds}$-V$_{ds}$ characteristics of the top-gate In$_2$O$_3$ transistor with L$_{ch}$ of 0.2 μm in Fig. 5.

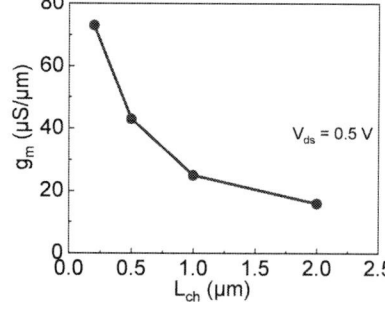

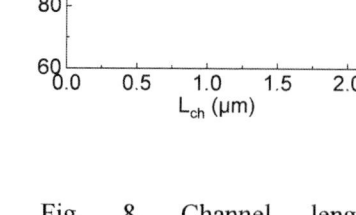

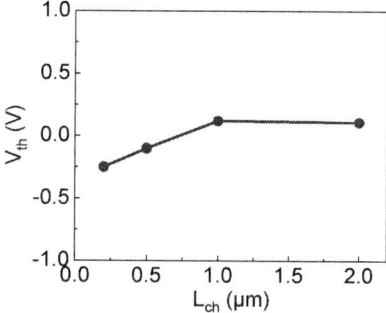

Fig. 7. Channel length dependent g$_m$ at V$_{ds}$ = 0.5 V for top-gate In$_2$O$_3$ transistors.

Fig. 8. Channel length dependent SS at V$_{ds}$ = 0.5 V for top-gate In$_2$O$_3$ transistors.

Fig. 9. Channel length dependent V$_{th}$ at V$_{ds}$ = 0.5 V for top-gate In$_2$O$_3$ transistors.

979-8-3503-9164-0/24 $31.00 © 2024 IEEE